Supercomputing, Collision Processes, and Applications

PHYSICS OF ATOMS AND MOLECULES

Series Editors

P. G. Burke, *The Queen's University of Belfast, Northern Ireland*
H. Kleinpoppen, *Atomic Physics Laboratory, University of Stirling, Scotland*

Editorial Advisory Board

R. B. Bernstein (*New York, U.S.A.*)
J. C. Cohen-Tannoudji (*Paris, France*)
R. W. Crompton (*Canberra, Australia*)
Y. N. Demkov (*St. Petersburg, Russia*)
C. J. Joachain (*Brussels, Belgium*)

W. E. Lamb, Jr. (*Tucson, U.S.A.*)
P.-O. Löwdin (*Gainesville, U.S.A.*)
H. O. Lutz (*Bielefeld, Germany*)
M. C. Standage (*Brisbane, Australia*)
K. Takayanagi (*Tokyo, Japan*)

Recent volumes in this series:

COINCIDENCE STUDIES OF ELECTRON AND PHOTON IMPACT IONIZATION
Edited by Colm T. Whelan and H. R. J. Walters

DENSITY MATRIX THEORY AND APPLICATIONS, SECOND EDITION
Karl Blum

ELECTRON MOMENTUM SPECTROSCOPY
Erich Weigold and Ian McCarthy

IMPACT SPECTROPOLARIMETRIC SENSING
S. A. Kazantsev, A. G. Petrashen, and N. M. Firstova

INTRODUCTION TO THE THEORY OF X-RAY AND ELECTRONIC SPECTRA OF FREE ATOMS
Roman Karazjia

PHOTON AND ELECTRON COLLISION WITH ATOMS AND MOLECULES
Edited by Philip G. Burke and Charles J. Joachain

PRACTICAL SPECTROSCOPY OF HIGH-FREQUENCY DISCHARGES
Sergei A. Kazantsev, Vyacheslav I. Khutorshchikov, Günter H. Guthöhrlein, and Laurentius Windholz

SELECTED TOPICS ON ELECTRON PHYSICS
Edited by D. Murray Campbell and Hans Kleinpoppen

SUPERCOMPUTING, COLLISION PROCESSES, AND APPLICATIONS
Edited by Bell, Berrington, Crothers, Hibbert, and Taylor

THEORY OF ELECTRON-ATOM COLLISIONS, PART I: POTENTIAL SCATTERING
Philip G. Burke and Charles J. Joachain

VUV AND SOFT-X-RAY PHOTOIONIZATION
Edited by Uwe Becker and David A. Shirley

A Chronological Listing of Volumes in this series appears at the back of this volume.

A Continuation Order Plan is available for this series. A continuation order will bring delivery of each new volume immediately upon publication. Volumes are billed only upon actual shipment. For further information please contact the publisher.

Supercomputing, Collision Processes, and Applications

Edited by

Kenneth L. Bell, Keith A. Berrington,
Derrick S. F. Crothers, Alan Hibbert,
and Kenneth T. Taylor
The Queen's University of Belfast
Belfast, Northern Ireland

Kluwer Academic / Plenum Publishers
New York, Boston, Dordrecht, London, Moscow

Proceedings of an International Conference on Supercomputing, Collision Processes, and Applications, held September 14–16, 1998, in Belfast, Northern Ireland, to mark the occasion of the retirement of Professor Philip G. Burke, CBE, FRS

ISBN 0-306-46190-0

©1999 Kluwer Academic/Plenum Publishers, New York
233 Spring Street, New York, N.Y. 10013

10 9 8 7 6 5 4 3 2 1

A C.I.P. record for this book is available from the Library of Congress.

All rights reserved

No part of this book may be reproduced, stored in a retrieval system, or transmitted in any form
or by any means, electronic, mechanical, photocopying, microfilming, recording,
or otherwise, without written permission from the Publisher

Printed in the United States of America

Preface

Professor Philip G. Burke, CBE, FRS formally retired on 30 September 1998. To recognise this occasion some of his colleagues, friends, and former students decided to hold a conference in his honour and to present this volume as a dedication to his enormous contribution to the theoretical atomic physics community. The conference and this volume of the invited talks reflect very closely those areas with which he has mostly been associated and his influence internationally on the development of atomic physics coupled with a parallel growth in supercomputing.

Phil's wide range of interests include electron-atom/molecule collisions, scattering of photons and electrons by molecules adsorbed on surfaces, collisions involving oriented and chiral molecules, and the development of non-perturbative methods for studying multiphoton processes. His development of the theory associated with such processes has enabled important advances to be made in our understanding of the associated physics, the interpretation of experimental data, has been invaluable in application to fusion processes, and the study of astrophysical plasmas (observed by both ground- and space-based telescopes).

We therefore offer this volume as our token of affection and respect to Philip G. Burke, with the hope that it may also fill a gap in the literature in these important fields.

K. L. Bell
K. A. Berrington
D. S. F. Crothers
A. Hibbert
K. T. Taylor

Contents

1. Fifty Years of Atomic and Molecular Collision Theory 1
   A. Dalgarno

2. Electron-Atom Resonances 9
   F. H. Read

3. Recent Progress in Electron-Atom Scattering 15
   I. Bray

4. Benchmark Studies in Electron-Impact Excitation of Atoms 33
   K. Bartschat

5. A Model Adiabatic Potential to Study Shapes and Locations of Single Particle Resonances: The Case of Electron-Ozone Scattering 51
   F. A. Gianturco, R. R. Lucchese, and N. Sanna

6. Aspects of an *Ab Initio* Approach to Electron Scattering by Small Molecules 67
   T. N. Rescigno

7. Atoms in Intense Laser Fields 77
   C. J. Joachain

8. Manipulating Small Molecules with Intense Laser Fields 105
   J. H. Posthumus, K. Codling, and L. J. Frasinski

9. Matrix Methods 119
   I. S. Duff

10.	Elastic Electron Collision with Chiral and Oriented Molecules .... K. Blum, M. Musigmann, and D. Thompson	137
11.	Positron Scattering by Atoms .... B. H. Brandsen	155
12.	From Positron to Positronium Scattering .... G. Laricchia, A. J. Garner, and K. Paludan	171
13.	Embedding and R-Matrix Methods at Surfaces .... J. E. Inglesfield	183
14.	Negative Ion Resonance of Molecules on Surfaces: From Spectroscopy to Dynamics .... L. Šiller and R. E. Palmer	197
15.	BERTHA—4-Component Relativistic Molecular Quantum Mechanics .... I. P. Grant	213
16.	JET Applications of Atomic Collisions .... H. P. Summers, R. W. P. McWhirter, H. Anderson, C. F. Maggi, and M. G. O'Mullane	225
17.	Radiation Pressure and Element Diffusion in Stellar Interiors M. J. Seaton	249
18.	Atomic Physics of Muon-Catalyzed Fusion .... I. Shimamura	269
Index		279

FIFTY YEARS OF ATOMIC AND MOLECULAR COLLISION THEORY

A. Dalgarno

Harvard-Smithsonian Center for Astrophysics
60 Garden Street
Cambridge, MA 02138
USA

Atomic, Molecular and Optical (AMO) Physics is about photons, electrons and positrons, positive and negative ions, atoms and molecules and their mutual interactions. There are two major strands of inquiry in AMO Physics: studies of spectra and studies of collisions. In spectroscopic measurements, ions, atoms and molecules emit or absorb electromagnetic radiation. Electromagnetic radiation consists of light waves and waves have wavelengths that can be measured by spectroscopic instruments. Our eyes are spectroscopic instruments with which we can distinguish colours increasing in wavelength from the violet to the red. Butterflies see in the ultraviolet which lies at the short wavelength side of the violet. At much shorter wavelengths are X-rays and γ-rays. At wavelengths longer than red are the infrared, millimetre and radio waves. The major impetus for the revolution in scientific thought that culminated in the formulation of quantum mechanics was driven by measurements of the spectrum of radiation emitted by hydrogen atoms. To explain the observation that narrow features appeared at specific wavelengths, Niels Bohr in 1913 gave up classical mechanics and in its place he invoked the idea that the electron in a hydrogen atom occupies only certain discrete orbits in its motion about the central nucleus, in contrast to classical mechanics according to which all orbits are allowed. Then radiative transitions between these stationary states of the electron produce spectral lines at specific wavelengths as observed. Earlier, Planck's theory of the thermal radiation emitted by a hot object had invoked the idea of discrete quantized energy modes but no one, least of all Planck, envisaged the theory as more than a temporary empirical fix that was needed to reproduce the measurements that had been made at long wavelengths. However, Planck's quantum theory of radiation is correct and it helped to establish a fundamental concept of quantum mechanics, the duality of particles and waves; in the case of electromagnetic waves, the particles are photons.

Spectroscopy is the major diagnostic probe in science. The wavelength of a spectral line is a unique property of the emitting or absorbing ion, atom or molecule and its presence in the spectrum identifies definitively the species. By measuring the spectra of an object, near or remote, the elements of which the object is made can be determined. The Astrophysical Journal was launched in 1895 with the subtitle "An International Review of Spectroscopy and Astronomical Physics".

The spectral line positions depend on the velocity of the emitter or absorber. Measurements of spectral lines emitted by stars in external galaxies reveal shifts in the wavelengths. The lines are red-shifted from which we conclude that all galaxies are moving away from us and the space of the Universe is expanding.

The other major strand of research in AMO Physics and indeed other disciplines of Physics is the study of the collisions of beams of particles. One of the most famous experiments in physics was the measurement by Rutherford of the angular distribution of a beam of α-particles scattered by a thin foil. It demonstrated that at the centre of an atom lies a heavy nucleus. Once we had learnt to control beams of particles—for charged particles like electrons and positive ions, electric and magnetic fields could be used—it became possible to explore the properties of atoms and molecules by using them as targets and measuring what happens when the beams collide with them. The quantization of the electron orbits in atoms, indicated by the spectroscopic data, was confirmed by the measurement of discrete energy losses in collisions of an electron beam with atoms. The duality of particles and waves was confirmed by experiments in which electrons showed wave-like properties of interference and diffraction. With continued technical improvements and a better understanding of the behaviour of particle beams, measurements of the collision of one particle beam with another could be performed. Particle beams have provided a general exploratory tool of enormous power and versatility that has greatly extended our knowledge of physical processes and has led to new technologies. Today collisions at extremely high energies generated in accelerators are the principal experimental tool for investigations of elementary particles and the fundamental laws and symmetries of physics. Led by Brian Gilbody, Queen's University has housed a distinguished research program, studying collisions of high energy multicharged ion beams.

Spectroscopy and collisions are not independent and each may be intimately involved in the other. Collisions taking place in the presence of radiation fields can be modified by them, raising the possibility of controlling the outcome of reactions by judicious use of photon beams. Photoionization, a process in which photons are absorbed and electrons are ejected and photodissociation, a process in which photons are absorbed and the molecules are broken into their constituent atoms, are examples of collision processes initiated by radiation, of particular interest today with the advent of laser sources that generate intense electromagnetic fields.

Spectroscopy and collisions are combined in investigations of ionized gases which are gases in which some of the atoms have lost electrons and exist as positive ions and some may have gained an electron and exist as negative ions. Spectral lines can be used to establish which of the ionization stages are present. Ionized gases are plasmas and they range from weakly ionized of the kind investigated at Queen's by Bill Graham and his collaborators to very nearly fully ionized as in fusion plasmas. The Universe can be regarded as ionized plasma, embedded in which are galaxies, stars and planets.

Historically the study of plasmas began with the exploration of electrical discharges in gases which led Thompson to the discovery of the electron one hundred years ago. Today plasma processing technologies comprise a substantial industrial activity of major economic importance.

Plasmas are extraordinarily complex and interesting phenomena and we are far from achieving a fully quantitative picture of them. They exist or can be produced in a wide range of physical conditions of composition, temperature, density and radiation fields. To model their behaviour involves the identification of the multitude of atomic, molecular and optical processes that together are responsible. The efficiency with which the processes occur must then be determined quantitatively by experimental measurement or by theoretical calculations.

AMO Physics has changed dramatically over the past fifty years. In 1948, Volumes 60 and 61 together occupying 1000 pages were published in the Proceedings of the Physical Society of London. They dealt with all of Physics The articles were mostly concerned with measurements of the properties of materials in solid and liquid form. An example is a measurement of the dielectric constant of water at centimetre wavelengths reported by C.H. Collie, J. B. Hasted and D.M. Ritson. I mention it in particular because John Hasted was one of the experimenters who led in the resurgence of AMO physics that has taken place in the last half century. Volume 60 of the Proceedings contained short papers on the spectra of the alkali metal pairs Li_2, K_2 and NaK, on the dissociation energy of nitric oxide and on the structure of carbon monoxide and a long paper on collision-broadening of ammonia in gases. There were three theoretical papers by Charles Coulson and his collaborators on momentum distributions in atoms and on the electronic structure of the conjugated molecules napthalene, anthracene, coronene and pirene, which happen to be of current astrophysical interest because of the possible existence of polycyclic aromatic hydrocarbons (PAHs) in the interstellar space of our Galaxy. The volume included one paper on gas discharges by L.B. Loeb, discussing the mechanism of spark breakdown. Setting aside the discussions of PAHs as chemistry, other than gas discharge physics AMO Physics occupied 36 pages of volume 60. Volume 61 is noteworthy because of a paper by R. H. Sloane and C. S. Watt presenting some intriguing results on the negative ions emitted from oxide-coated cathodes. The measurements were carried out in the Department of Physics here. Both Sloane and Watt became long term members of the faculty at Queen's. Volume 61 included a paper on the initiation of low pressure glow discharges by J. D. Craggs and J. H. Meek. These two papers took 32 pages. The only other paper in AMO Physics is a Letter of two pages on the spectrum of silver iodide by R. F. Barrow and M. F. R. Mulcahy. Thus in 1948, the Proceedings of the Physical Society devoted a total of 38 pages out of 1000 to AMO physics, if I exclude papers on gas discharges, 70 if I include them.

In 1949, it was decided to split the journal into parts A and B. Part A included quantum theory, atomic physics, molecules and spectra. The areas of application of AMO physics, electric discharges, astrophysics, radio, geo- and ionosphere physics and solar physics were to be in Part B. The total AMO content in 1949 in Part A was 38 pages again, mostly on measurements of molecular spectra and quantum mechanical calculations of momentum wave functions. Part B, though ostensibly not dealing with atomic physics or spectra, contained a paper on spectroscopic observations of pyrotechnic glows and a paper describing experiments on photoionization in gases. Including them, AMO physics occupied 55 pages of Volume 62.

Ten years later, in 1958, it was decided that the separation of the subject matter of Physics into two sections had led to an undesirable fragmentation and Parts A and B were reunited into a single publication. Volumes 71 and 72 of the Proceedings of the Physical Society were published that year. The number of pages was 2200 pages of which 312 were AMO Physics. The number of pages may not seem large but for AMO Physics it represented an increase by a factor of six in ten years. It signaled perhaps the end of an era in which it had been possible for an individual to pick up an issue of the journal and expect to read all the papers it contained on AMO Physics. I note in passing that amongst the authors of papers in Volumes 71 and 72 those forty years ago were Brian Bransden, Benno Moisewitsch, Michael Seaton, Arthur Kingston and myself, all present at this conference.

The growth continued and in 1968 the editors of the Proceedings gave up the unequal struggle and the Proceedings was again separated into sections. The Journal of Physics B emerged as the section devoted to Atomic and Molecular Physics. It contained 1250 pages, an increase in ten years by a factor of four. In 1978, there were 5000 pages. The number of pages remained at about 5000 through to 1988. In 1988, the title was enlarged to Journal of Physics B (Atomic, Molecular and Optical Physics) to reflect the changing distribution of research brought about by the use of lasers and other sophisticated optical techniques. The number of pages then increased steadily and in 1998, we can expect 7000 pages. The comparable journal in the United States, Physical Review A, will contain about 10,000 pages, but the pages are larger. There has also occurred a proliferation of more specialised journals. We now have time barely to read the titles of papers. The expansion in AMO Physics is surely not over. The fundamental discoveries and the technological applications that will arise from explorations of coherence and control, ultracold atoms and molecules, Bose-Einstein condensation, hollow atoms, atom interferometry, atom manipulation, atom lasers, X-ray lasers, intense fields, quantum computation and plasma manipulation and not least enhancements in computer power ensure the continued growth of AMO Physics.

Although physics and indeed most of science is driven by experiments, theory has a central role to play in providing a simplifying, logical framework, into which the results of different experiments can be incorporated and which provides a prescription for the prediction of physical phenomena. The theory is expressed as laws or principles, written in the form of mathematical equations. It is the responsibility of theory not only to write down the equations but also to solve them. Only by comparing the solutions with measurements can we be sure the equations are in fact correct and complete. Indeed, many advances in physics have stemmed from small differences found between theoretical predictions and experimental measurements. Theory is needed also for practical reasons. The data base of AMO processes required for applications to other sciences and technologies is so vast it cannot be supplied by experiments. Experiments provide crucial tests of the accuracy of theoretical predictions, but theory must be the principal future source of AMO data.

Fifty years ago saw the stirrings of the revolution that was to be brought about by the realisation of the electronic digital computer. Before and during the war, calculations were done mainly by pencil and paper aided by tables of mathematical functions and by a manual analogue computing device called a slide rule. Much effort had to be given to the ongoing detection and correction of errors as they inevitably occurred, because of the heavy price that would be paid if an error were to propagate deeply into the serial computation. The entire calculation from the point at which the error occurred would have to be repeated, hand operation by hand operation.

Slide rules were often used. They had the advantage of being readily transportable. In a biographical memoir, Sir David Bates mentions Sir Harrie Massey's custom of travelling with a slide rule and a notepad, computing as he went.

Electric analogue computers in the form of differential analysers existed but they were complicated to assemble and unreliable in operation. They met with some success in solving differential equations. Sir Harrie Massey and Sir David Bates used a differential analyzer built in the Physics Department at Queen's in 1938 by a highly skilled technician, John Wyler, to solve equations arising in the theory of the photoionization of atomic oxygen.

By 1948, manually operated mechanical calculating machines were widely available. The more fortunate individuals had access to versions with electric motors but computation remained an extraordinarily tedious activity. But it had to be done. Experiments are ultimately presented as numbers and numbers are what theory must provide for comparison.

To inject some interest into the process, students would sometimes indulge in races to see who could, for example, solve correctly a selected differential equation in the shortest time. In these competitions, the error rate increased rapidly with computing speed and the race often went to the slowest. Fortunately for the sanity of theorists, the period between electric and electronic machines was mercifully short, or so it seems in retrospect.

The emphasis in theoretical research in AMO Physics was on the invention of approximations and methods that whilst preserving the physical content of the problem yielded mathematical expressions and procedures that were computationally tractable. In a sense this is still the emphasis and still the challenge, but the speed, scale and complexity of what is practicable have been enormously enhanced with no end yet in sight. There has occurred a corresponding increase in the rapidity and flexibility of communication. Computers can be accessed remotely and computers at different locations can be brought to bear simultaneously on a common research problem. National and international networks, both formal and informal, have become a powerful unifying force in scientific research.

The challenges in responding to these dramatic changes in the computing environment and exerting control over its direction have been formidable. In AMO Physics, many have contributed but we would all agree that Philip Burke has been an intellectual leader at the many fronts of the enterprise. He has exhibited sophisticated skills and creativity in mathematical and numerical analysis, designing new mathematical approaches and numerical procedures in a mutual relationship that has been uniquely productive. Philip began his career with pioneering studies of the scattering of nuclei but soon took up what was to be his major interest, the much more versatile field of electron and photon collisions with atoms, ions and molecules.

His work is measured not only by his publications, extensive though they are, but also by the personal influence he has had on the research of his fellow scientists throughout the world and on the training of the next generation. I leave aside Philip's services in administrative capacities and his participation as a member of advisory committees, except to remark that his wise counsel been of the utmost importance in ensuring scientific access in the United Kingdom to the developing computer power as it came into being.

The history of the theory of scattering is one of increasingly elaborate formulations in which more and more of the physical mechanisms were included, but it had to be done in such a way that the resulting mathematical expressions were accessible to numerical computation. It began with the Born approximation. Scattering is characterised by cross sections which depend on the initial and final states of the target and which depend on the velocity of the projectile electron. In the Born approximation for excitation or ionization of the target, it is assumed that the transition is driven by a single direct interaction of the electron as it passes the target. There is a change in the momentum of the electron but its path is not disturbed by the target. If the electron is moving rapidly, the physical description, represented mathematically by the Born approximation, is accurate. Given the initial and final target wavefunctions, the Born approximation cross sections can be obtained by numerical quadrature. For elastic scattering there is no change in the electron momentum during the collision. The Born approximation again reduces to a simple quadrature. Quadrature is a relatively simple numerical exercise.

Electrons are fermions and in scattering, the projectile electron and any one of the electrons of the target can interchange places. Antisymmetrising the total wave function yields the Born-Oppenheimer approximation. The resulting expression can still be evaluated by quadrature but the formula has severe defects and the results are unreliable. There is a modification called the Born-Ochkur approximation that leads to some improvement and is easier to calculate. The Born approximation often failed at low collision energies, yielding cross sections that were much too large. The worst of its excesses could be restrained by an extension that preserved the unitarity of the scattering matrix, as noted by Michael Seaton and Valerie Burke, or by the use of the eikonal Born series as developed by Fred Byron and Charles Joachain.

More of the physics is contained in another approximation, the distorted wave approximation. It takes account of the modification in the path of the electron arising from its interaction with the electrostatic distribution of the target. In its antisymmetrised version, it becomes the exchange distorted wave approximation. It is an extension of the single configuration Hartree-Fock method for bound states to the continuum. The mathematics can be reduced to ordinary second order differential equations which can with some effort be solved by numerical integration. Variational methods could be used in place of direct numerical integration and they offered the prospect of a more elaborate description of the physics whilst being somewhat easier to implement.

A significant improvement to the theory for electron scattering by complex atoms was the extension by Michael Seaton of multiconfiguration Hartree-Fock theory. However, Hartree-Fock omits some physics that is important at low energies. The slowly moving electron polarises the target and the polarised target acts back on the motion of the electron, producing an attractive long range polarization force. The inclusion in the physics of polarization and exchange forces led to the successful explanation of the mysterious Ramsauer effect in which the scattering cross section is observed to pass through a minimum as the electron energy is increased.

The distorted wave approximation works well for weak interactions but it does not allow for the transition to reverse during the transit of the electron. That requires a balanced approximation in which the initial and final states are treated with equal accuracy. Studies of the coupled equations in limiting circumstances had been reported by Massey and Mohr and by Seaton in the Proceedings of the Physical Society, which demonstrated the importance of coupling in the excitation mechanism, but numerically we are faced with a demanding task. Still more demanding is the problem if we take the obvious conceptual step of expanding to a multi-state description which allows physically for the possibility of excitation by virtual transitions into and out of many states.

Progress would have been limited except for the advent of electronic digital computers. Responding to the new opportunity, in 1960 and 1961 Philip Burke, Kenneth Smith, Valerie Burke, Ian Percival, Ronald McCarroll and Harry Schey used the early electronic digital computers, the IBM 704 and the English Electric DEUCE machines, to solve by direct numerical integration the close-coupled set of equations that arises in the multistate formulation of the scattering of electrons by hydrogen atoms. The results were published in the Proceedings of the Physical Society and the Physical Review. This achievement transformed the theoretical study of electron and photon collisions with atoms and ions into a different kind of activity. In place of an effort to understand the principal mechanism at work in a particular process, a systematic investigation could be undertaken of the possibilities by exploring numerically the effects of different choices of basis sets. The calculations raised the hope that theory could have a predictive capacity, a hope that has been amply demonstrated by Michael Seaton's Opacity Project and by the talks presented at this conference.

Theory can also predict new phenomena. The close-coupling equations quickly revealed the presence of structure in cross sections that could be attributed to long-lived resonance states formed during the collision by the capture of the electron in temporary bound states that subsequently decayed. Thus numerical studies, rather than being simply a support of established theory, stimulated the discovery of new concepts needed for the physical interpretation of the features revealed by the calculations. With the extension of the close-coupling methods to complex ions and atoms, the dominating role of these resonance states in electron impact excitation was demonstrated. Resonance states lead to the processes of autoionization and dielectronic recombination which affect the ionization balance in plasmas.

The close-coupling method was soon extended by Philip Burke and Sydney Geltman by the addition of pseudo-states to the multi-state expansion. Suitably chosen pseudo-states provide a correct account of correlation and polarization. However because the final state has two continuum electrons, electron impact ionization poses a more severe theoretical problem which is still under active investigation.

Setting up the close-coupled equations for a complex system, especially when relativistic effects are included, is a demanding mathematical exercise, but still greater difficulties attend the construction of a reliable convergent procedure for the numerical integration of the equations. A method of great power and capable of generalization to a wide range of atomic and molecular problems is the R-matrix method introduced into AMO Physics by Philip Burke, Alan Hibbert and Derek Robb. A much greater variety of processes can occur when photons and electrons interact with molecules. Molecules lack spherical symmetry, the nuclei also move, and the theoretical problems are formidable. Nevertheless, the R-matrix method has been extended to molecular systems and substantial progress has been made in developing numerical codes that combine methods of scattering theory and quantum chemistry.

The power of the R-matrix method and the super power of supercomputers open up new areas of investigation to direct numerical attack. The R-matrix method has been extended to treat the behaviour of atoms and molecules in intense radiation fields and to include the additional dimension of time. Time-dependent studies in which systems can be followed as they evolve from different initial conditions will enlarge and deepen our understanding of the physics of AMO processes and open new pathways of research.

AMO Physics is integral to a broad range of scientific and technological disciplines and its growth in the past fifty years has been matched by its increased utility as a diagnostic probe and as an instrument for predicting the response of materials to disturbances induced by electric fields, radiation fields and dynamical interactions.

Rather than attempt a necessarily superficial summary of the vast array of the applications of AMO Physics, I will take a specific example from my own particular interests and consider the physics of the supernova 1987a. On February 23, 1987, the core of a blue supergiant star in the Large Magellanic Cloud collapsed and the star exploded. It was the brightest supernova since the invention of the telescope, the first from which neutrinos were detected and the first for which molecules were found in the ejecta. The progenitor star had also been observed. The event was a major happening in astronomy. Supernovae are believed to be the source of the elements heavier than iron found on Earth and the source of the energetic cosmic rays that bombard the Earth. SN 1987a enabled a decisive test to be made of theories of nucleosynthesis in stellar explosions.

Other than the neutrinos, all our information about the supernova comes from the electromagnetic radiation emitted by the object and the supernova can be regarded as an evolving laboratory of atomic and molecular physics. The light curve, which is the total energy emitted in the optical and infrared regions of the spectra, established that the principal source of luminous energy is the decay of radioactive ^{56}Co which is itself a product of the decay of ^{56}Ni. The decay time of ^{56}Co is 111.26 days which closely matches the decay of the light curve. At times beyond 1200 days, it is replaced by ^{57}Co which has a lifetime of 1.07 years. After 1500 days, the decay of radioactive ^{44}Ti which has a lifetime of 58±10 years takes over as the main source of energy. The decay products are γ-rays and positrons. The γ-rays scatter in collisions with the free and the bound electrons, producing X-rays and fast electrons. The electrons and the positrons collide with the ions, atoms and molecules in the ejecta, producing the light that is observed. Several years into the evolution of the ejecta, the steadily diminishing particle density allows the γ-rays to escape and the main energy source is the positrons, which slow down in collisions and annihilate, producing the 511 keV line. Spectral analysis of the light reveals features that can be identified with specific elements. Thus H, He, Ni, Co, Fe, O, Si, Mg, Na and Ca have all been identified in SN 1987a, the detection of Ni and Co confirming their role in energising the ejecta.

Of unique value is the evolutionary nature of the object. Changes in the emission spectrum can be followed over long periods of time. Because of the expansion of the ejecta and the loss of energy by radiation, the material cools with age and there has been a shift of the thermal part of the emission to longer wavelengths. The emission line of FeII at a wavelength of 715.5 nm, strong in the beginning, is now no longer detectable.

The spectrum shows that the elements are present as neutral atoms and singly ionized. The high ionization stages produced in the initial blast must have been removed by radiative and dielectronic recombination and by charge transfer recombination, once some neutral material had formed.

The ionization of the atoms arises from collisions with the fast electrons and positrons and the relative abundances of neutral and ionized material is determined by the ionization across sections. The excitation of the levels emitting the spectra that are measured has three sources. Low-lying levels are populated by thermal collisions. High-lying levels are populated by electron recombination and by direct impact excitation by the electrons and positrons. Given reliable cross sections we can infer the element abundances and test the models of nucleosynthesis. The physics has similarities to the physics of terrestrial auroras. Thus the oxygen doublet at 630.0 nm and 636.4 nm is a characteristic feature of the spectrum and it is produced mostly by electron impact. The doublet of ionized oxygen near 372.7 nm, produced in auroras by simultaneous excitation and ionization of neutral oxygen, is also seen in the supernova. A striking example of the importance of reliable AMO data is provided by the observation that the sodium doublet at 589 nm has a strength comparable to the oxygen red lines. Yet the nuclear models suggest that oxygen will be two orders of magnitude more abundant than sodium. The resolution must be that the excitation cross section for electron impact excitation of sodium is one hundred times larger than that for oxygen.

The spectra contain information about the progenitor star, the supernova explosion, the nucleosynthesis of the elements, the chemistry of molecule and dust formation, the dynamics of the expanding ejecta and possibly the contribution to the energy supply from the neutron star left behind by the explosion. AMO Physics, particularly of the kind discussed at this conference, is central to finding the answers to the many questions that remain.

ELECTRON-ATOM RESONANCES

Frank H Read

Department of Physics and Astronomy
University of Manchester
Manchester M13 9PL, UK

THE FIRST OBSERVATIONS OF ELECTRON-ATOM RESONANCES

Prof. Philip Burke has been a leader in the subject of electron-atom scattering resonances since the beginning of the subject in 1961. He provided (with Schey) the first clear theoretical evidence[1] for the existence of such resonances, as shown in Fig. 1. Using a close-coupling approximation it was found that the s-wave phase shift for scattering of electrons by hydrogen atoms increases steeply just below the energy of the first excited state. This was interpreted as being due to a narrow 'Breit-Wigner' resonance. About three months earlier Smith et al[2] had published a less conclusive result -the single critical point that showed a significant increase in the elastic scattering cross section below the first excited state had been slow to converge (taking $10hrs$ of computing time).

The first published experimental evidence for the existence of an electron-atom resonance is that of Schulz[3], reproduced in Fig. 2. A clear dip in the elastic scattering cross section, followed by a peak, is observed at an energy of approximately $19.4eV$, below the energy of the first excited state at $19.8eV$. Soon afterwards Fleming and Higginson[4] published a less precise observation of this resonance, using a simpler transmission experiment.

The asymmetric shape of the e-He resonance structure is analogous to that observed earlier by Silverman and Lassettre (although not published openly until 1964[5]) in electron impact excitation of autoionizing states of helium, which had been interpreted by Fano[6] as the interference of a resonant and a direct (ie non-resonant) scattering mechanism, giving the now-famous 'Fano profile'.

The first theoretical and experimental observations of electron-atom resonances all appear to have been 'accidental', in the sense that they were not being looked for, although scattering resonances were at that time well known in nuclear physics and also the existence of broad peaks in the scattering of low-energy electrons by molecules such as N_2 had been known for many years. It is told for example that Schulz observed the e-He resonance when he was looking for a cusp at the energy of the first excited state, which he was unable to see (and much later Cvejanovic et al[7] accidentally obtained the first clear observation of this cusp when they were looking at the resonance).

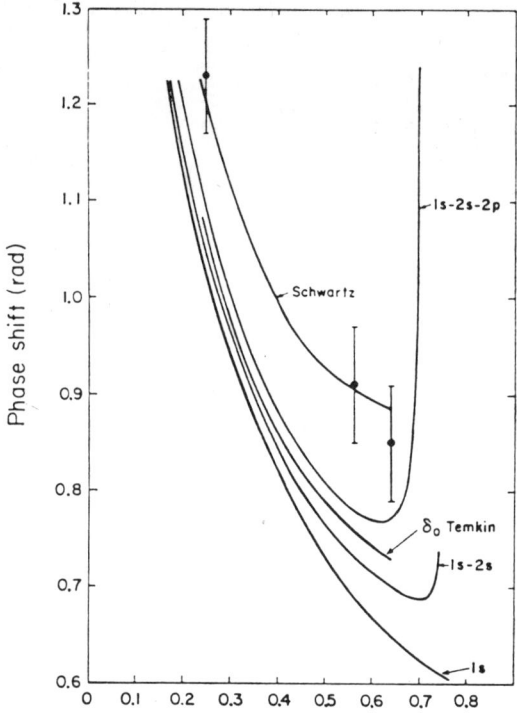

Figure 1. Calculated singlet S phase shifts for e-H scattering, as a function of k^2 (in atomic units). Reproduced from Burke and Schey 1962[1]. Also shown are curves that represent earlier calculations of Schwartz and Temkin, and points with associated errors, calculated by Temkin.

RECENT RESULTS

The subject continues to be well studied, and the Belfast group continues to maintain a leading position[8,9]. A new generation of high resolution studies has started in which laser excitation of atomic negative ions is used, for example in helium to reach narrow quartet resonance states from a metastable quartet state[10]. In the Manchester group we have recently completed a thorough exploration of the He doublet states in neutral helium that lie below the He$^+$ (N=3) threshold, by inelastic electron scattering from the He ground state. There is considerable overlapping of these states and so to deduce their individual energies and widths it is necessary to use a range of different incident energies and scattering angles. A representative result is shown in Fig. 3. It has been possible to deduce the energies and widths of 13 states in this energy region, most of which have not previously been measured.

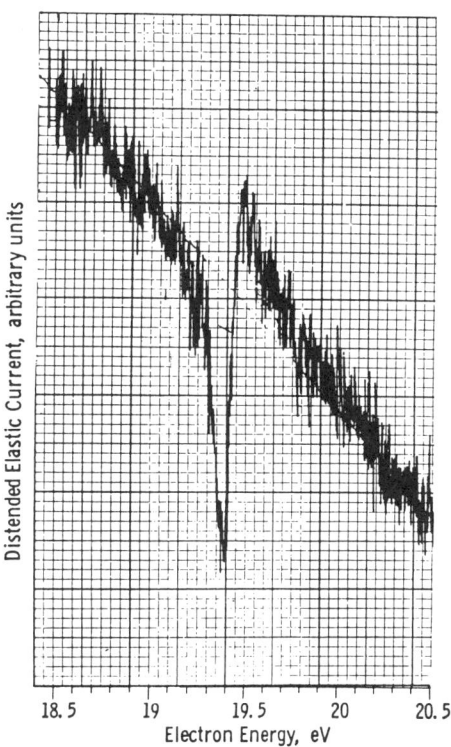

Figure 2. Observed yield of electrons elastically scattered through 72° by helium atoms, from Schulz[3].

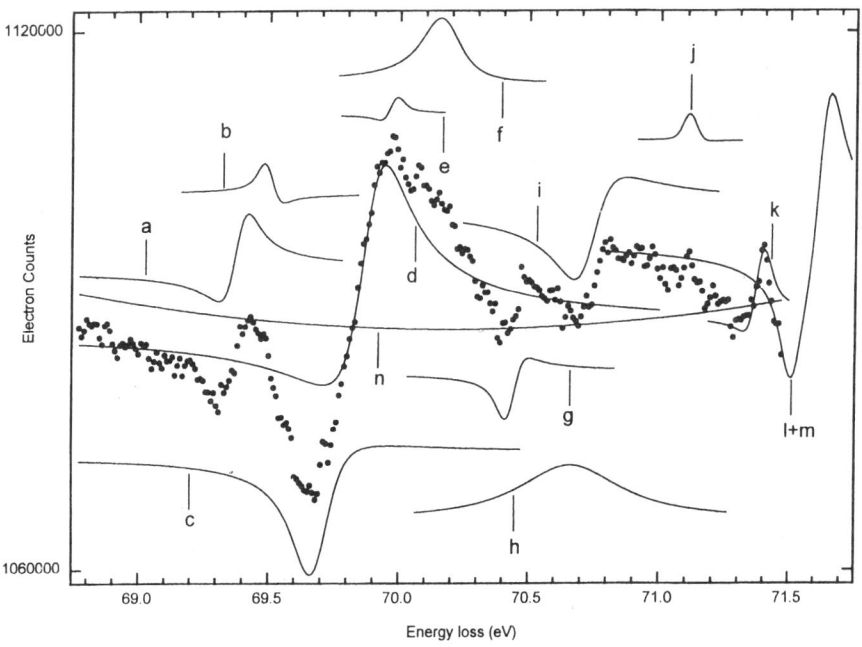

Figure 3. Electron energy loss spectrum for helium at a constant residual energy of $40 eV$ and a scattering angle of 20°, from Brotton et al[11]. The experimental results are shown as points, while the individual fits to the autoionizing states are shown by the displaced curves, which when taken together give a fit to the points.

A NEW TECHNIQUE FOR OBSERVING RESONANCES IN THE BACKWARD DIRECTION.

The most recent work on electron-atom resonances by the Manchester group has involved the use of a newly developed 'magnetic angle changing' technique[12,13] for making measurements in the previously inaccessible backward direction. In this technique a localised magnetic field is produced by sets of solenoids that surround the interaction region of an otherwise conventional spectrometer. The magnetic field is symmetric about an axis perpendicular to the scattering plane. The incident electron beam is deflected but the symmetry of the field ensures that when the beam is initially directed towards the interaction centre it will cross the centre after having changed direction. Similarly scattered electrons are deflected by the magnetic field but when they are at a sufficient distance from the interaction centre they move in a radial direction. The deflections of the incident and scattered electrons make it possible to physically separate the incoming electrons from those that are scattered through 180°, thus allowing measurements to be made at and around 180°.

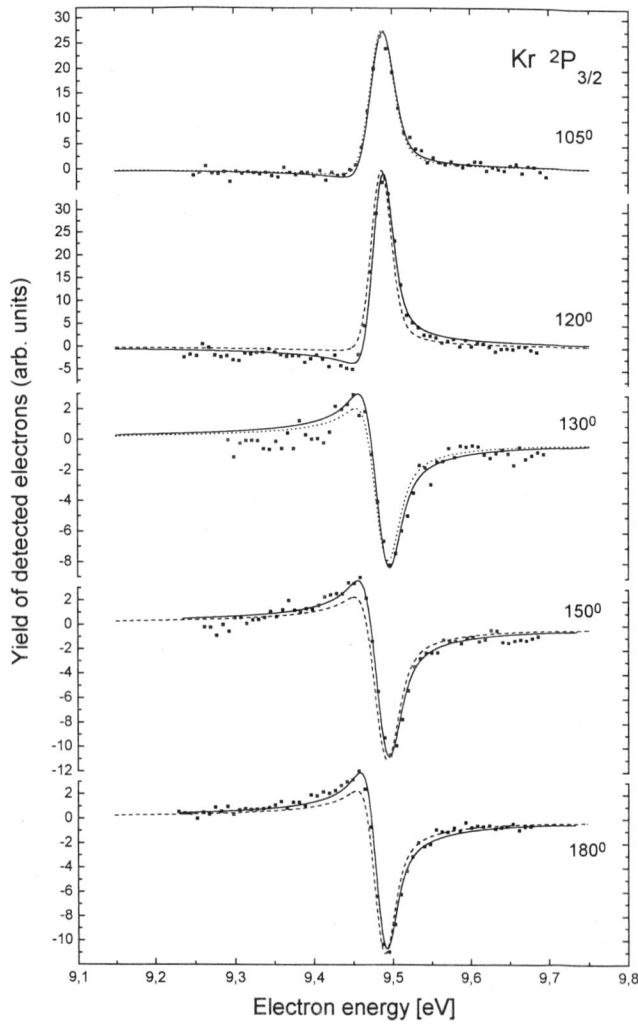

Figure 4. Yield of electrons elastically scattered form krypton at angles in the backward hemisphere, from Zubek et al[14]. The broken line is given by the calculated phase shifts of Sienkiewicz and Baylis[17], while the full lines represent an empirical fit.

Measurements of elastic electron scattering at large scattering angles are important in the determination of the phase shifts of the partial waves involved. At 180° all the partial waves that contribute (of which there are more for the heavier target atoms) have the same magnitude (modulus) of unity for their Legendre polynomials. Moreover the polarisation and exchange contributions to the interaction are usually expected to be more significant than the electric dipole contribution for backward scattering.

As an example of results obtained recently[14] using this technique, Fig. 4 shows features that correspond to the $4p^5(^2P_{3/2})5s^2$ resonance in e-Kr elastic scattering at a range of angles in the backward hemisphere, including 180°. High resolution measurements have previously only been published by Weingartshofer et al[15], in the angular range from 10° to 138°. The energy of the resonance has been determined by observing its energy positions with respect to the 'benchmark' energy[16] of $11.103 eV$ for the $^2P_{3/2}$ resonance in argon, using a mixture of the gases, and is found to be $9.490 \pm 0.012 eV$. The broken line in Fig. 4 has been obtained using the most accurate of the available calculated phase shifts[17]. The full line corresponds to an analysis in which the η_1^+ phase shift is varied while the difference between the two relevant phase shifts, $(\eta_1^+ - \eta_1^-)$, is kept constant. This analysis also gives the width of the resonance, $8 meV$.

FUTURE PROSPECTS

Much remains to be done. The use of laser excitation of metastable negative ions, briefly mentioned above, for high resolution studies has only just begun. Greatly increased target densities will be achieved through the trapping or storing of target atoms or ions by optical or electromagnetic means, particularly when combined with cooling techniques. Considerable improvements in electron guns can also be expected. Photoelectron sources, used mostly until now for electron-molecule studies, are an obvious possibilty for this, and extensions to shorter wavelengths or muliphoton ionization are intriguing prospects. The resolution of electron energy analyzers is also surely capable of improvement, to approach the values obtained in surface studies. These technical improvements will result in a wealth of new information on resonance states of negative ions and neutral atoms. This in turn will greatly assist in the important but difficult theoretical task of modelling states that are highly excited and highly correlated, the basis of which has been so firmly laid by Philip Burke.

ACKNOWLEDGEMENT

The author is grateful to Prof. Burke for the willing help and guidance given over many years.

REFERENCES

1. P. G. Burke and H. M. Schey, *Phys. Rev.* 126, 147 (1962).
2. K. Smith, R. P. McEachran and P. A. Fraser, *Phys. Rev.* 125, 553 (1962).
3. G. J. Schulz, *Phys. Rev. Lett.* 10, 104 (1963).
4. R. J. Fleming and G. S. Higginson, *Proc. Phys Soc.* 81, 974 (1963).
5. S. M. Silverman and E. N. Lassettre, *J. Chem. Phys.* 40, 1265 (1964).
6. U. Fano, *Phys Rev.* 124, 1866 (1961).
7. S. Cvejanovic, J. Comer and F. H, Read, *J. Phys B* 7, 468 (1974).
8. B. R. Odgers, M. P. Scott and P. G. Burke, *J. Phys B* 29, 4320 (1996).
9. E. T. Hudson, K. Bartschat, M. P. Scott, P. G. Burke and V. M. Burke, *J. Phys B* 29, 5513 (1996).

10. A. E. Klinkmuller, G. Haeffler, D. Hanstorp, I. Y. Kiyan, U. Berzinsh and D. J. Pegg, *J. Phys B* 31, 2549 (1998).
11. S. J. Brotton, S Cvejanovic, F. J. Currell, N. J. Bowring and F. H. Read, *Phys. Rev. A* 55, 318 (1997).
12. F. H. Read and J. M. Channing, *Rev. Sci. Instrum.* 67, 2372 (1996).
13. M. Zubek, N. Gulley, G. C. King and F. H. Read, *J. Phys B* 29, L239 (1996).
14. M. Zubek, B. Mielewska, J. Channing, G. C. King and F. H. Read, to be published.
15. A. Weingartshofer, K. Willmann and E. M. Clarke, *J. Phys B* 7, 79 (1974).
16. P. Hammond, *J. Phys B* 29, L231 (1996).
17. J. E. Sienkiewicz and W. E. Baylis, *J. Phys B* 24, 1739 (1991).

RECENT PROGRESS IN ELECTRON-ATOM SCATTERING

Igor Bray

Physics Department, Flinders University, GPO Box 2100, Adelaide 5001, Australia

INTRODUCTION

It is a great pleasure for me to make a contribution to this volume in honour of Prof. P. G. Burke. His visit to Adelaide, Australia in the mid 1980s was pivotal in my changing fields from gravitation lenses to atomic collision theory. Though the gravitational lens field boomed soon after my departure I have no regrets having been a part of the recent substantial progress in the field of electron-atom collision theory.

We entered this decade with some uncertainty. Experimental evidence suggested that even the most sophisticated electron-atom scattering theories were unable to correctly describe 2P excitation in the most fundamental electron-hydrogen atom scattering problem. Our own work was very much in response to this single issue. In contrast, we leave this decade with great certainty in our ability to calculate even the more complicated excitation processes for a large range of atoms, including hydrogen, at all energies. Furthermore, even electron-impact ionization (e,2e) is able to be described relatively accurately at all energies and excess energy sharing.

The primary reason for the computational progress has to be the incredible growth in the power of modern computers. The speed with which arithmetic operations proceed has increased by nearly two orders of magnitude during the last decade. The price of random access memory (RAM) has dropped immensely. At the beginning of the decade we performed our calculations on 16-megabyte machines. Currently we use 2-gigabyte machines routinely. Furthermore, there has been a constant progress in the improvement of compilers and the multiprocessing software and hardware. Together, these advances have made possible what so recently was even beyond contemplation. The greater computational resources allow not only for speedier throughput, but also to ask questions that previously were not able to be answered. We will see that this has happened in our case with some surprising results.

The structure of this paper is as follows. We first review some of the recent progress in the field demonstrating the success of the close-coupling approach to the calculation of electron-atom scattering. Then we give some brief detail of the close-coupling theory and show how it may be extended to calculate ionization processes. This will be followed by an example of a detailed study of in- and out-of-plane ionization of the ground state of helium by 44.6 eV electrons with the two outgoing electrons sharing the excess 20 eV energy equally.

RECENT PROGRESS IN ELECTRON-ATOM SCATTERING

Electron-Hydrogen Scattering

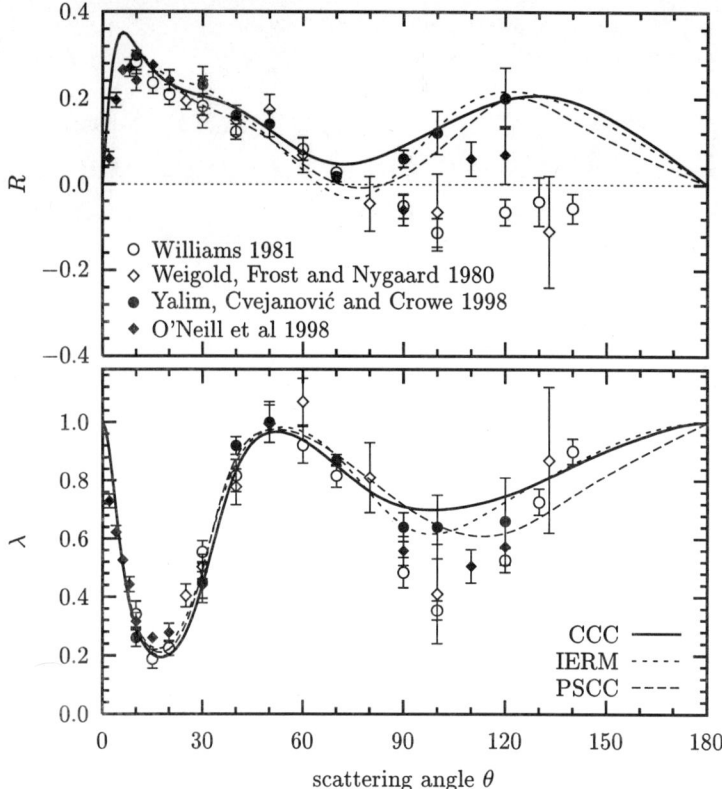

Figure 1. Electron-impact 2P excitation of atomic hydrogen angular correlation parameters R and λ. The measurements are due to Weigold et al.[1], Williams[2], Yalim et al.[3] and O'Neill et al.[4]. The CCC calculation is due to Bray and Stelbovics[5]. The pseudostate close-coupling (PSCC) calculation is due to van Wyngaarden and Walters[6]. The presented intermediate-energy R-matrix (IERM) calculation was obtained using the $L \leq 4$ partial-wave results of Scholz et al.[7] and topped up by the CCC higher partial waves (see text).

To our mind the most important problem in the field has been the discrepancy between experiment and theory for the 2P electron-impact excitation of atomic hydrogen at 54.4 eV. Immense theoretical effort has been put towards this problem without resolution. So much so that many theorists began to doubt the accuracy of the two experiments[1, 2], which showed good consistency with each other. Theoretical success for other more complicated systems gave renewed motivation to revisit this problem experimentally. New measurements were recently performed and found to be in much better agreement with the more reliable theories. The present status of this problem is illustrated in figure 1. We see that the measurements of Yalim et al.[3] are in excellent agreement with the convergent close-coupling (CCC) theory[5], the pseudostate close-coupling (PSCC) approach[6] and the intermediate-energy R-matrix (IERM) method[7]. Note that the originally presented IERM results were combined with higher partial waves from a pseudostate calculation. These results showed some unexpected oscillation. For this reason we took the published IERM matrix elements and incorporated higher partial waves using the CCC theory[5]. The curves denoted by IERM in figure 1 have been obtained this way. Our conclusion arising from the figure is that the primary difficulty in obtaining accurate e-H angular correlation parameters at large scattering angles is more of a problem for experiment than theory. The experimental difficulties have been addressed recently by Williams[8], who also performed new measurements, but at lower energies where the 2P cross section is not as low at the backward angles as it is at 54.4 eV. Good agreement with theory was found.

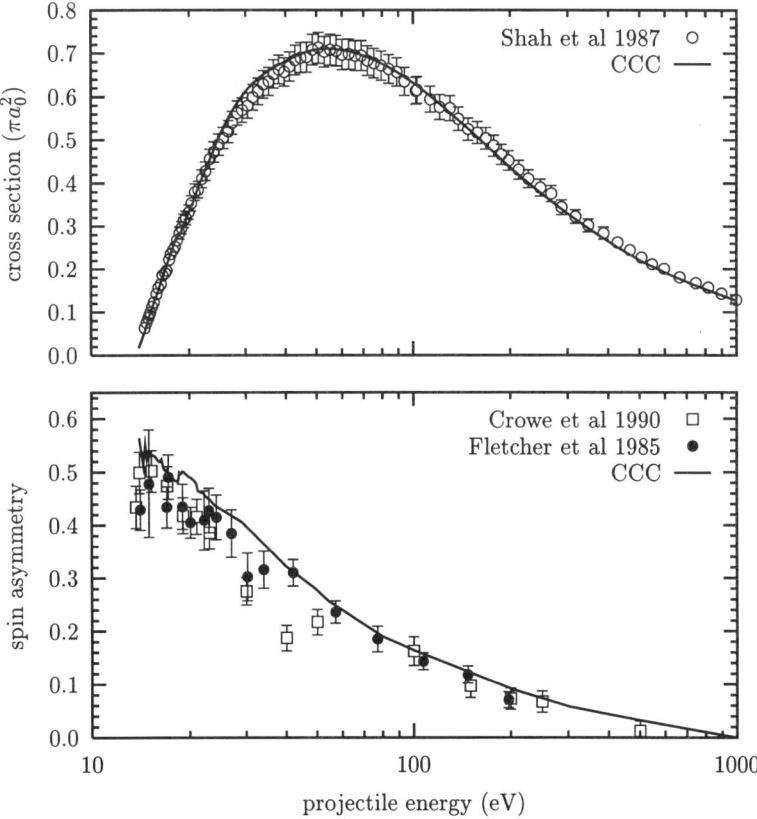

Figure 2. Electron-impact ionization of atomic hydrogen cross sections and spin asymmetry. The measurements are due to Crowe et al.[9] and Fletcher et al.[10]. The CCC calculations are due to Bray and Stelbovics[11].

The CCC theory was developed in response to the abovementioned discrepancies. It was to our great disappointment, at that time, that it was unable to resolve this discrepancy. However, other successes soon became clear. The first of these was the reproduction of the highly accurate e-H total ionization cross section[11]. It was the first ab initio theory to do so and gave the first hint that the close-coupling formalism may also provide for an accurate treatment of electron-impact ionization at all energies. In figure 2 we show this result in addition with the associated spin asymmetry. Since this demonstration other close-coupling-based theories also showed the ability to do this. These include the hyper-spherical method of Kato and Watanabe[12], the R-matrix with pseudostates (RMPS) and IERM methods[13].

Electron-Lithium Scattering

We next progress to the lithium atom. Though this atom has three electrons the frozen-core Hartree-Fock approximation yields a very accurate description of the atomic wavefunctions. Thus, the valence electron effectively moves in the fixed potential formed by the two 1s electrons. In this model the electron-lithium scattering problem looks very much like electron-hydrogen scattering. The only difference is that the frozen-core Hartree-Fock potential replaces the simple Coulomb core potential of hydrogen. Therefore, the calculation of electron-lithium scattering is only marginally more difficult than the electron-hydrogen case.

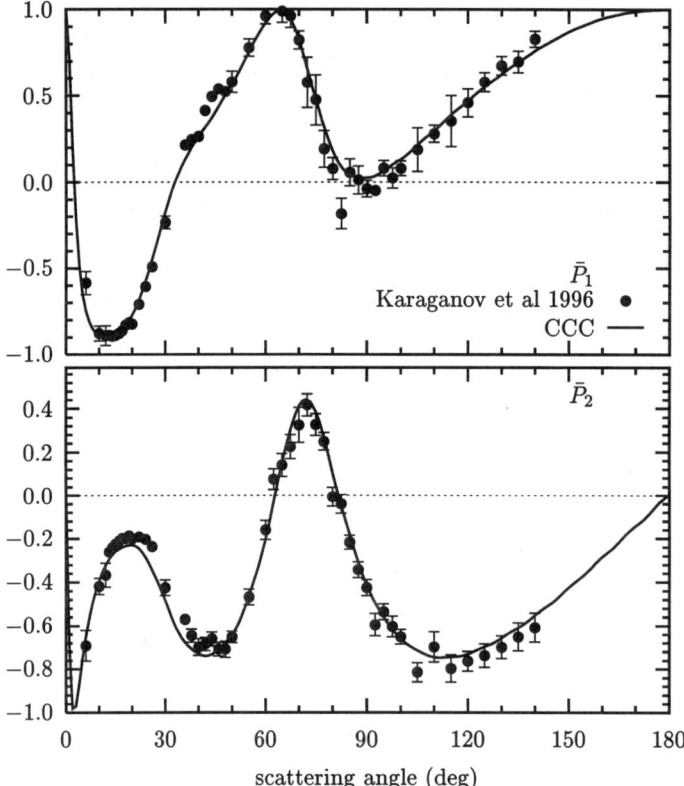

Figure 3. Electron-impact 2P excitation of lithium reduced Stokes parameters at 22 eV. The measurements and calculations are from Karaganov et al.[14].

For experiment, on the other hand, lithium has some advantages over atomic hydrogen. Atomic beams are readily produced and, most importantly, the highly accurate super-elastic scattering technique may be applied to obtain estimates of the angular correlation parameters. Whereas as for atomic hydrogen a coincidence experiment is performed involving simultaneous detection of the outgoing electron and the 2p photon, in lithium the 2P state may be optically pumped and the deexcitation to the ground 2S state measured as a function of the polarization of the laser light and electron scattering angle. Such measurements typically determine three reduced Stokes parameters $\bar{P}_n$. They are trivially related to the angular correlation parameters, e.g. $\bar{P}_1 = 2\lambda - 1$ and $\bar{P}_2 = -2\sqrt{2}R$.

In figure 3 we look at the comparison of the measurements and the CCC theory in the case of 22 eV energy relative to the ground state. We see spectacular agreement between theory and experiment. The choice of energy was such as to make the e-Li system as similar as possible to the problematic e-H case with the incident energy being approximately four times the ionization threshold. Karaganov et al.[14] argued that the agreement for the above-mentioned e-H case should be as good as for the presented e-Li case. Both Yalim et al.[3] and O'Neill et al.[4] used the presented e-Li results as motivation to revisit the e-H problem.

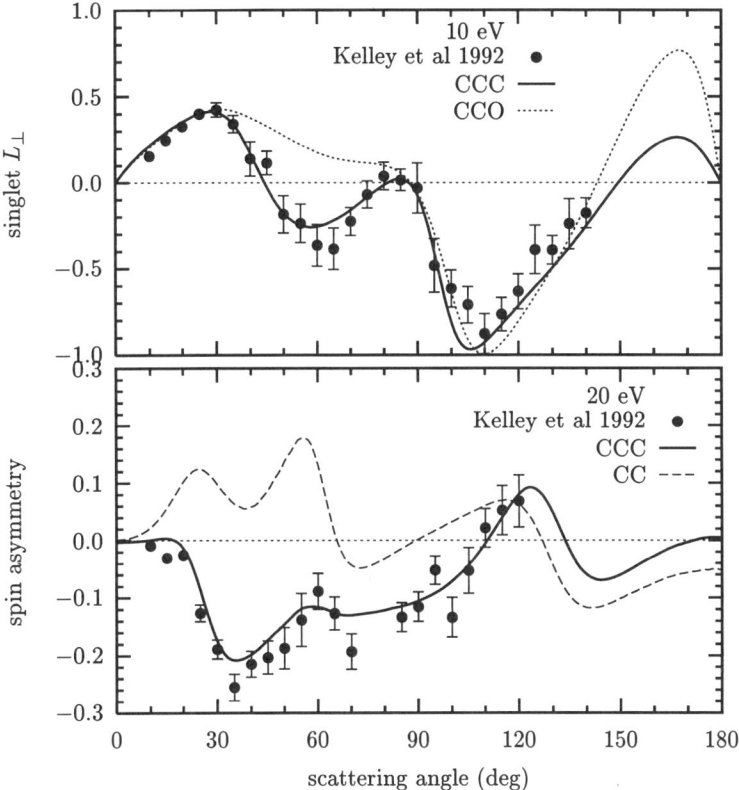

Figure 4. Spin-resolved electron-impact 3P excitation of sodium The measurements are due to Kelley et al.[15] and the calculations are from Bray[16].

Some of the best experiments performed in the electron-atom scattering field, in our view, were at NIST where spin-polarized electrons were incident on spin-polarized sodium atoms in the ground or the 3P state, see Kelley et al.[15] and references therein. The technique allowed for spin-resolved scattering observables to be obtained with high accuracy on a broad range of energies. Apart from having intrinsic scientific merit, collectively these measurements proved an excellent test of theory. In figure 4 we look at two cases. At 10 eV the singlet $L_\perp$ is best reproduced by the CCC theory. At that time the coupled-channels optical (CCO) method[17] yielded best agreement with experiment, with the presented case showing the poorest agreement. The main approximation in the CCO theory is that of weak coupling in the continuum. The good agreement of the CCC theory with the experiment indicates that strong coupling in the continuum must be allowed for when calculating the sodium 3P excitation. This is a very much unexpected result. The fact that the CCC theory does so well here indicates that it does treat the ionization channels very accurately and was motivation for application of the CCC method directly to ionization. The 20 eV part of the figure shows the stark contrast between a close-coupling (CC) calculation that ignores the continuum and a CCC one which does not. The sign of the spin asymmetry is opposite for the two calculations over a considerable part of the angular range. The reason for the substantial difference is that the CC model does not allow for any electron flux to the continuum which in turn affects the calculation of the 3P amplitudes[18].

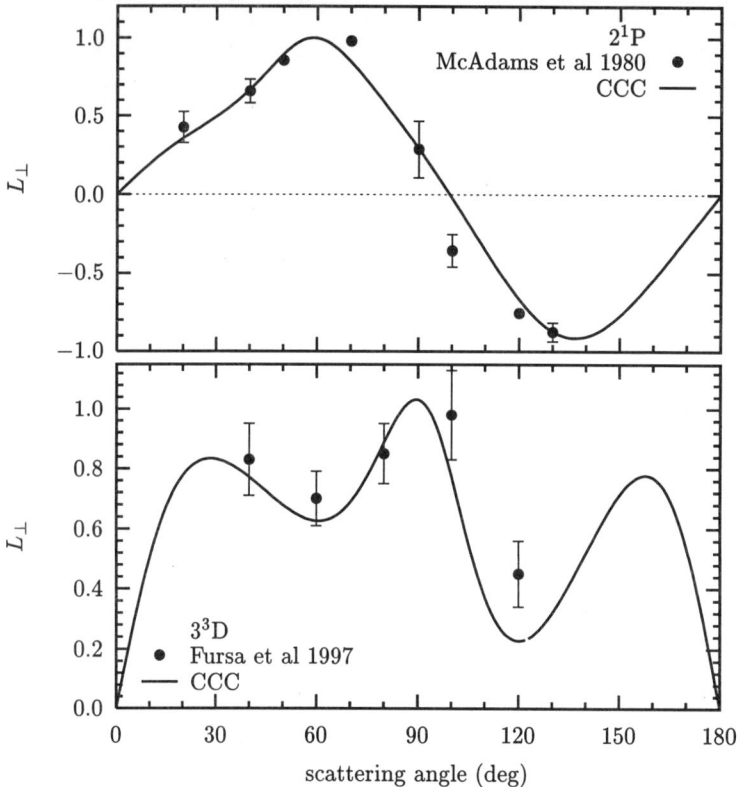

Figure 5. Electron-impact excitation of helium. The 2^1P measurements and calculations are due to McAdams et al.[19] and Fursa and Bray[20], respectively. The 3^3D measurements and calculations have been reported by Fursa et al.[21].

Thus far we have discussed atomic targets that are readily treated by the frozen-core Hartree-Fock approximation. These then become hydrogen-like and the corresponding CCC theory has been given by Bray[22]. The next step was to extend the CCC theory to helium and helium-like targets. This was done by Fursa and Bray[20]. In figure 5 we give just a couple of examples showing that the CCC theory works as well for helium excitation involving a simple state like 2^1P or a more complicated one like 3^3D. The many applications of the CCC theory to the excitation of the ground state of helium show good agreement with experiment[20] and subsequent RMPS calculations[23]. Not so in the case of scattering from the metastable states[24]. This remains an unresolved problem with the recent RMPS calculations of Bartschat[25] in support of the CCC theory.

With overwhelming success of the CCC theory for the lighter targets it was extended to incorporate helium-like targets[26], light or heavy, and applied to e-Ba excitation[27]. The incorporation of the heavy targets was done by the use of phenomenological one- and two-electron core-polarization potentials while still remaining within the L-S coupling scheme. In figure 6 we give an example which shows that the non-relativistic CCC method still works well at least for the spin-preserving transitions, and better than an implementation of a relativistic distorted-wave (RDW) approximation[31].

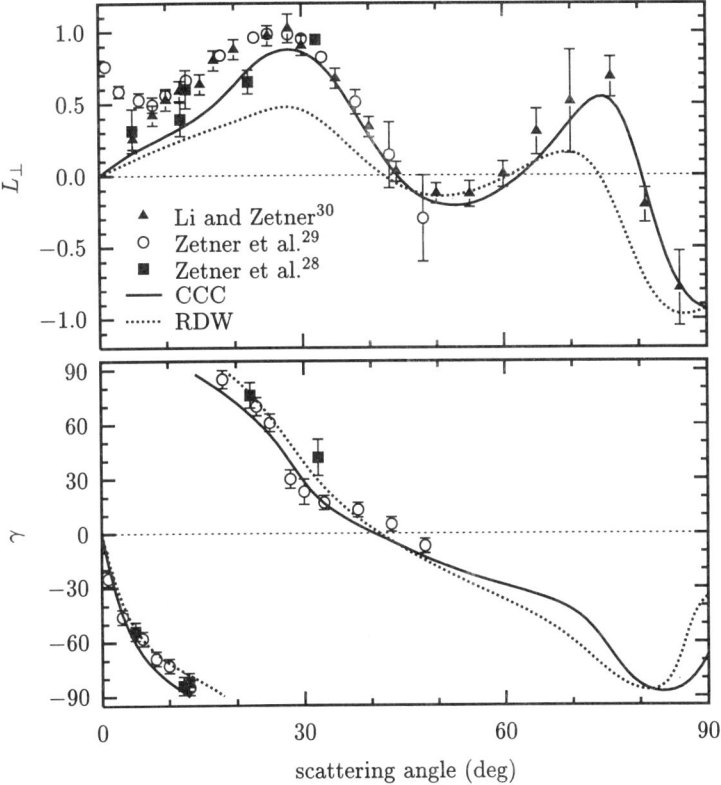

Figure 6. Electron-impact 6^1P excitation of barium. The CCC calculations are due to Fursa and Bray[27]. The relativistic distorted-wave approximation (RDWA) calculations are due to Srivastava et al.[31].

Close-Coupling Theory

The close-coupling theory began with Massey and Mohr[32], and presently has many numerical implementations such as the R-matrix approach of Burke and Robb[33], the variational approach of Callaway[34], the intermediate-energy R-matrix method of Burke et al.[35], the eigenchannel R-matrix method of Greene[36], the R-matrix method with pseudostates (RMPS) of Bartschat et al.[37] and Badnell and Gorczyca[38], the pseudostate approach of van Wyngaarden and Walters[6], the hyper-spherical close-coupling method of Watanabe et al.[39], the CCC method of Bray and Stelbovics[5], and others. We will concentrate on the CCC implementation, but the ideas should be applicable to all close-coupling approaches.

In the close-coupling approximation the total electron-atom scattering wave function is written as an antisymmetric expansion using N square-integrable states $\phi_n^{(N)}$

$$|\Psi_i^{(+)}\rangle = \mathcal{A}|\psi_i^{(+)}\rangle \approx \mathcal{A}|\psi_i^{(N)(+)}\rangle = \mathcal{A}\sum_{n=1}^{N}|\phi_n^{(N)} F_{ni}^{(N)}\rangle, \qquad (1)$$

where the unknown one-electron functions $|F_{ni}^{(N)}\rangle = \langle\phi_n^{(N)}|\psi_i^{(+)}\rangle$, $\mathcal{A}$ is the antisymmetrization operator, and typically the states $\phi_n^{(N)}$ are obtained by diagonalising the target Hamiltonian H_T in some basis

$$\langle\phi_f^{(N)}|H_T|\phi_i^{(N)}\rangle = \epsilon_f^{(N)}\langle\phi_f^{(N)}|\phi_i^{(N)}\rangle = \epsilon_f^{(N)}\delta_{fi}. \qquad (2)$$

The electron-atom scattering boundary conditions are simply $|\phi_n^{(N)}\mathbf{k}_n\rangle$, where $|\mathbf{k}_n\rangle$ is a plane (for a neutral target) wave of energy (a.u.) $k^2/2$.

In the CCC theory[5] we use a truncated Laguerre basis thereby ensuring that

$$\lim_{N\to\infty}\sum_{n=1}^{N}|\phi_n^{(N)}\rangle\langle\phi_n^{(N)}| = \lim_{N\to\infty} I^{(N)} = I, \qquad (3)$$

where I is the target-space identity operator. The close-coupling equations, at total energy E, are formed in momentum space for the K matrix

$$\langle \mathbf{k}_f\phi_f^{(N)}|K|\phi_i^{(N)}\mathbf{k}_i\rangle = \langle \mathbf{k}_f\phi_f^{(N)}|V|\psi_i^{(N)}\rangle, \qquad (4)$$

where V is the effective potential (incorporates $\mathcal{A}$). The K matrix is evaluated via

$$\langle \mathbf{k}_f\phi_f^{(N)}|K|\phi_i^{(N)}\mathbf{k}_i\rangle = \langle \mathbf{k}_f\phi_f^{(N)}|V|\phi_i^{(N)}\mathbf{k}_i\rangle$$
$$+ \sum_{n=1}^{N}\mathcal{P}\int d^3k \frac{\langle \mathbf{k}_f\phi_f^{(N)}|V|\phi_n^{(N)}\mathbf{k}\rangle\langle \mathbf{k}\phi_n^{(N)}|K|\phi_i^{(N)}\mathbf{k}_i\rangle}{E - \epsilon_n^{(N)} - k^2/2}. \qquad (5)$$

To obtain the physical T matrix involves a simple set of linear equations

$$\langle \mathbf{k}_f\phi_f^{(N)}|K|\phi_i^{(N)}\mathbf{k}_i\rangle = \sum_{n:\epsilon_n^{(N)}\leq E}\langle \mathbf{k}_f\phi_f^{(N)}|T|\phi_n^{(N)}\mathbf{k}_n\rangle$$
$$\times\left(\delta_{ni} + i\pi k_n\langle \mathbf{k}_n\phi_n^{(N)}|K|\phi_i^{(N)}\mathbf{k}_i\rangle\right). \qquad (6)$$

In principle, it should not matter which numerical implementation of the close-coupling theory is chosen. If the same states $\phi_n^{(N)}$ are taken then different numerical techniques should yield the same T-matrix elements.

In the case of discrete excitation with $\epsilon_f^{(N)} = \epsilon_f < 0$ we use the calculated T matrix to define the scattering amplitude

$$f_{fi}^{(N)} = \langle \mathbf{k}_f\phi_f^{(N)}|T|\phi_i^{(N)}\mathbf{k}_i\rangle. \qquad (7)$$

For positive-energy states we multiply (7) by the overlap of $|\phi_f^{(N)}\rangle$ and $\langle \mathbf{q}_f^{(-)}|$ the true target continuum wave of energy $q_f^2/2 = \epsilon_f^{(N)}$

$$f_{fi}^{(N)}(\mathbf{q}_f) = \langle \mathbf{q}_f^{(-)}|\phi_f^{(N)}\rangle\langle \mathbf{k}_f\phi_f^{(N)}|T|\phi_i^{(N)}\mathbf{k}_i\rangle. \qquad (8)$$

This has the effect of transforming the unity normalisation of the states $|\phi_f^{(N)}\rangle$ to that of the continuum, and introduces the Coulomb one-electron phase. We use (8) to define fully differential ionization cross sections[40].

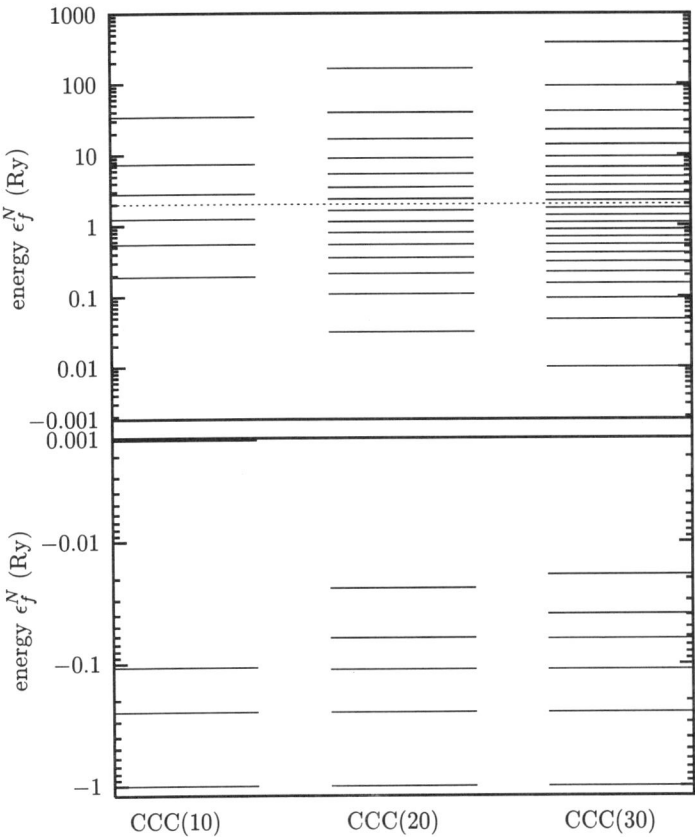

Figure 7. Energy levels arising in the CCC(10), CCC(20) and CCC(30) calculations of the Temkin-Poet (S-wave) model of e-H scattering. The dotted line indicates the total energy $E=2$ Ry.

Now, in a series of publications Heller, Yamani, Reinhardt and Broad[41, 42, 43] showed that the summation in (3) forms a quadrature rule for both the infinite sum over the discrete bound states and the integration over the true target continuum, that comprise the identity operator I. This summation is evident in (5), but in (6) it is truncated to include only those states $|\phi_n^{(N)}\rangle$ whose energy $\epsilon_n^{(N)} \leq E$ (open channels). What then does this truncated sum represent? At energies below the ionization threshold ($E < 0$) we simply have a finite sum over discrete eigenstates (assuming large enough N). For $E > 0$ the sum represents the infinite sum over the true discrete eigenstates and an integral over the true continuum energy range $[0, E]$. However, the quadrature rule defined by the solution of (2) knows nothing about E and so there is no guarantee that the truncation in (6) is consistent with the quadrature rule. If, however, we know the endpoint of integration for each n then, given a single fixed E, we can adjust the basis parameters such that the last $\epsilon_n^{(N)} < E$ had E as the endpoint of the integration. This has ramifications for the RMPS approaches[37, 38] since the R-matrix method is used to calculate the T matrix on a fine energy range of E though a fixed set of states $|\phi_n^{(N)}\rangle$ is used. Using the Laguerre basis the endpoints of integration for each $\epsilon_n^{(N)} > 0$

with $n < N$ happen to be $(\epsilon_n^{(N)} + \epsilon_{n+1}^{(N)})/2$. For a long time it has been known that if an energy $\epsilon_n^{(N)} > 0$ is near E then a large pseudoresonance may appear in the T matrix. In the CCC method we have a formal requirement that E should always be inbetween $\epsilon_n^{(N)}$ and $\epsilon_{n+1}^{(N)}$ for some n. A detailed numerical discussion of these issues has been given by Bray and Clare[44].

We use the quadrature rule notion in the CCC method and so expect that the error in the calculated scattering amplitudes decreases with increasing N. Let us consider a simple numerical example at a total energy $E = 2$ Ry of the Temkin-Poet model e-H scattering problem, where only states of zero orbital angular momentum are included[45, 46]. We shall consider calculations for $N = 10$, 20 and 30. In figure 7 we present the three sets of energy levels. We see that with increasing N the negative energies converge to the true eigenvalues, while the positive energies simply yield a more dense discretization of the continuum. The Laguerre basis parameter $\lambda \approx 2$ has been varied to ensure that E was inbetween two consecutive $\epsilon_n^{(N)}$ for some n.

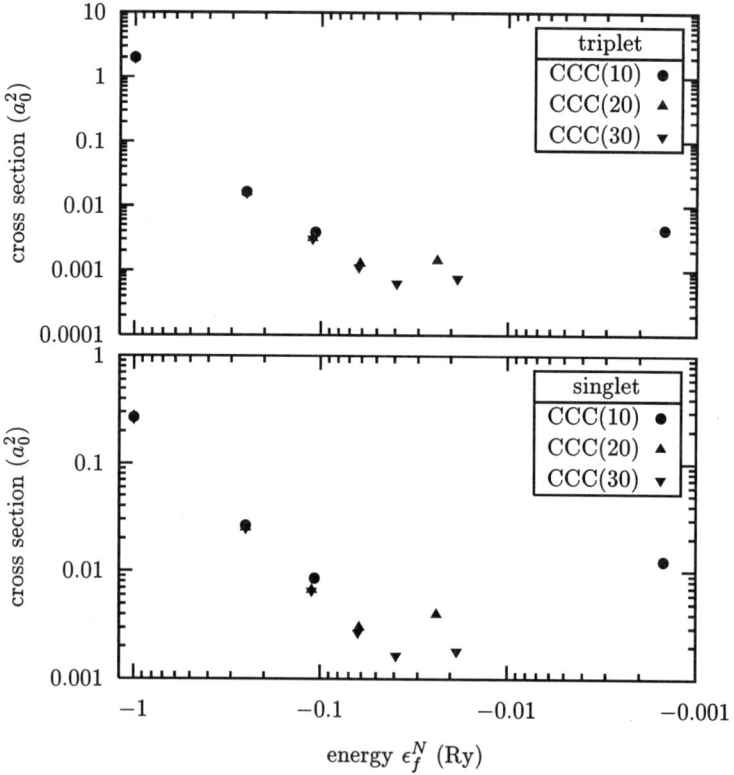

Figure 8. Electron-impact cross sections for excitation of the negative-energy states arising in the CCC(N) calculations of the Temkin-Poet (S-wave) model of e-H scattering at $E=2$ Ry using the indicated values of N.

In figure 8 we give the singlet and triplet cross sections ($|f_{n1}|^2$) for the negative-energy states. We see that for all three calculations there is good convergence in the elastic and the $n = 2$ excitation cross sections, with the larger calculations being necessary for accurate larger n results. Note also how the last cross section for each N is higher than the previous. This is an indication of how the last points are attempting to take into account the infinite true discrete series, and not a violation of the expected $1/n^3$ scaling of the cross sections.

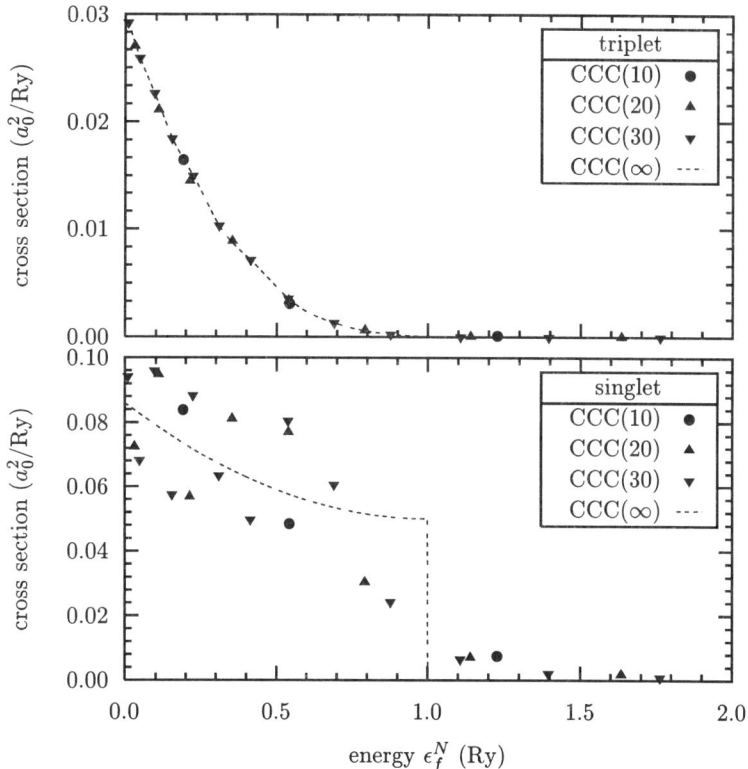

Figure 9. Electron-impact cross sections for excitation of the positive-energy states arising in the CCC(N) calculations of the Temkin-Poet (S-wave) model of e-H scattering at $E=2$ Ry using the indicated values of N.

We next look at the cross sections ($|f_{n1}(q_n)|^2$) for the excitation of the positive-energy states, given in figure 9. These define the singly differential cross section (SDCS). In the triplet case we find excellent convergence since all points lie on the same curve denoted by CCC(∞). The cross section on the energy range $[E/2, E]$ is effectively zero. At first glance this is surprising because the physics on either side of $E/2$ should be the same. However, the close-coupling theory, which has antisymmetry of the total wavefunction fully implemented, distinguishes between the two energy ranges on either side of $E/2$ due to the choice of boundary conditions. In both cases only the true continuum electron is allowed to escape to true infinity. On the energy range $[E/2, E]$ this leads to the unphysical case of a slow "continuum" electron being shielded completely from the residual ion by a fast "L^2" electron. The presented numerical result is in fact ideal since the unitary close-coupling theory cannot have double counting problems and should describe all possible physics with the energy range of the "L^2" electron being restricted to $[0, E/2]$. Turning to the singlet case, we see that the situation is much less satisfactory. There is no convergence in the region $[0, E/2]$ and the cross sections on $[E/2, E]$ are non-zero, though small. In fact, we suspect that any finite N calculation will not yield the correct result as this should be a step-function[47], which we guess to be

25

the curve denoted by CCC(∞). The suggestion is that the scattering amplitude for the excitation of the state whose energy $\epsilon_f^{(N)} > k_f^2/2$ converges to zero with increasing N, irrespective of total spin S or energy E. This idea also explains why the RMPS techniques work reasonably well for positive E. The step-function idea says that it should not matter whether there is a quadrature point $\epsilon_f^{(N)}$ near E since the integrand at this point should be zero. In practice, pseudoresonances rapidly diminish with increasing N, due to the smallness of the cross sections on the $[E/2, E]$ energy range.

Equal energy-sharing 44.6 eV electron-impact ionization of helium

Faced with the results of figure 9 how do we interpret them? For either the singlet or triplet case we have two independent estimates of ionization on either side of $E/2$. Since in forming the total ionization cross section we simply sum the excitation cross sections for positive-energy states, equivalent to performing integration from 0 to E in figure 9, we must combine these as cross sections. To see how this works in a realistic calculation we present the results of two 149-state 44.6 eV e-He CCC calculations. The first we label by the pair (17,4) indicating that the maximum target orbital angular momentum $l_{max} = 4$ and the number of states for each l is $N_l = 17 - l$ (have only 16 ^{3}S states). The other is labeled by (15,5) indicating $l_{max} = 5$ and $N_l = 15 - l$. The results of the (17,4) calculation have been recently presented by Rioual et al.[48].

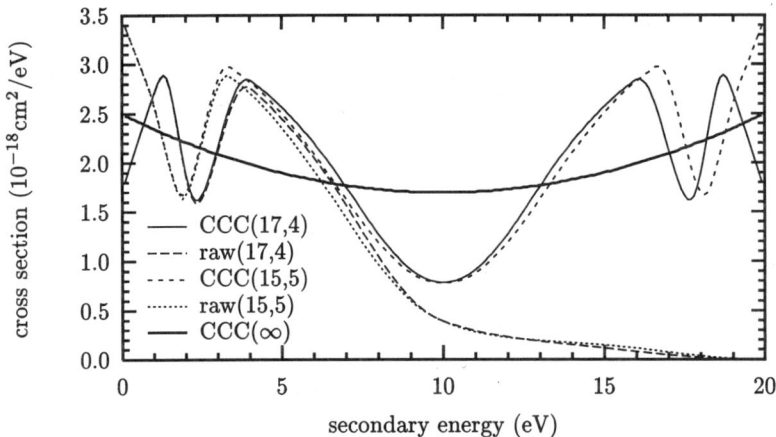

Figure 10. The singly differential cross sections arising in the two 149-state (see text) calculations.

In figure 10 we give the results for the SDCS obtained by summing the "raw" results on either side of $E/2$. The integral of the CCC results from 0 to $E/2$ is the same as the integral of the "raw" results from 0 to E, and is in excellent agreement with experiment[49]. However, there is no convergence in the SDCS with oscillation being most evident for the asymmetric energy sharing. The curve given by CCC(∞) is an estimate of what the true SDCS should be[48]. We make this estimate so that we may rescale any CCC-calculated ionization angular distribution to yield, upon integration over the angular variables, the SDCS result indicated by CCC(∞). For this reason the following CCC angular distributions, for two 10 eV outgoing electrons, have been multiplied by 2.2.

The first geometry we consider, presented in figure 11, is the "fixed θ_A" where one detector is fixed at angle θ_A and the other scanned from $-180°$ to $180°$. We see good convergence in the angular distributions and good agreement with experiment after appropriate rescaling. The choice of 0.8 for the experiment was from best overall fit to all of the data presented by Rioual et al.[48]. If the CCC theory is able to reproduce all of the possible in- and out-of-plane angular distributions, after multiplication by a single factor, then the rescaling of the experiment is made consistent with the estimate of the true SDCS given in figure 10. Note, the experimental normalisation of Rioual et al.[48] is ±25%.

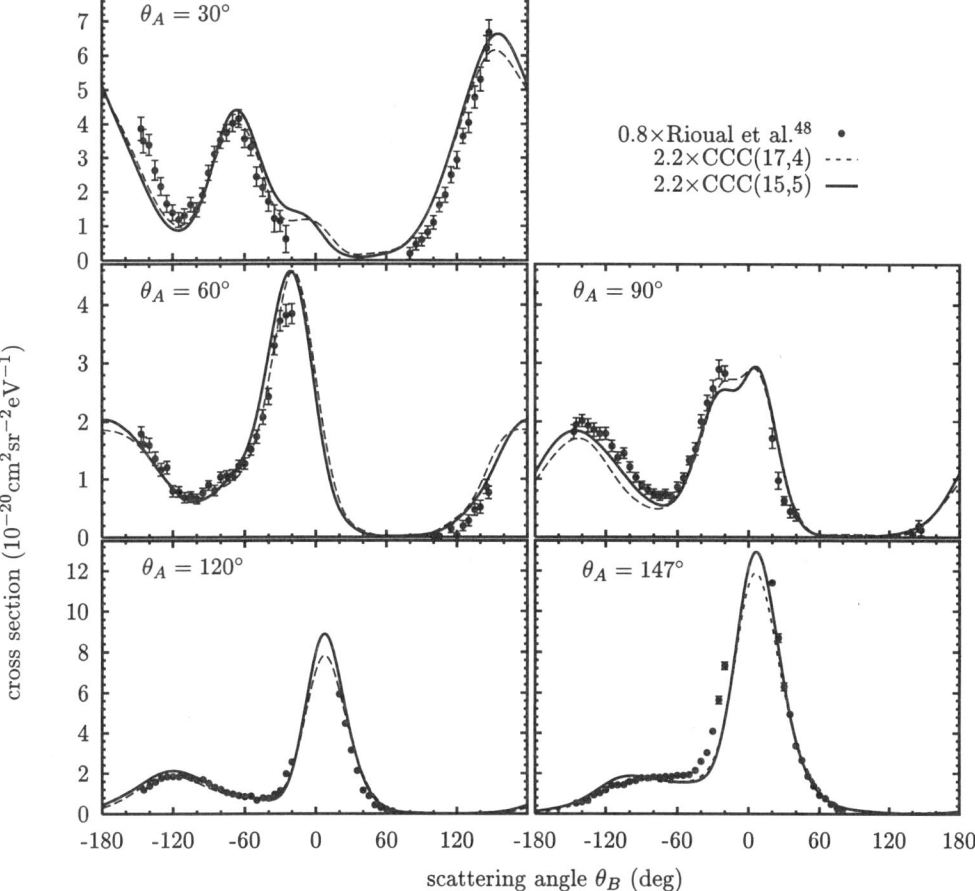

Figure 11. Electron-impact ionization of helium triply differential cross sections for 44.6 eV incident energy with 10 eV outgoing electrons for the coplanar fixed θ_A geometry. The experiment and the 149-state CCC(17,4) calculation (see text) have been presented by Rioual et al.[48]. The CCC(15,5) (see text) calculation has not been previously published.

Some shoulder-like structure is visible around $-20°$ for $\theta_A = 30°$ and $\theta_A = 90°$. This is a convergence problem and is not likely to be physical. It may also be coming from the fact that in the case of helium we are unable to ensure that for all target symmetries, considered in the close-coupling expansion, there was always some pseudostate of exactly 10 eV energy. As a result some interpolation is required of the scattering amplitudes resulting in some uncertainty[40].

In figure 12 we consider the "fixed θ_{AB}" geometry where the angle between the two detectors is kept fixed. Again, good agreement is found in the angular distributions and magnitudes after rescaling.

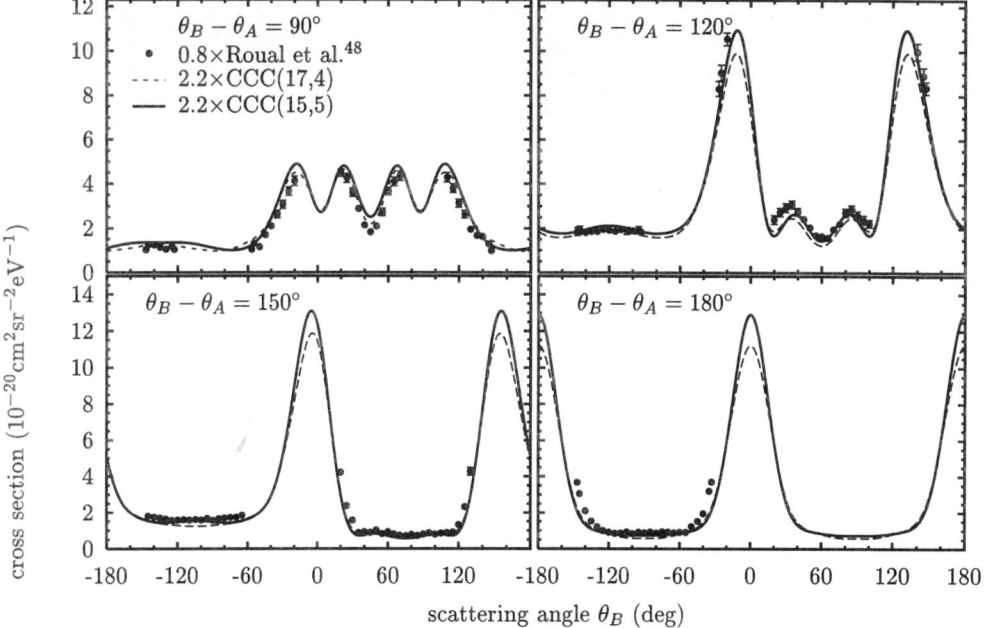

Figure 12. Electron-impact ionization of helium triply differential cross sections for 44.6 eV incident energy with 10 eV outgoing electrons for the coplanar fixed $\theta_B - \theta_A$ geometry. The theory and experiment are as for figure 11.

Lastly, in figure 13, we consider the symmetric out-of-plane geometry where the two outgoing electrons are measured on either side of the z-axis with the incident electron being at angle ψ to the plane formed by the two outgoing electrons. At $\psi = 0$ we have the standard symmetric coplanar geometry. For the out-of-plane geometry we have the earlier measurements of Murray et al.[50] with both sets available for $\psi = 0$. We see now that there is substantial difference between the two calculations with the (15,5) one yielding better agreement with experiment. This indicates the difficulty in obtaining results which have converged to a high accuracy, but also indicates that the CCC formalism is able to yield correct ionization angular distributions if more and more states are taken.

Generally, the smaller the cross sections the bigger are the required calculations. The out-of-plane cross sections are of substantially smaller magnitude than the corresponding coplanar ones. Both of the presented calculations require immense, by even today's standards, computational resources. The core memory requirements were approximately 1.5G and 2G for the CCC(17,4) and CCC(15,5) calculations, respectively. The latter requires more memory because the larger l_{max} leads to more channels and hence bigger matrices.

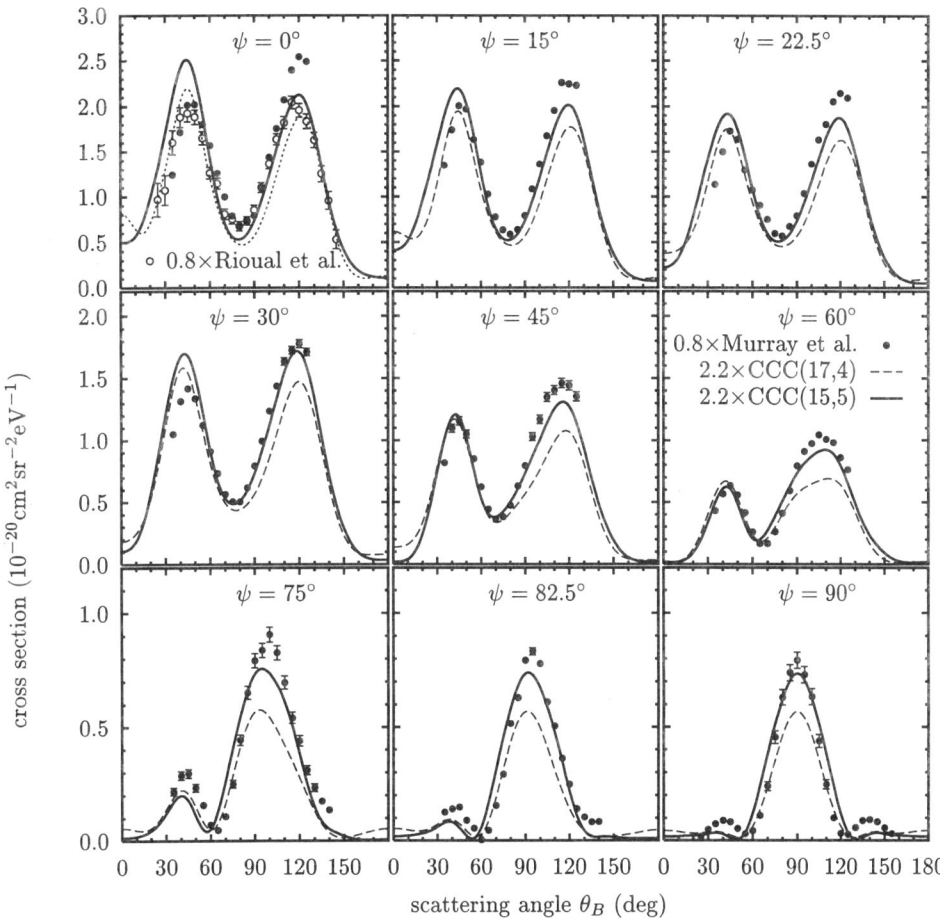

Figure 13. Electron-impact ionization of helium triply differential cross sections for 44.6 eV incident energy with 10 eV outgoing electrons for the symmetric out-of-plane geometry. The theory is as for figure 11. The experiment is due to Murray et al.[50].

Conclusions

We have briefly glanced at some of the progress made during this decade in the field of electron-atom scattering and ionization calculations. To our mind the close-coupling formalism is adequate in obtaining highly accurate scattering amplitudes for the discrete transitions at all energies. Though the theory was designed specifically for such transitions it may be readily applied to ionization processes with substantial, but not complete success. We suspect that the close-coupling boundary conditions result in the SDCS converging to a step-function with increasing N. For finite calculations the CCC(N) results may be rescaled to the true SDCS, if known, yielding accurate magnitudes and angular distributions.

The future for the close-coupling formalism is very bright. With recent experimental predictions confirming the accuracy of the theory the foundations are very solid and considerable effort may be applied with confidence to extend the formalism to incorporate ever more complicated targets of value to science and industry.

REFERENCES

1. E. Weigold, L. Frost, and K.J. Nygaard, Large-angle electron-photon coincidence experiment in atomic hydrogen, *Phys.Rev.A* 21:1950–1954 (1980).
2. J.F. Williams, Electron-photon angular correlations from the electron impact excitation of the 2s and 2p electronic configurations of atomic hydrogen, *J.Phys.B* 14:1197–1217 (1981).
3. H. Yalim, D. Cvejanovic, and A. Crowe, 1s-2p excitation of atomic hydrogen by electron impact studied using the angular correlation technique, *Phys.Rev.Lett.* 79:2951–2954 (1997); private communication (1998).
4. R.W. O'Neill, P.J.M. van der Burgt, D. Dziczek, P. Bowe, S. Chwirot, and J.A. Slevin, Polarization correlation measurements of electron impact excitation of H(2p) at 54.4 eV, *Phys.Rev.Lett.* 80:1630–1633 (1998).
5. I. Bray and A.T. Stelbovics, Convergent close-coupling calculations of electron-hydrogen scattering, *Phys.Rev.A* 46:6995–7011 (1992).
6. W.L. van Wyngaarden and H.R.J. Walters, Elastic scattering and excitation of the 1s→2s and 1s→2p transitions in atomic hydrogen by electrons at medium to high energies, *J.Phys.B* 19:929–968 (1986).
7. T.T. Scholz, H.R.J. Walters, P.G. Burke, and M.P. Scott, Electron scattering by atomic hydrogen at intermediate energies: II. Differential elastic, 1s-2s and 1s-2p cross sections and 1s-2p electron-photon coincidence parameters, *J.Phys.B* 24:2097–2126 (1991).
8. J.F. Williams, New and old measurements of electron scattering in atomic hydrogen, *Aust.J.Phys.* 51:633–643 (1998).
9. D.M. Crowe, X.Q. Guo, M.S. Lubell, J. Slevin, and M. Eminyan, Spin-tagged electron-hyrogen scattering: new measurements of ionisation asymmetries from threshold to 500 eV, *J.Phys.B* 23:L325–L331 (1990).
10. G.D. Fletcher, M.J. Alguard, T.J. Gay, P.F. Wainwright, M.S. Lubell, W. Raith, and V.W. Hughes, Experimental study of spin-exchange effects in elastic and ionizing collisions of polarized electrons with polarized hydrogen atoms, *Phys.Rev.A* 31:2854–2883 (1985).
11. I. Bray and A.T. Stelbovics, Calculation of the total ionization cross section and spin asymmetry in electron-hydrogen scattering, *Phys.Rev.Lett.* 70:746–749 (1993).
12. D. Kato and S. Watanabe, Two-electron correlations in $e + H \to e + e + p$ near threshold, *Phys.Rev.Lett.* 74:2443–2446 (1995).
13. M.P. Scott, P.G. Burke, K. Bartschat, and I. Bray, Near-threshold electron impact ionisation of atomic hydrogen, *J.Phys.B* 30:L309–L315 (1997).
14. V. Karaganov, I. Bray, P.J.O. Teubner, and P. Farrell, Super-elastic electron scattering on lithium, *Phys.Rev.A* 54:R9–R12 (1996).
15. M.H. Kelley, J.J. McClelland, S.R. Lorentz, R.E. Scholten, and R.J. Celotta, Scattering of polarized electrons from optically pumped sodium atoms, in: *Correlations and Polarization in Electronic and Atomic Collisions and (e,2e) Reactions*, P.J.O. Teubner and E. Weigold, editors, number 122, Institute of Physics Conference Series, pages 23–32, IOP, London, (1992).
16. I. Bray, Convergent close-coupling calculation of electron-sodium scattering, *Phys.Rev.A* 49:R1–R4 (1994).
17. I. Bray and I.E. McCarthy, Spin-dependent observables in electron-sodium scattering calculated using the coupled-channel optical method, *Phys.Rev.A* 47:317–326 (1993).
18. I. Bray, Calculation of spin asymmetries in electron-alkali scattering, *Z.Phys.D* 30:99–103 (1994).
19. R. McAdams, M.T. Hollywood, A. Crowe, and J.F. Williams, Alignment and orientation of He(2^1P), *J.Phys.B* 13:3691–3701 (1980).
20. D.V. Fursa and I. Bray, Calculation of electron-helium scattering, *Phys.Rev.A* 52:1279–1298 (1995).
21. D. Fursa, I. Bray, B.P. Donnelly, D.T. McLaughlin, and A. Crowe, Electron impact excitation of the 3D states of helium: comparison between experiment and theory at 30 eV, *J.Phys.B* 30:3459–3473 (1997).
22. I. Bray, Convergent close-coupling method for the calculation of electron scattering on hydrogen-like targets, *Phys.Rev.A* 49:1066–1082 (1994).
23. K. Bartschat, E.T. Hudson, M.P. Scott, P.G. Burke, and V.M. Burke, Differential cross sections and electron-impact coherence parameters for electron scattering from helium atoms, *J.Phys.B* 29:2875–2885 (1996).
24. D.V. Fursa and I. Bray, Topical review: Convergent close-coupling calculations of electron-helium scattering, *J.Phys.B* 30:757–785 (1997).

25. K. **Bartschat**, Electron-impact excitation of helium from the 1^1S and 2^3S states, *J.Phys.B* 31:L469–L476 (1998).
26. D.V. Fursa and I. Bray, Convergent close-coupling calculations of electron scattering on helium-like atoms and ions; electron-beryllium scattering, *J.Phys.B* 30:5895–5913 (1997).
27. D.V. Fursa and I. Bray, Calculation of electron scattering from the ground state of barium, *Phys.Rev.A* 57 (1998).
28. R. Srivastava, T. Zuo, R.P. McEachran, and A.D. Stauffer, Excitation of the $^{1,3}P_1$ states of calcium, stroncium and barium in the relativistic distorted-wave approximation, *J.Phys.B* 25:3709–3720 (1992).
29. P.W. Zetner, Y. Li, and S. Trajmar, Measurement of electron impact coherence parameters by superelastic scattering form laser-excited ^{138}Ba(... 6s6p)$^1P_1$, *J.Phys.B* 25:3187–3199 (1992).
30. P.W. Zetner, Y. Li, and S. Trajmar, Charge-cloud alignment in the electron-impact excitation of ^{138}Ba(... 6s6p)$^1P_1$, *Phys.Rev.A* 48:495–504 (1993).
31. Y. Li and P.W. Zetner, Transfered orbital angular momentum in the excitation of ^{138}Ba(... 6s6p)$^1P_1$ by electron impact, *Phys.Rev.A* 49:950–955 (1994).
32. H.S.W. Massey and C.B.O. Mohr, The collision of slow electrons with atoms, *Proc. Roy. Soc. A* 136:289–311 (1932).
33. P.G. Burke and W.D. Robb, The R-matrix theory of atomic processess, *Adv.Atom.Mol.Phys.* 11:143–214 (1975).
34. J. Callaway, The variational method in atomic scattering, *Phys. Rep.* 45:89 (1978).
35. P.G. Burke, C.J. Noble, and M.P. Scott, R-matrix theory of electron scattering at intermediate energies, *Proc.R.Soc.A* 410:289–310 (1987).
36. C.H. Greene, Variational calculation of channel interaction parameters, in: *Fundamental Processes of Atomic Dynamics*, J.S. Briggs, H. Kleinpoppen, and H.O. Lutz, editors, pages 105–127, Plenum Publishing Corporation, New York (1988).
37. K. Bartschat, E.T. Hudson, M.P. Scott, P.G. Burke, and V.M. Burke, Electron-atom scattering at low and intermediate energies using a pseudo-state/R-matrix basis, *J.Phys.B* 29:115–123 (1996).
38. N.R. Badnell and T.W. Gorczyca, Iso-electronic trends for continuum coupling effects on the electron-impact excitation of H-like ions, *J.Phys.B* 30:2011–2019 (1997).
39. S. Watanabe, Y. Hosada, and D. Kato, Hyperspherical close-coupling method extended to the two-electron continuum region: test on the S-wave model for e-H scattering, *J.Phys.B* 26:L495–L501 (1993).
40. I. Bray and D.V. Fursa, Calculation of ionization within the close-coupling formalism, *Phys.Rev.A* 54:2991–3004 (1996).
41. E.J. Heller and H.A. Yamani, New L^2 approach to quantum scattering: Theory, *Phys.Rev.A* 9:1201–1208 (1974).
42. H.A. Yamani and W.P. Reinhardt, L^2 discretizations of the continuum: radial kinetic energy and Coulomb Hamiltonian, *Phys.Rev.A* 11:1144–1156 (1975).
43. J.T. Broad, Gauss quadrature generated by diagonalization of H in finite L^2 bases, *Phys.Rev.A* 18:1012–1027 (1978).
44. I. Bray and B. Clare, S-wave model for H-like ions, *Phys.Rev.A* 56:R1694–R1696 (1997).
45. A. Temkin, Nonadiabatic theory of electron-hydrogen scattering, *Phys. Rev.* 126:130–142 (1962).
46. R. Poet, The exact solution for a simplified model of electron scattering by hydrogen atoms, *J.Phys.B* 11:3081–3094 (1978).
47. I. Bray, Close-coupling theory of ionization: successes and failures, *Phys.Rev.Lett.* 78:4721–4724 (1997).
48. S. Rioual, J. Röder, B. Rouvellou, H. Ehrhardt, A. Pochat, I. Bray, and D.V. Fursa, Absolute (e,2e) cross sections for the electron-impact ionization of helium in energy sharing kinematics at 44.6 eV, *J.Phys.B* 31:3117–3127 (1998).
49. M.B. Shah, D.S. Elliot, P. McCallion, and H.B. Gilbody, Single and double ionization of helium by electron impact, *J.Phys.B* 21:2751–2761 (1988).
50. A.J. Murray, F.H. Read, and N.J. Bowring, (e,2e) collisions at intermediate energies, *Journal de Physique* 3:51–58 (1993).

BENCHMARK STUDIES IN ELECTRON-IMPACT EXCITATION OF ATOMS

Klaus Bartschat

Department of Physics and Astronomy
Drake University
Des Moines, IA 50311, U.S.A.

INTRODUCTION

Electron collisions with atoms and ions, involving elastic scattering, excitation, and ionization, are of tremendous importance not only as a fundamental branch of atomic physics, but also because of the urgent practical need for accurate data in many applications. These data serve as input for modeling processes in air pollution research, astronomy, electrical discharges, laser developments and applications, magnetically confined thermonuclear fusion devices, planetary atmospheres, and surface science, to name just a few.

In light of the importance of these processes, the need for benchmark comparisons between experimental data and theoretical predictions is obvious. Such comparisons can be made at various levels of detail, ranging from rather global observables such as rate coefficients, i.e., total, angle-integrated cross sections that are further integrated over a range of collision energies, to the very detailed parameters measured in "complete experiments" (Bederson, 1969) which resolve the scattering angle of the projectile, its spin, and even the polarization of the light emitted from excited targets in possibly spin-resolved electron–photon coincidence experiments.

In this article, we present some key examples to illustrate the current status of theory and experiment in such benchmark comparisons for electron–atom collisions. We begin by introducing a basic experimental setup and demonstrating the need to perform experiments which test, by investigating independent sets of observables, as many aspects as possible of the theoretical predictions. This is followed by a summary of the key ideas behind two currently very promising theoretical approaches to describe electron–atom collisions, namely the "convergent close-coupling" (CCC) and the "R-matrix with pseudo-states" (RMPS) methods. Predictions from these and other methods are then compared with a variety of experimental results to allow for a critical assessment of the present situation and to provide direction for future developments.

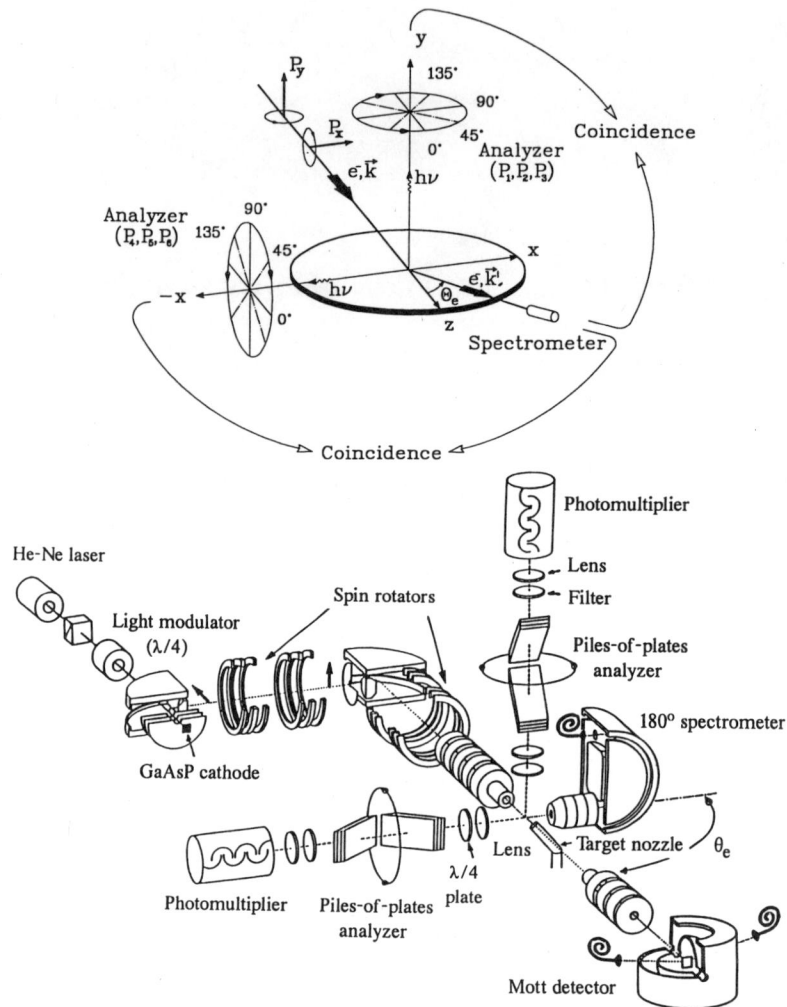

Figure 1. Schematic diagram (top) and actual experimental setup (bottom) of the Münster experiment (Sohn and Hanne, 1992) in which spin-polarized electrons are scattered inelastically from atoms at an angle θ. The photons emitted during the decay of the atom in a specific direction are polarization analyzed and detected in coincidence with the scattered electrons without further spin analysis.

AN EXPERIMENTAL SETUP AND EXAMPLE RESULTS

Figure 1 shows an experimental setup, developed in the Münster group (Sohn and Hanne, 1992), to perform scattered-electron–polarized-photon coincidence studies. The main difference compared to similar experiments with unpolarized electrons is the GaAs source of spin-polarized electrons which replaces a standard electron gun. After impact excitation of the target, the scattered electrons and the emitted light are observed in coincidence, and the three Stokes parameters,

$$P_1 \equiv \frac{I(0°) - I(90°)}{I(0°) + I(90°)}, \quad P_2 \equiv \frac{I(45°) - I(135°)}{I(45°) + I(135°)}, \quad P_3 \equiv \frac{I(RHC) - I(LHC)}{I(RHC) + I(LHC)}, \quad (1)$$

are determined. Here $I(\beta)$ is the light intensity transmitted by a linear polarization analyzer oriented at an angle β with respect to a previously defined direction (Blum,

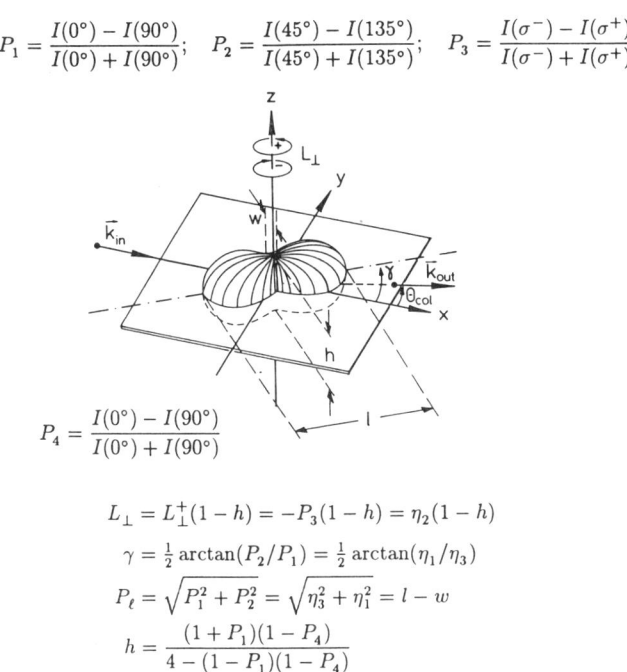

Figure 2. Physical interpretation of Stokes parameter measurements for excitation of atomic P states. The set $(\eta_3, \eta_1, -\eta_2)$ is a frequently used alternative notation to (P_1, P_2, P_3).

1996; Andersen and Bartschat, 1996; Andersen et al.,1997), while $I(RHC)$ and $I(LHC)$ are the intensities transmitted by filters for right-circularly and left-circularly polarized light, respectively. Note that the result of the polarization measurement depends upon the direction of light observation. While various notations have been used to describe the basic cases, the essence of such experiments lies in the determination of the so-called "generalized Stokes parameters" introduced by Andersen and Bartschat (1994).

Figure 2 shows the physical interpretation of the results for excitation of an atomic P-state (Andersen et al., 1988). The alignment angle γ gives the principal axis of the charge cloud relative to the incident beam direction, while the degree of linear polarization P_ℓ indicates the relative length and width; $P_\ell = 0$ thus corresponds to a circular state with the maximum value of the angular momentum transfer $L_\perp$. In some cases, the charge cloud may also exhibit a finite height h that can be measured directly through a $P_1(\equiv P_4)$ measurement with photon observation *in* the scattering plane.

There are several modifications that can be made to the above apparatus. For example, one might use unpolarized electrons or refrain from a coincidence study and only analyze the light polarization. The latter measurement determines the so-called "integrated Stokes parameters" whose physical importance, especially for polarized incident electron beams, was analyzed by Bartschat and Blum (1982). Alternatively, the same information (and sometimes more) as in electron-photon coincidence experiments can be obtained in "time-reversed" setups that were pioneered by the NIST group (McClelland et al., 1985,1989) and involve de-excitation of a laser-prepared target by the incident electron beam.

Furthermore, one can analyze the polarization of initially polarized electrons after scattering from unpolarized or even spin-polarized atomic targets. The latter experiment has not been performed to date, but the "generalized *STU* parameters"

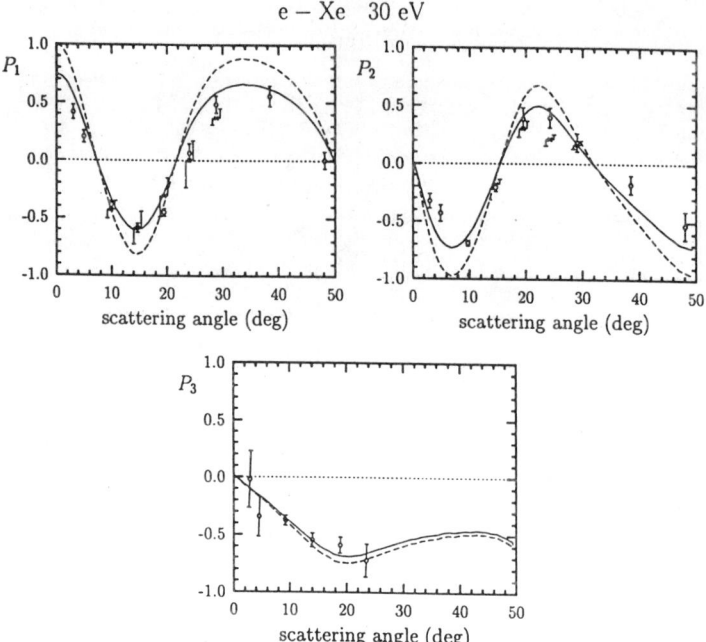

Figure 3. Stokes parameters (P_1, P_2, P_3) for electron-impact excitation of the $(5p^5[3/2]6s)$ ("$^3P_1$") state in xenon at an incident energy of 30 eV. The experimental data of Nishimura et al. (1986) and of Corr et al. (1990) are compared with semi-relativistic DWBA results of Bartschat and Madison (1987). The solid line includes the depolarization effects due to the hyperfine structure.

(Bartschat, 1989) that fully describe the change of an arbitrary initial electron polarization through scattering from unpolarized target atoms, have been determined for some cases by the Münster group (Berger and Kessler, 1986; Müller and Kessler, 1994). Besides the *absolute* differential cross section σ_u for the scattering of unpolarized electrons, there are seven *relative* generalized STU parameters that completely describe the reduced spin density matrix of the scattered projectiles. In this article, we will restrict the discussion to just one of those parameters, namely the asymmetry function S_A that determines the left-right asymmetry in the differential cross section for scattering of spin-polarized projectiles (Kessler, 1985).

As the first example of such experiments and a comparison of their results with theoretical predictions, we show in Fig. 3 the set of Stokes parameters (P_1, P_2, P_3) for electron-impact excitation of the $(5p^5[3/2]6s)$ ("$^3P_1$") state in xenon at an incident energy of 30 eV. (The inverted commas indicate that the LS notation for this state is only an approximation, since it is heavily mixed and must be described by an intermediate coupling scheme. The number in square brackets denotes the angular momentum of the Xe^+ core.) The incident electron beam is unpolarized, and the radiation is observed perpendicular to the scattering plane. After modifying the first-order distorted-wave (DWBA) results of Bartschat and Madison (1987) to account for the hyperfine-structure depolarization of the radiation, there is excellent agreement between theory and experiment (Nishimura et al., 1986; Corr et al., 1990) for the set of Stokes parameters. Hence, one might conclude that the relatively simple, semi-relativistic theoretical approach is sufficient to describe this collision process.

However, this assessment clearly needs to be revised after examining Fig. 4 which

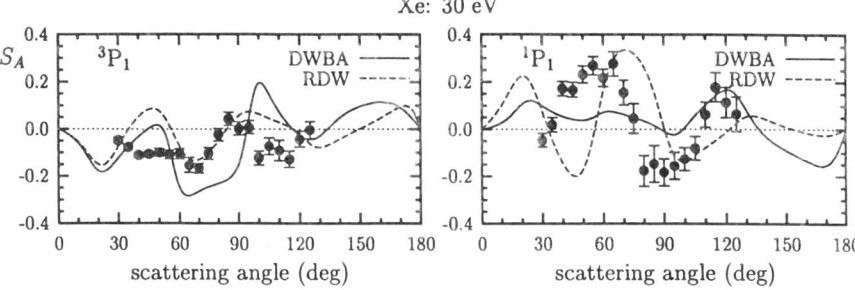

Figure 4. Spin asymmetry function S_A for electron-impact excitation of the $(5p^5[3/2]6s)$ ($"^3P_1"$) and $(5p^5[1/2]6s)$ ($"^1P_1"$) states in xenon at an incident energy of 30 eV. The experimental data of Dümmler et al. (1995) are compared with semi-relativistic (DWBA) and full-relativistic (RDW) first-order distorted-wave results of Bartschat and Madison (1987) and of Zuo et al. (1992).

displays results for the spin-asymmetry function S_A for impact excitation of the same state as well as the $(5p^5[1/2]6s)$ ($"^1P_1"$) state. Investigating an independent observable at the same collision energy apparently reveals major deficiencies in the theoretical model, and the discrepancies between theory and experiment become even more visible due to the wide range of scattering angles for which the experiment was performed. We also note that predictions from a full-relativistic distorted-wave calculation by the Toronto group (Zuo et al., 1992) agrees neither with the semi-relativistic DWBA results nor with experiment over a large range of energies for these and also for other excited states (see Dümmler et al., 1995, for more details).

NUMERICAL METHODS

The above example clearly shows the need for the development of sophisticated theoretical methods, stimulated by numerous such benchmark experiments that were performed over the past 25 years. We therefore summarize the principal ideas behind two recently developed close-coupling methods that account for the effect of the target continuum states in the expansion and keep the coupling between all the physical discrete as well as the bound and continuum pseudo-channels. As will be demonstrated below, the success of these methods for simple target systems such as helium and sodium provides an excellent basis for further improvement. Alternatively, the use of more direct numerical approaches, such as a time-dependent close-coupling approach (Pindzola et al., 1997), has become very popular, particularly in light of the recent trends towards massively parallel supercomputers. However, such methods have not yet been used in calculating angle-differential parameters for electron–atom collisions.

The convergent close-coupling method (CCC)

The details of the CCC theory have been given by Bray and Stelbovics (1995). The method can be thought of as a standard close-coupling approach where, in addition to the discrete target states, the target continuum is treated with the aid of positive-energy, square-integrable pseudo-states. All states are obtained by diagonalising the target Hamiltonian in an orthogonal Laguerre basis. The usage of such a basis ensures that "completeness" is, in principle, approached by increasing the basis size until the results of interest become sufficiently stable.

A key feature of the formalism, as developed by Bray and Stelbovics, lies in the fact that the coupled equations are formulated in momentum space, where they take the form of coupled Lippmann-Schwinger equations for the T-matrix. These are solved separately, upon partial-wave expansion, for each collision energy of interest. As such the method is not ideally suited for the study of detailed scattering behavior as a function of incident energy, as needed, for example, in the vicinity of Rydberg resonance series. However, modern computational resources involving large clusters of workstations allow for the parallel execution of the CCC program at many energies, thereby somewhat reducing this problem. Furthermore, so far the method has only been applied to (quasi-)one-electron and (quasi-)two-electron targets in a non-relativistic framework, although the formalism, as presented by Fursa and Bray (1995), is more general.

The R-matrix with pseudo-states method (RMPS)

The low-energy R-matrix method (for details, see Burke and Robb, 1975) is another method to solve the close-coupling equations, this time in coordinate space. The important difference to the standard formulation is the division of configuration space into two regions, $r \leq a$ and $r > a$, where the R-matrix radius a is chosen in such a way that exchange effects between the projectile and the target electrons can be neglected in the external region. Here the coupled equations (without exchange) are solved for each collision energy and matched, at the boundary $r = a$, to the solution in the inner region. However, instead of solving a set of coupled integro-differential equations in the internal region for each collision energy, the $(N+1)$-electron wavefunction at energy E is expanded in terms of an energy-independent basis set, ψ_k, as

$$\Psi_E = \sum_k A_{Ek} \psi_k. \tag{2}$$

The basis states ψ_k are constructed as

$$\psi_k = \mathcal{A} \sum_{ij} \Phi_i(1,\ldots,N) \frac{u_j(N+1)}{r_{N+1}} a_{ijk} + \sum_i \chi_i(1,\ldots,N+1) b_{ik}, \tag{3}$$

where the Φ_i are N-electron target states, the u_j are members of a complete set of numerical continuum orbitals used to describe the motion of the scattered electron inside the box, and the χ_i are $(N+1)$-electron configurations. The latter are formed from the one-electron orbital basis used to describe the N-electron target. They are included to allow for electron correlation effects when the scattered electron is close to the nucleus, and also to ensure completeness of the trial wavefunction if the continuum orbitals are constructed orthogonal to the bound orbitals.

Each of the N-electron target states is expanded in terms of a sum of orthonormal configurations

$$\Phi_i = \sum_{j=1}^m c_{ij} \phi_j. \tag{4}$$

The configurations ϕ_j are built up from one-electron orbitals, coupled together to give a function which is completely antisymmetric with respect to the interchange of the space and spin coordinates of any two electrons.

The one-electron orbitals used in constructing the ϕ_j configurations may be of two types: i) physical (Hartree-Fock) orbitals and ii) suitably chosen non-physical pseudo-orbitals. In standard R-matrix calculations, performed over many years by the Belfast group and their collaborators worldwide (Burke and Berrington, 1993), the number of pseudo-orbitals is very small (one or at most two per angular momentum), and generally their sole purpose is to improve the description of the discrete target spectrum.

The R-matrix with pseudo-states (RMPS) method was introduced by Bartschat et al. (1996a) and described in detail by Bartschat (1998a). It is an alternative to the "intermediate energy R-matrix method" (IERM) introduced by Burke et al. (1987). The principal idea behind both of these methods is the same as in the CCC approach, i.e., one includes a large set of pseudo-orbitals which can then be used both (i) to improve the description of the physical target states and (ii) to approximate the effects of the infinite number of discrete and continuum states of the target atom or ion that cannot be included explicitly in the calculation. As in the CCC method, these effects are only represented accurately in a certain region of configuration space (usually the R-matrix box), due to the finite range of the pseudo-orbitals. However, this is expected to be sufficient if one is interested in transitions between discrete states whose range is restricted to within this box. Like the CCC approach, the methods also work for ionization (Bartschat and Bray, 1997; Scott et al., 1997) if the box is made sufficiently large and some averaging over effective box sizes, determined by different ranges of the pseudo-orbitals, is included.

The RMPS method has recently been implemented in updated versions of the general Belfast R-matrix codes RMATRX I (Berrington et al., 1995) and RMATRX II (Burke et al., 1994). Consequently, it may already be used for complex targets and, in principle, relativistic effects can be accounted for through the inclusion of the one-electron terms in the Breit-Pauli Hamiltonian. Due to the computational demands of such a project, no semi-relativistic RMPS calculation has yet been performed, but work in this direction is currently in progress. Ultimately, an RMPS version of the full-relativistic DARC program (Norrington and Grant, 1987; Wijesundera et al., 1991), based upon the Dirac-Breit Hamiltonian, would be highly desirable.

BENCHMARK CONPARISONS

We now present a few selected benchmark comparisons between experimental and theoretical results for angle-integrated sections, angle-integrated Stokes parameters, and angle-differential Stokes as well as spin polarization parameters in electron–atom collisions. We begin with angle-integrated cases in which the scattered electrons are not observed and the counting statistics are generally much better than in the angle-differential measurements discussed further below.

Angle-Integrated Cross Sections

A problem of great interest for many fields of physics is electron-impact excitation of helium. Predictions from a recent RMPS calculation (Bartschat, 1998b) for excitation of the $n=3$ states from the He ground state $(1s^2)^1S$ are shown in Fig. 5, compared with the 75-state CCC results of Fursa and Bray (1995) and also with some selected experimental data. The overall agreement between RMPS and CCC(75) is very satisfactory, and the remaining differences in the theoretical predictions are often smaller than the experimental uncertainties.

Results for excitation of the $n=2$ and $n=3$ states from the $(1s2s)^3S$ metastable state are presented in Fig. 6. The agreement between CCC(75) and the RMPS results is once again very satisfactory and generally even better than that for excitation from the ground state. In fact, this is not surprising since the excited $n=2,3$ states are very well described by a frozen-core model. We also note the good agreement between RMPS,

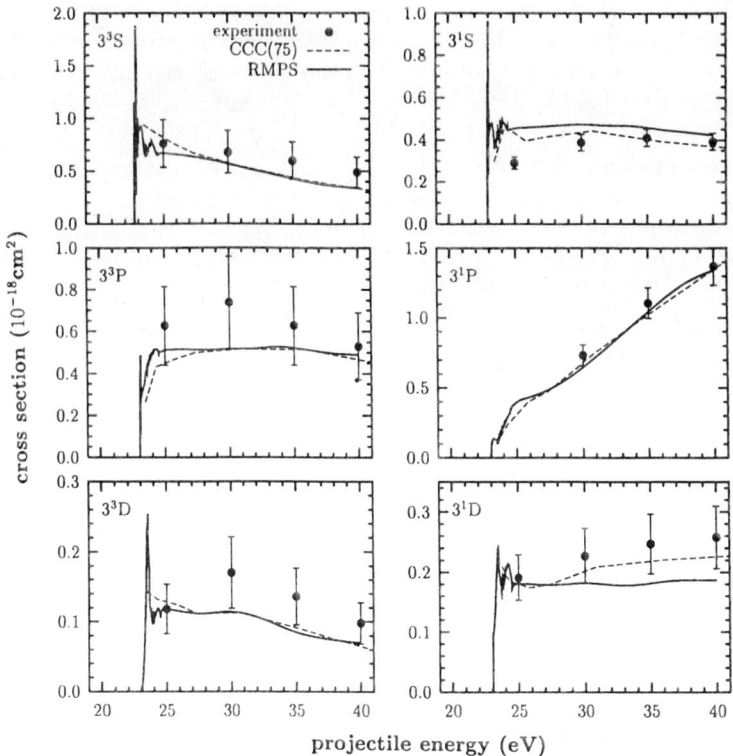

Figure 5. Angle-integrated cross sections for electron impact excitation of helium from the ground state 1^1S to the $n=3$ states as a function of the projectile energy. The RMPS results (Bartschat, 1998b) are compared with the 75-state CCC predictions of Fursa and Bray (1995) and with experimental data recommended by de Heer et al. (1992).

CCC and the 10-state Eikonal Approximation (EA10) of Flannery and McCann (1975) for the $(1s2s)^3S - (1s2p)^3P^o$ cross section, particularly at a projectile energy of 10 eV. Again, this is not unexpected, since 10 eV is already a fairly high energy relative to the excitation threshold from the $(1s2s)^3S$ state.

As a consequence of the good agreement between the predictions from these very sophisticated theoretical models, the disagreement of the CCC/RMPS predictions with the experimental results of the Wisconsin group (Lagus et al., 1996; Piech et al., 1997,1998) remains unresolved. While we note the better agreement of the five-state R-matrix results of Fon et al. (1981) with the results of Lagus et al. (1996) for the $(1s2s)^3S - (1s2p)^3P^o$ transition, and of the 11-state R-matrix results of Berrington et al. (1985) and the First-Born results of Briggs and Kim (1971) for excitation of the $n=3$ triplet states, it should be pointed out that these simple models would generally not be expected to yield reliable results for most of the transitions in this energy regime.

The difficulties associated with the absolute experimental normalization procedures (note, for example, the revision by nearly a factor of two between the results of Piech et al., 1997, and Piech et al., 1998) further emphasizes the need for reliable numerical calculations. It is worth noting that the recommended data for use in modeling programs for excitation of helium from the ground state have recently been revised, with dominant weight given to the CCC and RMPS predictions (deHeer, 1998).

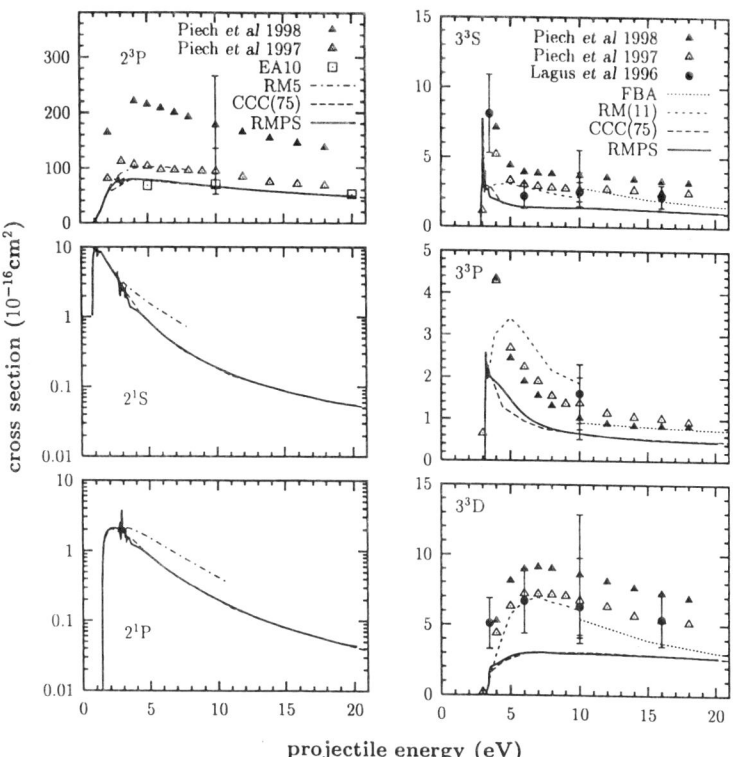

Figure 6. Angle-integrated cross sections for electron impact excitation of helium from the metastable 2^3S state to the $n=2$ and the triplet $n=3$ states as a function of the projectile energy. Theoretical results from the RMPS and CCC models are shown together with predictions from a 10-state Eikonal approximation (EA10) of Flannery and McCann (1972), a 5-state R-matrix calculation (RM5) of Fon et al. (1981), and an 11-state R-matrix calculation (RM11) of Berrington et al. (1985). The experimental data are from Lagus et al. (1996) and Piech et al. (1997,1998), with characteristic total (statistics, cascade contributions, plus absolute normalization) uncertainties indicated at 10 eV.

Angle-Integrated Stokes Parameters

Figure 7 compares recent results for the Stokes parameters (P_1, P_2, P_3) observed by the Perth group (Yu et al., 1997) after impact excitation of two states in the $(2p^53p)$ manifold of neon by a transversely spin-polarized electron beam with polarization P_e as a function of the incident electron energy. The light is observed in a direction perpendicular to the incident beam but parallel to the spin-polarization vector. In this case, P_1 is independent of the electron polarization, whereas P_2 and P_3 are directly proportional to P_e (Bartschat and Blum, 1982).

In the near-threshold regime, there is certainly very satisfactory, though not perfect agreement between the experimental data and theoretical predictions from a 31-state (discrete states only) Breit-Pauli R-matrix calculation of Zeman and Bartschat (1997). The very narrow resonance structure predicted by theory was apparently unresolved in the experiment. We also note that cascade effects, which become substantial for incident energies above approximately 20 eV, generally lead to a depolarization of the emitted radiation and may thus explain why theory overestimates the magnitude of the polarization components at higher energies.

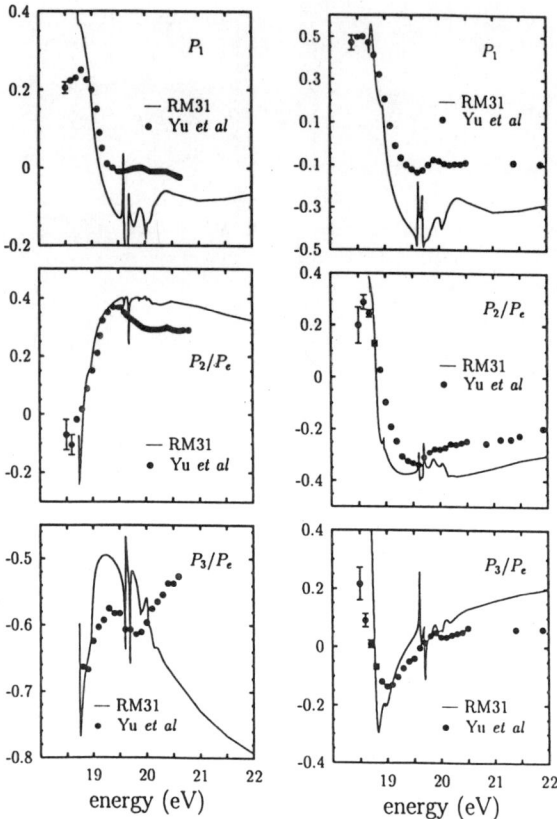

Figure 7. Stokes parameters (P_1, P_2, P_3) for electron impact excitation of the $2p^53p[1/2]_1$ (left) and the $2p^53p[3/2]_1$ (right) states in neon and optical decay to $2p^53s[1/2]_1$, plotted as a function of the incident electron energy. The experimental results of Yu et al. (1997) (P_2 and P_3 normalized to 100% incident electron polarization) are compared with theoretical predictions based upon a Breit-Pauli R-matrix approach (Zeman and Bartschat, 1997).

Probably the most interesting parameter from a physical point of view is the linear light polarization P_2. While electron exchange, together with some spin-orbit interaction within the target, can lead to non-vanishing values of the circular light polarization P_3, these effects cannot lead to non-zero values of P_2. In other words, spin-polarization of the electron beam can easily be transferred into angular momentum orientation of the excited atomic ensemble, but not into alignment. In fact, it was shown by Bartschat and Blum (1982) that LS-coupling must be violated already *during the collision* for a non-zero P_2 to occur. The two principal mechanisms that have been proposed to cause such a result are:

- The spin-orbit interaction within the target, combined with configuration mixing, makes it impossible to describe the excitation process in LS-coupling. Instead, an intermediate coupling scheme, i.e., combinations of LS-states with different values of L and S but the same value of the total electronic angular momentum J, must be used.

- Coupling of the continuum electron spin to the target angular momenta is so strong that it can effectively rotate the charge cloud (see also Fig. 2). This mechanism represents the optical analogue to Mott scattering, but this time affecting the *target* without even detecting a specific electron scattering angle.

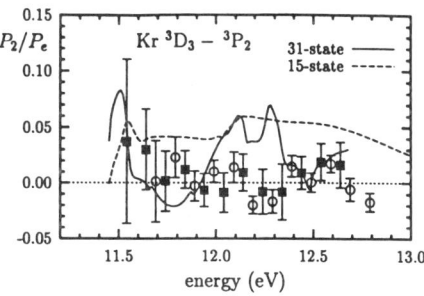

Figure 8. Stokes parameter P_2 (normalized to 100% incident electron polarization) for impact excitation of the $(4p^55p)^3D_3 \to (4p^55s)^3P_2$ transition in krypton, plotted as a function of the incident electron energy. The experimental results of Gay's group [41] are compared with theoretical predictions based upon a Breit-Pauli R-matrix approach (Zeman and Bartschat, 1997; Zeman et al., 1997).

As in the first experimental verification of a non-zero P_2 in an angle-integrated experiment with spin-polarized electrons, carried out by Bartschat et al. (1982) for electron-impact excitation of the $(6s6p)^3P_1$ state in mercury, the results in Fig. 7 can be understood in terms of the intermediate coupling nature of the excited states. Despite several attempts on different targets (Furst et al., 1992), however, the optical detection of Mott scattering has not been successful to date. Figure 8 shows the most recent results of Gay's group (Al-Khateeb et al., 1998) for the linear polarization P_2 in the $(4p^55p)^3D_3 \to (4p^55s)^3P_2$ transition in krypton, after impact excitation by spin-polarized electrons. Although krypton is a fairly heavy target ($Z = 36$), the states of interest are well LS-coupled, and hence the intermediate-coupling mechanism is not expected to be relevant. Note that two different ways of determining P_2 did not yield an experimental result that could be interpreted as non-zero in a statistically meaningful way. On the other hand, theoretical predictions from two semi-relativistic 15-state and 31-state Breit-Pauli R-matrix calculations suggest positive values of P_2/P_e in the order of a few percent over an energy range that is free of cascades and several times wider than the energy width of the electron beam. Although the differences between the two theoretical curves clearly indicate a potential lack of convergence with respect to the number of states included in the calculation, the qualitative disagreement between theory and experiment regarding the P_2 parameter is somewhat surprising, since the theoretical model was otherwise very successful in describing near-threshold electron-impact excitation of the noble gases (Zeman and Bartschat, 1997; Zeman et al., 1997; Brunger et al., 1998). The reason for the discrepancy is currently unknown, and hence this problem represents an important open question in the field of polarized electron physics.

Angle-Differential Stokes Parameters

Figure 9 shows results for the angular momentum transfer $L_\perp$ and the alignment angle γ or electron impact excitation of the $(1s2p)^1P$ state of helium from the ground state $(1s^2)^1S$ at an incident energy of 50 eV. There is excellent agreement between the CCC (Fursa and Bray, 1995) and RMPS (Bartschat et al., 1996b) results, and also with various experimental data. For this optically allowed transition, first-order theories, such as a distorted-wave (Madison and Shelton, 1971) or a many-body (FOMBT) (Cartwright and Csanak, 1988) approach, are generally in reasonably good agreement with experiment as well, especially at small scattering angles and sufficiently high energies. In fact, only the *large-angle measurements* of electron-impact coherence para-

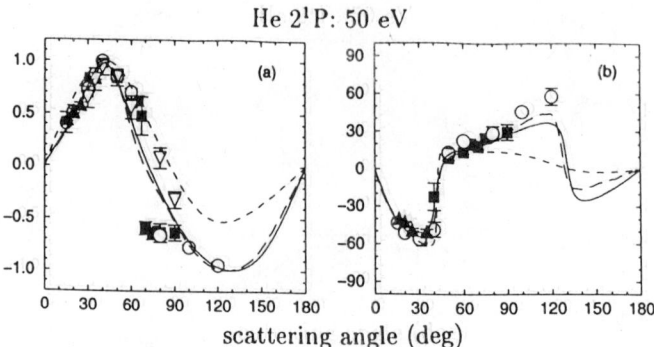

Figure 9. Angular momentum transfer $L_\perp$ (a) and alignment angle γ (b) for electron impact excitation of the $(1s2p)^1P$ state of helium at an incident energy of 50 eV. The theoretical predictions from RMPS (solid line, Bartschat et al., 1996b), CCC (long-dashed line, Fursa and Bray, 1997), and DWBA (short-dashed line, Beijers et al., 1987) calculations are compared with experimental data of Eminyan et al. (1974) (full triangles), McAdams et al. (1980) (open circles), Khakoo et al. (1986) (open triangles), and Beijers et al. (1987) (full squares).

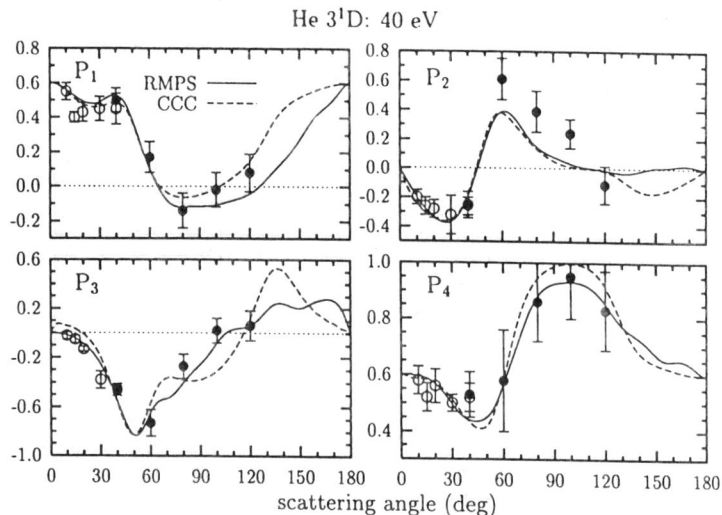

Figure 10. Stokes parameters (P_1, P_2, P_3, P_4) for electron impact excitation of the $(1s3d)^1D$ state of helium at an incident energy of 40 eV. The theoretical predictions from RMPS (solid line, Bartschat, 1998c) and CCC (dashed line, Fursa and Bray, 1997) calculations are compared with experimental data from McLaughlin et al. (1994) and Mikosza et al. (1994).

meters revealed problems with the above first-order theories for this particular collision system — problems that cannot be seen by comparing the predictions for the differential cross section (Bartschat et al., 1996b).

As shown by Crowe (1994, 1998), however, first-order theories are much less successful in the description of optically forbidden transitions, while CCC and RMPS calculations continue to reproduce the experimental data in a satisfactory way. This is demonstrated in Fig. 10 where the Stokes parameters (P_1, P_2, P_3, P_4) for electron impact excitation of the $(1s3d)^1D$ state in helium are compared with measurements from the Newcastle and Perth groups.

Figure 11 shows another example of the detailed information that can be obtained

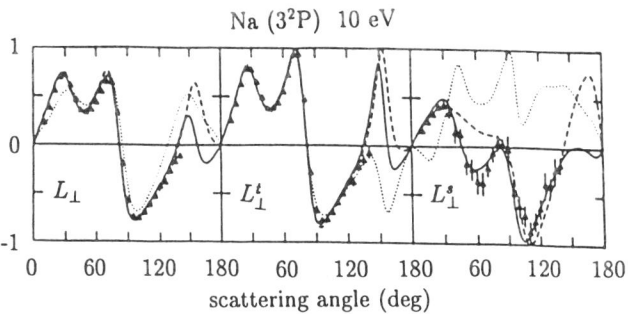

Figure 11. Angular momentum transfer $L_\perp$, spin-averaged and separated into contributions from the triplet (t) and singlet (s) spin channels for electron impact excitation of the $(3p)^2P$ state of sodium at an incident energy of 10 eV. The experimental data from the NIST group (Scholten et al., 1991) are compared with theoretical predictions from CCC (solid line, Bray, 1994), CCO (dashed line, Bray and McCarthy, 1993), and DWBA2 (dotted line, Madison et al., 1992) calculations.

in spin-resolved studies of atomic orientation parameters. In the experiment performed by the NIST group (Scholten et al., 1991), spin-polarized electrons were scattered from laser-prepared spin-polarized and angular-momentum-oriented sodium atoms in the $(3p)^2P$ state. By studying the superelastic signal, i.e., the collision-induced de-excitation to the $(3s)^2S$ state, it was not only possible to determine the corresponding angular momentum transfer $L_\perp$ for the time-reversed inelastic collision, but also the individual contributions from the singlet and triplet spin channels. As seen from Fig. 11, only the CCC method (Bray, 1994) was able to reproduce the experimental results for $L_\perp^s$ (the singlet spin channel) between 30° and 90°. On the other hand, the spin-averaged experimental results are also reproduced in a "close-coupling plus optical potential" (CCO) approach (Bray and McCarthy, 1993), and even by a second-order distorted-wave (DWBA2) approach (Madison et al., 1992) which one would generally not expect to be suitable at such a low energy of approximately twice the ionization threshold. In fact, the DWBA2 does very well for the triplet spin channel, but it fails completely for the singlet channel where electron-electron correlation effects are generally much more important than in the triplet channel.

As a final example for angle-differential generalized Stokes parameters and the spin-resolved electron-impact coherence parameters derived from their measurement, we show in Fig. 12 results for excitation of the $(6s6p)^3P_1$ state of mercury by a spin-polarized incident electron beam with an energy of 8 eV. For details of this work we refer to the paper by Andersen et al. (1996), but it is worthwhile to mention a few important points, namely:

- Based upon their definition, the parameters Q_{13}^z, Q_{23}^z, and Q_{33}^z are identical, but they can be determined experimentally in a variety of ways. Hence, the analysis in terms of generalized Stokes parameters revealed that the original experimental data were not completely consistent.

- The overall agreement with predictions from a 5-state Breit-Pauli R-matrix calculation (Scott et al., 1993; Bartschat et al., 1994; Andersen et al., 1996) is surprisingly good, although one might expect that internally consistent experimental data with smaller uncertainties will show the limits of the theoretical model. Such an experiment is currently in preparation (Hanne, 1998).

- The negative value of $L_\perp^{+\downarrow}$, i.e., the angular momentum transfer induced by electrons with spin down relative to the scattering plane, at small scattering angles is

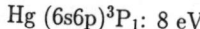

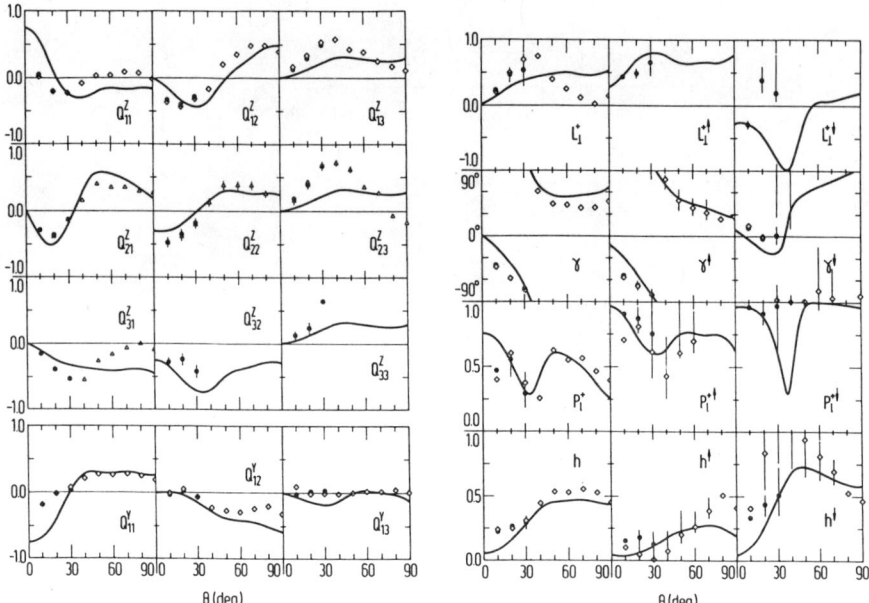

Figure 12. Generalized Stokes parameters and the corresponding spin-averaged and spin-resolved electron-impact coherence parameters for electron impact excitation of the $(6s6p)^1P_1$ state of mercury at an incident energy of 8 eV. The theoretical predictions from a 5-state Breit-Pauli R-matrix calculation are compared with experimental data from the Münster group (Andersen et al., 1996).

a violation of well-established, though empirical, propensity rules found for spin-averaged angular momentum transfers in S → P excitations. For more details, we refer to the discussion by Andersen (1998).

- Measurement of the generalized Stokes parameters allows for the determination of all relative sizes and four out of five relative phases between the six (complex) scattering amplitudes that fully determine this excitation process. Together with the differential cross section (for the absolute size) and a generalized *STU* parameter measurement (for the missing phase as well as for consistency checks), a *complete experiment* for this complicated process is thus within reach of current technology.

Angle-differential spin-polarization parameters

We finish by presenting two further examples to demonstrate that methods based upon the close-coupling method indeed promise good agreement between theory and experiment even for observables that are very sensitive to the details of the theoretical models. Figure 13 compares results for the spin-asymmetry function S_A (see also Fig. 5) for electron-impact excitation of two krypton states at an incident energy of 15 eV. While there are again major differences between the semi-relativistic (DWBA) and full-relativistic (RDW) predictions and experiment (Dümmler et al., 1995), a 31-state Breit-Pauli R-matrix model produces very satisfactory results. However, other comparisons show that RMPS-type calculations would likely be necessary to obtain such good agreement on a routine basis, particularly for slightly higher excitation energies.

Finally, we show results for the two left-right spin asymmetries A_1 and A_2 that oc-

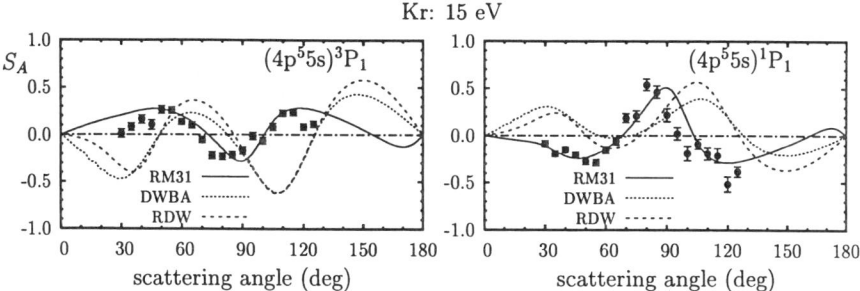

Figure 13. Spin asymmetry function S_A for electron-impact excitation of the $(4p^5[3/2]5s)$ ("$^3P_1$") and $(4p^5[1/2]5s)$ ("$^1P_1$") states in krypton at an incident energy of 15 eV. The experimental data of Dümmler et al. (1995) are compared with semi-relativistic (DWBA) and full-relativistic (RDW) first-order distorted-wave results of Bartschat and Madison (1987) and of Zuo et al. (1992), and with predictions from a 31-state Breit-Pauli R-matrix calculation.

cur in elastic scattering of unpolarized electrons from spin-polarized cesium atoms (A_1) and spin-polarized electrons from unpolarized cesium atoms ($A_2 \equiv S_A$), respectively. After a thorough analysis of the general aspects of this collision problem by Burke and Mitchell (1974), it was predicted by Farago (1974) that the asymmetry A_1 should only occur if both electron exchange and the spin-orbit interaction between the projectile electron and the target nucleus are important in the process — either mechanism alone is not sufficient.

Recently, the Bielefeld group produced benchmark results for elastic e–Cs collisions at an energy of 3 eV for various spin asymmetries (Baum et al., 1998). The results for A_1 and A_2 are shown in Fig. 14 and compared with theoretical predictions from 8-state Breit-Pauli (Bartschat, 1993) as well as 8-state Dirac-Breit R-matrix (Ait-Tahar et al., 1997) calculations. (Note that the non-relativistic result for both asymmetries is exactly zero.) Nearly perfect agreement exists between experiment and the Breit-Pauli results for A_2; only the data points for the two largest angles lie significantly below the prediction. The results of the Dirac treatment, however, deviate considerably from the Breit-Pauli curve and thus disagree with experiment, especially in the angular ranges from 30° to 50° and from 100° to 130°. This is likely due to some deficiencies in the structure part of the (all-electron) Dirac model, which is apparently inferior to the model-potential approach used in the Breit-Pauli calculation.

As one might expect, the spin asymmetry A_1 is the most sensitive parameter with respect to the details of the theoretical model, as it is small to begin with and is affected by both exchange and relativistic effects. Here the two theoretical treatments predict distinctly different results. Both show a sharp peak of the asymmetry in the forward direction, but of considerably different size and location, and structure in the backward scattering region, although with opposing angular dependence and even different signs of the asymmetry. The measurements show reasonable agreement with the predictions in the limited angular range from 60° to 105°, but deviate below and above this range. Note that in this case the experimental data near the forward direction lie significantly above the Breit-Pauli predictions after convoluting the latter with the experimental angular resolution.

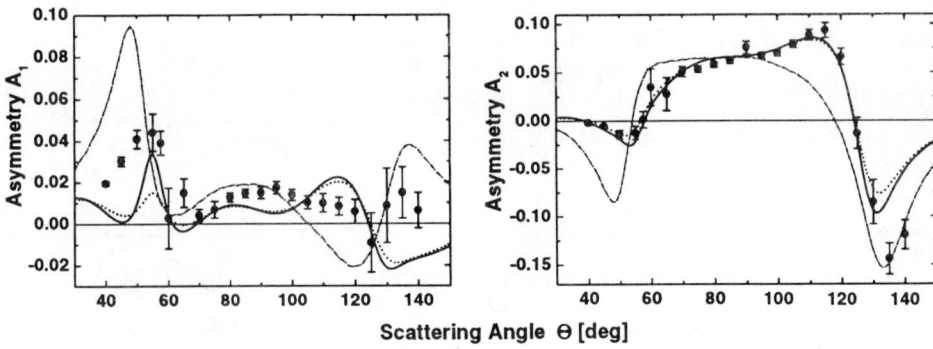

Figure 14. Spin asymmetry functions A_1 and A_2 for elastic electron scattering from cesium atoms at an incident energy of 3 eV. The experimental data of the Bielefeld group (Baum et al., 1998) are compared with predictions from semi-relativistic (solid line, Bartschat, 1983) and full-relativistic (dash-dotted line, Ait-Tahar et al., 1997) 8-state R-matrix calculations. The dotted line was obtained after convoluting the Breit-Pauli results with the experimental angular resolution of 8.5° (FWHM).

CONCLUSIONS

During the past 25 years, the field of electron–atom collisions, including quantum mechanically complete experiments, has developed to considerable maturity. Data for a few elastic and inelastic scattering processes clearly have provided important benchmarks to test state-of-the art scattering theory.

Advancements in the formulation of the standard time-independent close-coupling method resulting, for example, in the CCC, RMPS, and IERM approaches, have achieved a break-through in describing electron collisions with simple target systems. However, the methods need to be further developed and tested for more complex systems and also for heavier targets, such as noble gases other than helium, where relativistic effects are likely to become very important. Reliable results for many of these systems are of crucial importance for modeling the physics of plasmas, lasers, stars, and planetary atmospheres. In light of the experimental difficulties, particularly with respect to absolute normalization of the cross sections, theoretical CCC and RMPS predictions are increasingly being used to revise the corresponding databases.

Regarding the general formulation of the theory of measurement, it has become evident that all cases can be discussed within a *common framework* by parameterizing the change of the scattered electron polarization by means of generalized *STU* parameters and, for excitation, by the evaluation of generalized Stokes parameters for full characterization of the radiation pattern (Andersen and Bartschat, 1996; Andersen et al., 1997). Systematic mapping of the various dimensionless parameters provides a much closer insight into the detailed collision dynamics than the differential cross section alone. Being the only *absolute* observable, however, a truly complete experiment cannot be achieved with relative measurements alone, and the need for accurate cross section measurements should not be underestimated.

From an experimental point of view, future progress in the field requires further development of sophisticated coincidence setups and the ability to handle very long data accumulation times under stable conditions. Alternatively, selected *angle-integrated experiments* may provide important benchmarks as well, and more progress is also expected from scattering and (de-)excitation studies involving *optically prepared states*.

Finally, a very promising theoretical alternative lies in the use of time-dependent approaches, particularly in light of the current developments towards massively parallel computing for which such methods are ideally suited. Although it has not yet been done for the electron–atom collision studies reported in this article, observing the time-development of atomic orientation and alignment in such collisions would undoubtedly provide new insights, perspectives, and further stimulation for this exciting field.

Acknowledgments

I would like to express my sincere thanks to Professor P.G. Burke for his expert guidance in the use of the R-matrix method since my first time as a visiting PhD student at the Queen's University of Belfast in 1982, and for the continuous fruitful collaboration that I have truly enjoyed since then with him and his group. I am also grateful to numerous other colleagues from Adelaide, Belfast, Bielefeld, Boulder, Brisbane, Gaithersburg, Lincoln, Münster, Newcastle, Perth, Rolla, Toronto, Windsor, and Drake University who have contributed to this work. This research was supported, in part, by the United States National Science Foundation under grants PHY-9318377 and PHY-9605124, by the Deutsche Forschungsgemeinschaft in SFB-216, and through visiting fellowships at Adelaide, Belfast, Boulder, and Münster.

REFERENCES

Ait-Tahar, S., Grant, I.P., and Norrington, P.H., 1997, *Phys. Rev. Lett.* 79:2955.
Al-Khateeb, H.M., Birdsey, B.G., Bowen, T.C., Johnson, M.L., and Gay, T.J., 1998, *Bull. Am. Phys. Soc.* 43:1273.
Andersen, N. (1999), Orientation and alignment in atomic collisions, in: "16th International Conference on Atomic Physics (ICAP'98)," G.W.F. Drake, ed., American Institute of Physics, New York.
Andersen, N., and Bartschat, K., 1994, *J. Phys. B* 27:318; corrigendum, 1996, *J. Phys. B* 29:1149.
Andersen, N., and Bartschat, K., 1996, *Adv. At. Mol. Opt. Phys.* 36:1.
Andersen, N., and Bartschat, K., 1997, *J. Phys. B* 30:5071.
Andersen, N., Bartschat, K., Broad, J.T., Hanne, G.F., and Uhrig, M., 1996, *Phys. Rev. Lett.* 76:208.
Andersen, N., Bartschat, K., Broad, J.T., and Hertel, I.V., 1997, *Phys. Rep.* 279:251.
Andersen, N., Gallagher, J.W., and Hertel, I.V., 1988, *Phys. Rep.* 165:1.
Bartschat, K., 1989, *Phys. Rep.* 180:1.
Bartschat, K., 1993, *J. Phys. B* 26:3995.
Bartschat, K., 1998a, The R-matrix with pseudo-states method: Theory and applications to electron scattering and photoionization, *Comp. Phys. Commun.* 114, in press.
Bartschat, K., 1998b, *J. Phys. B* 31:L469.
Bartschat, K., 1998c, R-matrix with pseudo-states calculations for coherence parameters in electron-impact excitation of helium, in preparation for submission to *J. Phys. B*.
Bartschat, K., and Madison, D.H., 1987, *J. Phys. B* 20:5839.
Bartschat, K., and Blum, K., 1982, *Z. Phys. A* 304:85.
Bartschat, K., Hanne, G.F., and Wolcke, A., 1982, *Z. Phys. A* 304:89.
Bartschat, K., Hudson, E.T., Scott, M.P., Burke, P.G., and Burke, V.M., 1996a, *J. Phys. B* 29:115.
Bartschat, K., Hudson, E.T., Scott, M.P., Burke, P.G., and Burke, V.M., 1996b, *J. Phys. B* 29:2875.
Bartschat, K., and Bray, I., 1997, *J. Phys. B* 30:L571.
Bartschat, K., Scott, N.S., Blum, K., and Burke, P.G., 1984, *J. Phys. B* 17:269.
Baum, G., Raith, W., Roth, B., Tondera, M., Bartschat, K., Bray, I., Ait-Tahar, S., Grant, I.P., and Norrington, P.H., 1998, Spin asymmetries in low-energy electron scattering from cesium atoms, submitted to *Phys. Rev. Lett.*
Bederson, B., 1969, *Comments At. Mol. Phys.* 1:41 and 1:65.
Beijers, J.P., Madison, D.H., van Eck, J., and Heideman, H.G.M., 1987, *J. Phys. B* 20:167.
Berger, O., and Kessler, J., 1986, *J. Phys. B* 19:3539.
Berrington, K.A., Burke, P.G., Freitas, L.C.G., and Kingston, A.E., 1985, *J. Phys. B* 18:4135.
Berrington, K.A., Eissner, W., and Norrington, P.H., 1995, *Comp. Phys. Commun.* 92:290.

Blum, K., 1996, "Density Matrix Theory and Applications," 2nd edition, Plenum Press, New York.
Bray, I., 1994, *Phys. Rev. A* 49:1066.
Bray, I., and McCarthy, I.E., 1993, *Phys. Rev. A* 47:317.
Bray, I., and Stelbovics, A.T., 1995, *Adv. Atom. Mol. Phys.* 35:209.
Briggs, J.S., and Kim, Y.K., 1971, *Phys. Rev. A* 3:1342.
Brunger, M.J., Buckman, S.J., Teubner, P.J.O., Zeman, V., and Bartschat, K., 1998, *J. Phys. B* 31:L387.
Burke, P.G., and Berrington, K.A., 1993, "Atomic and Molecular Processes – An R-Matrix Approach," Institute of Physics, Bristol.
Burke, P.G., Burke, V.M., and Dunseath, K.M., 1994, *J. Phys. B* 27:5341.
Burke, P.G., and Mitchell, J.F.B., 1974, *J. Phys. B* 7:214.
Burke, P.G., Noble, C.J., and Scott, M.P., 1987, *Proc. R. Soc. A* 410:289.
Burke, P.G., and Robb, W.D., 1975, *Adv. At. Mol. Phys.* 11:143.
Cartwright, D.C., and Csanak, G., 1988, *Phys. Rev. A* 38:2740.
Corr, J.J., Plessis, P., and McConkey, J.W., 1990, *Phys. Rev. A* 42:5240.
Crowe, A., 1997, Electron–helium correlation studies, *in:* "Photon and Electron Collisions with Atoms and Molecules," P.G. Burke and C.J. Joachain, eds., Plenum, New York.
Crowe, A., 1999, Correlation Studies of Inelastic Electron Scattering from Simple Atoms: Recent Advances," *in:* "16th International Conference on Atomic Physics (ICAP'98)," G.W.F. Drake, ed., American Institute of Physics, New York.
de Heer, F.J., 1998, private communication.
de Heer, F.J., Hoekstra, R., Kingston, A.E., and Summers, H.P., 1992, *Supplement to Nuclear Fusion* 6:7.
Dümmler, M., Hanne, G.F., and Kessler, J., 1995, *J. Phys. B* 28:2985.
Eminyan, M., MacAdam, K.B., Slevin, J., and Kleinpoppen, H., 1974, *J. Phys. B* 7:1519.
Fursa, D.V., and Bray, I., 1995, *Phys. Rev. A* 52:1279.
Fursa, D.V., and Bray, I., 1997, *J. Phys. B* 30:757.
Farago, P.S., 1974, *J. Phys. B* 7:L28.
Flannery, M.R., and McCann, K., 1972, *Phys. Rev. A* 12:846.
Fon W.C., Berrington, K.A., Burke P.G., and Kingston A.E., 1981, *J. Phys. B* 14:2921.
Furst, J.E., Gay, T.J., Wijayaratna, W.M.K.P., Bartschat, K., Geesmann, H., Khakoo, M.A., and Madison, D.H., 1992, *J. Phys. B* 25:1089.
Hanne, G.F., 1998, private communication.
Kessler, J., 1985, "Polarized Electrons," 2nd edition, Springer, Berlin.
Khakoo, M.A., Becker, K., Forand, J.L., Madison, D.H., and McConkey, J.W., 1986, *J. Phys. B* 19:L209.
Lagus, M.E., Boffard, J.B., Anderson, L.W., and Chun, C.C., 1996, *Phys. Rev. A* 53:1505.
Madison, D.H., Bartschat, K., and McEachran, R.P., 1992, *J. Phys. B* 25:5199.
Madison, D.H., and Shelton, W.N., 1971, *Phys. Rev. A* 7:499.
McAdams, R., Hollywood, M.T., Crowe, A., and Williams, J.F., 1980, *J. Phys. B* **13**, 3691 (1980).
McClelland, J.J., Kelley, M.H., and Celotta, R.J., 1985, *Phys. Rev. Lett.* 55:688.
McClelland, J.J., Kelley, M.H., and Celotta, R.J., 1989, *Phys. Rev. A.* 40:2321.
McLaughlin, D.T., Donnelly, B.P., and Crowe, A., 1994, *Z. Phys. D* 29:259.
Mikosza, A.G., Hippler, R., Wang, J.B., and Williams, J.F., 1994, *Z. Phys. D* 30:129.
Müller, H., and Kessler, J., 1994, *J. Phys. B* 27:5993; corrigendum, 1995, *J. Phys. B* 28:911.
Nishimura, H., Danjo, A., and Takahashi, A., 1986, *J. Phys. B* 19:L167.
Norrington, P.H., and Grant, I.P., 1987, *J. Phys. B* 20:4869.
Piech, G.A., Chilton, J.E., Anderson, L.W., and Chun, C.C., 1998, *J. Phys. B* 31:859.
Piech, G.A., Lagus, M.E., Anderson, L.W., and Chun, C.C., 1997, *Phys. Rev. A* 55:2842.
Pindzola, M.S., Robicheaux, F., Badnell, N.R., and Gorczyca, T.W., 1997, *Phys. Rev. A* 56:1994.
Scholten, R.E., Lorentz, S.R., McClelland, J.J., Kelley, M.H., and Celotta, R.J., 1991, *J. Phys. B* 24:L653.
Scott, M.P., Burke, P.G., Bartschat, K., and Bray, I., 1997, *J. Phys. B* 30:L309.
Scott, N.S., Burke, P.G., and Bartschat, K., 1983, *J. Phys. B* 16:L361.
Sohn, M., and Hanne, G.F., 1992, *J. Phys. B* 25:4627.
Wijesundera, W.P., Parpia, F.A., Grant, I.P., and Norrington, P.H., 1991, *J. Phys. B* 24:1803.
Yu, D.H., Hayes, P.A., Furst, J.E., and Williams, J.F., 1997, *Phys. Rev. Lett.* 78:2724.
Zeman, V., and Bartschat, K., 1997, *J. Phys. B* 30:4609.
Zeman, V., Bartschat, K., Gay, T.J., and Trantham, K.W., 1997, *Phys. Rev. Lett.* 79:1825.
Zuo, T., McEachran, R.P., and Stauffer, A.D., 1992, *J. Phys. B* 25:3393.

A MODEL ADIABATIC POTENTIAL TO STUDY SHAPES AND LOCATIONS OF SINGLE PARTICLE RESONANCES: THE CASE OF ELECTRON-OZONE SCATTERING

F.A. Gianturco,[1] R.R. Lucchese,[2] and N. Sanna[3]

[1]Department of Chemistry, The University of Rome, Città Universitaria, 00185 Rome, Italy
[2]Department of Chemistry, Texas A & M University, College Station, Texas 77843-325, USA
[3]Supercomputing Center for University and Research, CASPUR, Piazzale A. Moro 5, 00185 Rome, Italy.

INTRODUCTION

The study of low-energy electron scattering from polyatomic molecules has revealed, through a rather wide variety of experimental observations, that one of the most useful features of the cross section behaviour in that energy range is provided by the appearance of marked structures at specific energy values that are strongly dependent on the molecular properties of the relevant target [1]. Such features are generally classed together as resonant features and are loosely described as due to temporary electron trapping in the vicinity of the molecular environment, leading in turn to enhancements of several of the final channels from the scattering event like molecular excitations, dissociative attachment and dissociative recombination probabilities, to cite but a few of the possible outgoing channels [2]. It is therefore of great current interest to try and find relatively simple methods of analysis for such features in order to be able to relate them more directly to specific molecular properties of the target system. This is usually a rather tall order since it has been made clear by extensive theoretical and experimental studies of the last thirty years or so, on both diatomic and polyatomic targets, that such resonant features come about from the rather complicated interplay between the direct and exchange forces acting on the incoming electron, and due to the electronuclear interaction with its target, and the response forces of the latter to the perturbation field caused by the impinging electron. Thus, to correctly describe the resonances, to locate them at the right energy positions and to extract from them information on their effects on scattering attributes, the theoretical treatments have strived to handle the above effects in as much balanced a way as possible and to include in the description of the physical forces as many contributions as possible. A direct result from this required sophistication has been that accurate and reliable data from theory have been fairly slow to come by, even for the simpler diatomic targets like

N_2 or CO and definitely much fewer and far apart in time for the smaller polyatomics like CO_2, H_2O or NH_3 [3,4]. For the larger molecular systems like benzene, SF_6 or C_{60} the existing calculations are unfortunately only at their beginning [5,6].

In order to efficiently balance the effort required to include the correct interactions needed to deal with molecular resonances (both single-particle and Feshbach-type resonances) with a realistic handling of their computational demands, we have recently developped the use of an effective adiabatic potential for polyatomic targets to carry out the analysis of the individual symmetry components of the single-particle resonances which may occur in the lower energy scattering region that extends from the lowest cross section thresholds around 0 eV up to about 30 eV [8]. The model potential that we have introduced can be, generally speaking, a rigorous potential developed from first principles. On the other hand, in order to make calculations more accessible for the larger systems, we have introduced some simplifying features in it that allow us, as we shall show below, to examine fairly complicated multielectron, non-linear targets and to provide microscopic explanations for various of the observed experimental features while performing a marked reduction of the required computational time.

The following Section therefore describes briefly our adiabatic potential model and the ensuing scattering equations, while Section III reports the existing situation on the possible resonant features of electron scattering from Ozone, a fairly complex molecule which we shall use here as an example for our analysis of the reliability of this model potential. Section IV presents our computed adiabatic potentials and the corresponding one-particle resonances obtained with them. Finally, Section V gives the present results for the resonant states, together with their corresponding wavefunctions and discusses their connection with what has been observed by the existing experiment. Our general conclusions are reported in Section VI.

THE MODEL ADIABATIC POTENTIAL

The main focus of this paper is to examine in some detail the mechanism and qualitative characteristics of low-energy, one-electron resonances. It requires a model which is simple enough to be clearly understood but includes sufficient details of the full scattering problem to accurately reproduce the essential features of the more accurate calculations. Thus, we wish to be able to study the resonances by employing some model, purely local potential that we shall call the Static Model Exchange Correlation Polarisation (SMECP) Potential, V_{SMECP}. Although it is certainly possible that the non-locality of the correct exchange interaction could have provided an additional mechanism for resonant trapping, the excellent qualitative agreement between results obtained using exact exchange and local model exchange potentials in analysing the isolated resonances in earlier systems [5,6,7] gives us confidence that the resonant trapping mechanism is likely to be the same for both potentials, although quantitative aspects are bound to be different.

The standard, symmetry-adapted angular momentum eigenstates, $X_{lh}^{p\mu}$, do not form the most compact angular basis set for the electron-molecule scattering problem. An alternative expansion basis set is the angular eigenfunctions obtained from diagonalizing the angular Hamiltonian at each radius r. The angular functions obtained in this fashion are referred to as the adiabatic angular functions $Z_k^{p\mu}(\theta, \phi, r)$ which are linear combinations of the symmetry adapted harmonics discussed before

$$Z_k^{p\mu}(\theta, \phi, r) = \sum_{lh} X_{lh}^{p\mu}(\theta, \phi) C_{lh,k}(r) \qquad (1)$$

where the expansion coefficients are solutions to the matrix eigenvalue equation

$$\sum_{lh} V_{l'h',lh}(r) C_{lh,k}(r) = V_k^A(r) C_{l'h',k}(r) \qquad (2)$$

The eigenvalues $V_k^A(r)$ then form an adiabatic radial potential for each index value k.

Solving the scattering equations using $V_k^A(r)$ can have several advantages. First the expansion of the scattering wavefunction in adiabatic angular states converges more rapidly than the corresponding expansion in angular momentum eigenstates. Thus, the number of coupled radial equations which must be solved with the adiabatic basis set is usually much smaller [6,7,8]. The second advantage is that the numerical instabilities found in the solution of the standard momentum eigenfunction expansion are greatly reduced. The third advantage is that often a single radial adiabatic potential can be responsible for the appearance of a shape resonance. In such cases the adiabatic potential can be used as a qualitative guide to better understand the main features of the resonant state. The spatial extent of the resonant wavefunction can be determined from the well and angular momentum barrier, and the physical mechanism for the resonant trapping is seen to be due to the slow rate of tunnelling through the potential barrier. One drawback to the adiabatic potential approach is that the non-adiabatic radial coupling introduces additional terms in the radial differential equations for which the standard integration method is not directly applicable.

In order to avoid the non-adiabatic coupling terms, we actually employed a piecewise diabatic (PD) representation for the potential [8]. In this approach the radial coordinate is divided into a number of regions so that region i is defined as $r_{i-1} < r < r_i$, with: $r_o=0$. In each radial region we averaged the coupling potential $V_{l'h',lh}(r)$ over r and the resulting averaged potential was diagonalized as in equation (2) to yield a set of angular functions $Z_{k,i}^{p\mu}(\theta,\phi)$. Then in region i the scattering potential was transformed into the new representation in which it was nearly diagonal. The resulting equations could then be solved using the full scattering potential in each region. A further approximation we made was to ignore the off-diagonal coupling in each region. If the regions were small enough this turned out to be a rather good approximation.

The key step in solving the radial equations using the PD approach was matching the radial functions and their derivatives at the boundary between radial regions. The transformation of the radial functions from one region to the next was given by the transformation matrix $U_{k,k'}^{(i+1 \leftarrow i)}$ defined by

$$U_{k,k'}^{(i+1 \leftarrow i)} = \sum_{lh} C_{lh,k}^{(i+1)} C_{lh,k'}^{(i)} \qquad (3)$$

When the size of the angular momentum eigenfunction basis used was larger than the size of the diabatic angular basis set, the transformation matrix $U_{k,k'}$ was not in general unitary. In order for the unitarity of the **S** matrix to be mantained, it was essential to modify $U_{k,k'}^{(i+1 \leftarrow i)}$ so that it was unitary. We accomplish the unitarization of $U_{k,k'}^{(i+1 \leftarrow i)}$ using simple Graham-Schmidt orthonormalization on the columns of $U_{k,k'}^{(i+1 \leftarrow i)}$.

A narrow and isolated resonance in a scattering process at an energy E_R and width Γ can be shown to be due to a pole in the **S** matrix which has been analytically continued into the complex plane at an energy $E = E_R - i(\Gamma/2)$. The **S** matrix is obtained from a scattering wavefunction by finding a solution with the usual asymptotic form that includes the appropriate Hankel functions. It is then possible to compute directly the **S** matrix at complex energy by computing the solution using the standard numerical procedures with a complex valued energy and by matching the solution to the Hankel functions with the appropriate complex argument. A resonance energy in the scattering process can be located by finding those energies at which $0 = 1/\det \mathbf{S}$.

Once the zeros of $1/\det \mathbf{S}$ in the region of interest have been located, the actual scattering state at that complex energy can be computed. We can then analyse such resonant scattering states in terms of their radial functions and as full three-dimensional spatial functions. One interesting observation which has been made previously is that the resonant states could be compared to the virtual orbitals from a much simpler MBS-SCF calculation, which uses a minimal basis expansion. We have in fact shown before that useful comparisons between the scattering resonant functions and the L^2 virtual orbitals can be obtained fom such simple calculations. The eigenphase sums obtained from the SECP calculations are also fitted, whenever possible, to a Breit-Wigner form to extract the resonance parameters using the familiar functional form [1].

$$\delta(E) = a + b(E - E_R) + c(E - E_R)^2 + \tan^{-1}\left[\frac{\Gamma}{2(E - E_R)}\right] \quad (4)$$

We will show in the following, and we have already extensively shown such features in previous systems, [6, 7, 8], that the generation of such potential 'curves' can guide us in the analysis of possible resonant states and of their spatial features in terms of molecular orbitals of the target molecule.

One simple approximation for the target wavefunction is to use a single Slater determinant constructed from a product of one-electron molecular orbitals. Using the Hartree-Fock (HF) wave function or a self-consistent field (SCF) approximate solution of the HF equations to represent the target, and truncating the expansion over target states of the total $(N + 1)$ electron wavefunction to one state only, leads to the exact-static-exchange (ESE) approximation for the potential to be employed in the scattering equations. For a target which has a closed shell electronic structure with n_{occ} doubly occupied orbitals, ϕ_i, this potential can be written as

$$V_{SE} = \sum_{\gamma=1}^{M} Z_i |r - R_\gamma|^{-1} + \sum_{i=1}^{n_{occ}} 2\hat{J}_i - \hat{K}_i \quad (5)$$

where $\hat{J}_i$ is the usual local static potential defined by

$$\hat{J}_i(r_1) = \int \frac{\phi_i^*(r_2)\psi_i(r_2)}{r_{12}} d^3r_2 \quad (6)$$

and $\hat{K}$ is the non-local exchange potential operator defined by

$$(\hat{K}_i \psi)(r_1) = \phi_i(r_1) \int \frac{\phi_i^*(r_2)\psi(r_2)}{r_{12}} d^3r_2 \quad (7)$$

Electron-molecule scattering cross-sections computed using the ESE potential are usually in fairly good agreement with experimental total scattering cross-sections. However, a major deficiency in the model for studying negative ion resonances is the lack of target response, i.e. electron correlation, in the scattering process. At higher energies, this is reflected in the fact that electronically inelastic processes are not obtained in the single-channel ESE approximation. At low energies, the lack of target response leads to the neglect of important polarization effects. This can significantly alter the energy and width of resonances which occur below a scattering energy of ~ 10 eV. This difficulty can be avoided by the inclusion of additional target states in the expansion of the wavefunction. However, this approach can significantly increase the computational difficulty of the problem.

The approach which we have used here attempts to include the effects of correlation and polarization through the addition of a local, energy-independent model potential $V_{CP}(r)$. [9, 10] Briefly, this potential contains a short-range correlation potential smoothly connected to a long-range polarization potential. The short-range correlation potential V_{CP} is obtained by defining an average correlation energy of a single particle, within the formalism of the Kohn and Sham variation theorem, and by obtaining the short-range correlation forces as a functional of the target electron density. The long-range polarization potential V_{CP} is then obtained by first constructing a model polarization potential which asymptotically agrees with the potential obtained from the static polarizability α of the molecule. This corresponds to including the first term in the second-order perturbation expansion of the polarization potential. This model potential can be either constructed assuming a single polarization center, or by partitioning the static polarizability to different centers.

In general the long-range potential does not exactly match the short-range potential V_{CP} at any given value of r. To select an appropriate matching radius, r_{match}, at which we wish to connect these two potentials, we first expand both V_P and V_C in a single-center expansion. Then we find the radii where the two $\ell=0$ radial potentials intersect. In all cases considered to date, these potentials cross at two values of r. Empirically we found that the best matching radius is the smaller of these two radii where the $\ell=0$ potentials cross. Then the final definition for the model correlation-polarization potential is

$$V_{CP}(r) = \begin{cases} V_C(r) & r < r_{match} \\ V_P(r) + \sum_{lm} C_{lm} r^{-\lambda(l)} Y_{lm}(\theta,\phi) & r > r_{match} \end{cases}$$

where the values of C_{lm} have been determined so that potential is continuous at r_{match} and where λ is a function of ℓ such that $\lambda=6,5,6$ for $\ell=0,1,2$ and $\lambda(\ell)=\ell+2$ for $\ell \geq 3$. The function $\lambda(l)$ was chosen so that the term added to V_P has the same functional form as the first term neglected in the perturbation expansion of V_P. The static-exchange correlation-polarization (SECP) scattering potential was then used for the full scattering problem, V_{SECP}, and was just the sum of the V_{ESE} defined in equation (7) and V_{CP} defined in equation (8).

Although we solved many times the scattering problem using V_{SECP}, [11] in order to effectively implement the present diabatic model we need to further modify the interaction and therefore the non-local exchange was replaced by an approximate local model exchange (ME) potential denoted by V_{ME}. For this purpose we have used the Hara free-electron-gas-exchange (HFEGE) potential [9]. In V_{ME}, the exchange interaction is approximated by an energy-dependent local function of the electron density of the target. We found that the energy dependence of the HFEGE potential was fairly weak so that scattering results over a given range of energies where the mean energy was used in the HFEGE potential were very similar to the results where the actual scattering energy was used to compute the HFEGE potential. The potential where the model exchange was used in place of the exact exchange will be referred to as the static-model-exchange-correlation-polarization (SMECP) potential V_{SMECP}.

When the scattering potential is purely a local potential, standard differential equation integrators can be used to obtain the wavefunction. The asymptotic form of the wavefunction is then analysed to obtain the appropriate scattering matrix from which total and differential cross-sections can be computed. When the potential given in equations (5) and (8) contain non-local terms an alternative approach must be used. We have developed a particularly powerful approach based on the Schwinger variational expression for the scattering matrix with Padé approximant corrections that was discussed several times in our earlier publications [11].

In the present study, however, we wish to illustrate the powerful pictorial value of the model diabatic potential discussed above and therefore we will present the results obtained not by solving the coupled scattering equations with the more correct non-local interaction but rather by using the diabatic potential curves discussed before, whereby a more limited number of channels is employed to obtain the low-lying metastable states (resonant scattering states) supported by the V_{SMECP} model potential in the case of the Ozone molecule.

RESONANT SCATTERING FROM OZONE

Having information on the general features of the scattering of electrons by Ozone is especially important to the studies of upper atmosphere processes, where the electronic quenching mechanisms and energy transfer efficiency involving O_3 lower-lying electronic states can increase the populations of its excited states thereby perturbing the local thermodynamic equilibrium. The consequences of such changes in local population can then appear as increased emission from infrared-active gases, thereby affecting radiative cooling and temperature structure of the atmosphere. On the other hand, perhaps because of its difficult preparation and handling and of its corrosive properties, fairly few experimental studies have been carried out on this system. Electron-energy-loss spectra (EELS) of Ozone have been reported by various authors and further energy-loss spectra and absolute differential cross sections (DCS) were also reported at energies between 3 and 20 eV and angles between 12° and 168° (ref. [12]). More recently, an experimental collaboration [13, 14] reported vibrational energy-loss spectra along a series of energies between 3.5 and 7.0 eV, over a range of scattering angle between 40° and 120°. Partial electron ionization cross sections for incident electron energies from 40 to 500 eV have also been determined by using time-of-flight mass spectrometry [15], while, most recently, the dissociative electron attachment in Ozone has been explored first for electron energies between 0 and 10 eV [16]. and by Allan et al. (ref. [12]), who also reported vibrationally inelastic differential cross sections in the resonance region.

The corresponding theoretical studies on the structure of Ozone in its ground electronic state and in its lower-lying electronic states have also been quite extensive over the years, as could be gleaned from the discussion and results of ref. [17]. On the other hand, the actual calculations for the scattering dynamics and scattering observables in the case of electron as projectiles have been rather few and fairly limited in scope. No calculations, however, had appeared at a more sophisticated level before the work of Okamoto and Itikawa [18], in which differential, integral and momentum transfer cross sections were computed for the vibrationally elastic process at collision energies between 5 and 20 eV. Further ab initio work at the exact static-exchange (ESE) level of approximation was carried out more recently by Sarpal et al. [19] using the polyatomic R-matrix method, where integral elastic scattering cross sections were evaluated and compared with the few existing experiments. Additional static-exchange calculation have been recently completed by making use of the Schwinger Variational Iterative method (SVIM) [20]. Our recent work on angular distributions and rotationally inelastic processes has employed the ESE potential with the addition of the V_{CP} potential discussed above [21].

The specific analyses of possible resonant states, although raised by several of the papers discussed above has not yet been completely discussed in our work. However, the presence of resonant behaviour was clearly pointed out by the experiments on the vibrational excitation process and, most recently, also surmised by dissociative attachment experiments observing the threshold behaviour of the attachment cross sections in the $O_2^- + O$ dissociative channel. We therefore intend to show below that the generation of the diabatic potentials V_{SMECP}, for each of the relevant symmetries of the C_{2v} molecule kept at its equilibrium geometry, is already capable of giving us some qualitative understanding of the molecular origins for the effects which indicate selective resonant excitations of different vibrational modes in the experiemntal findings. The present calculations employ the target wavefunction, the nuclear geometry and the polarisability value already discussed in our previous study on angular distributions from the same system.[21]

Symmetry	E_R (eV)	$\Gamma_R(eV)$
$^2A_1$	6.05	1.54
$^2A_2$	0.34	0.006
$^2B_1$	2.10	0.74
$^2B_2$	0.35	0.08
$^2B_2$	10.76	1.03

Table 1.: Computed resonance positions and widths using the V_{SMECP} diabatic potential discussed in the main text.

DIABATIC POTENTIALS AND SCATTERING RESONANCES

In Table 1 we are reporting the widths and positions of the metastable states obtained from the model sector-diabatic potentials descussed in the previous Section II. The numerical calculations using the V_{SMECP} were carried out using very small sectors for a total of about 40 sectors up to an outer distance of 20 a_o. The HFEGE model exchange was evaluated at a fixed energy value of 5.0 eV. For each of the I.R. considered in the present work, we have solved coupled-channel equations which include ten diabatic channels. The totally-symmetric representation a_1 is seen to produce a rather broad resonance (Γ_{a1}=1.5 eV) located around 6.05 eV on the energy scale of the FN calculations. The lower potential curves for each of the contributing partial waves (from the lowest curve for $l = 0$ up to $l = 3$) are shown in Figure 1. It is interesting to note there that the $l = 1$ component shows a very small barrier of a few meV at a distance of about $4a_o$ away from the center of mass of the C_{2v} molecule, while the stronger barriers appears for $l = 2$ and, much higher up in energy, for $l = 3$. It turns out that the dominant contribution to the a_1 resonance of Table 1 is indeed the $l = 2$ component and therefore the present calculations suggest that a metastable wavefunction for the continuum electron could have two nodal planes in the 3D space of the equilibrium geometry of the molecule. The spatial presentation of such wavefunction is given by Figure 2 where a cut of the full wavefunction is given across the molecular plane (yz). The projection on the plane containing the three atoms clearly shows that the additional resonant orbital is located in the interbond region where its density fills the area between the three bound atoms and also adds additional electronic density along the molecular bonds. In other words, the metastable state supported by the $l = 2$ effective potential of our calculations is seen to be able to couple the extra electron in it with those in the bonding region and in the region between bonds. Thus, we can expect that the formation of the metastable ion would cause for this particular symmetry equally strong vibronic couplings with the stretching and bending motions of the nuclei since it is seen to be capable of modifying the force constants for both modes.

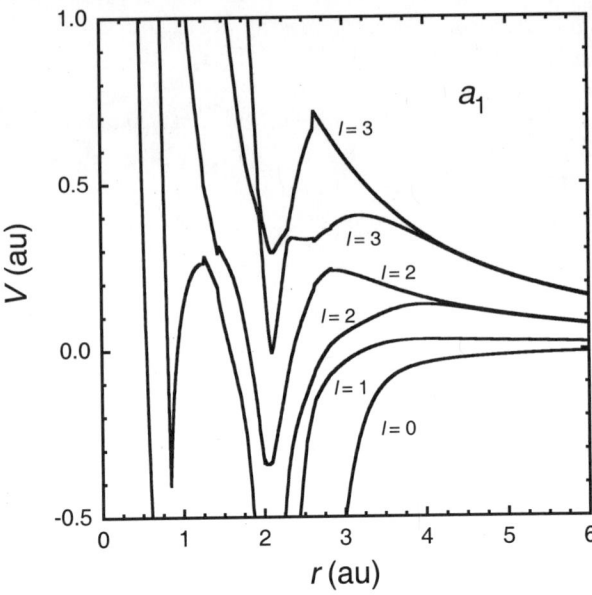

Figure 1.: Computed diabatic potential curves for the a_1 symmetry of the ozone molecule.

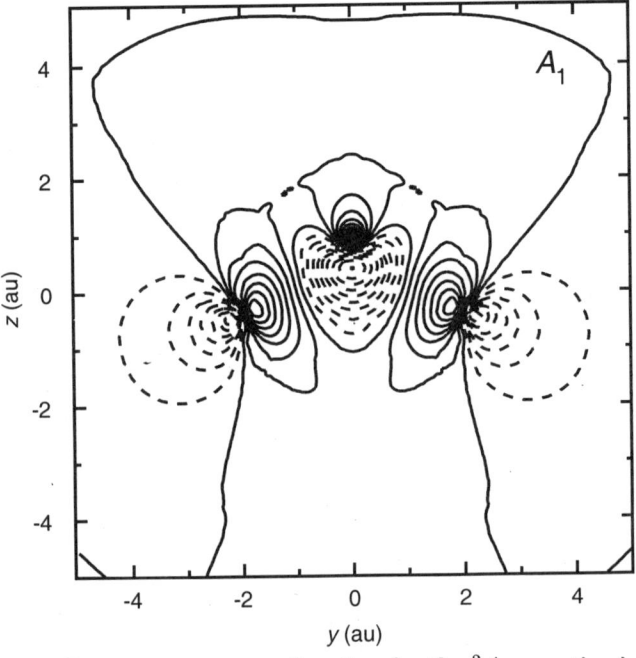

Figure 2.: Resonant state wavefunction for the $^2A_1$ negative ion of O_3. The plane shown is the one of the molecule (yz) with the origin at the center of mass. The dashed density curves correspond to negative values and help to identify nodal planes.

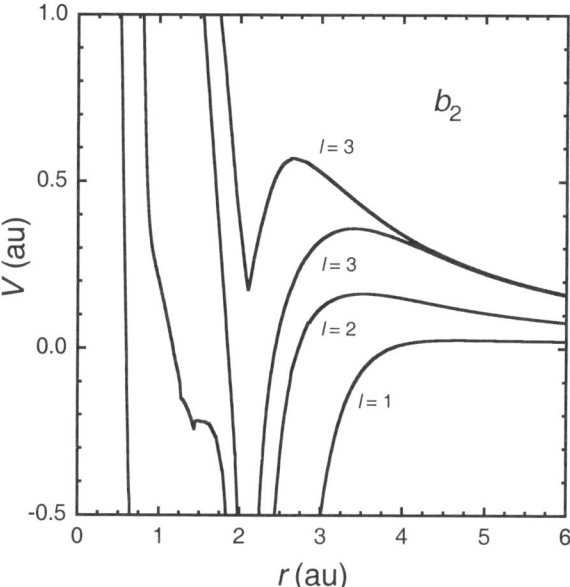

Figure 3.: Computed diabatic potential curves for the b_2 symmetry of the ozone molecule.

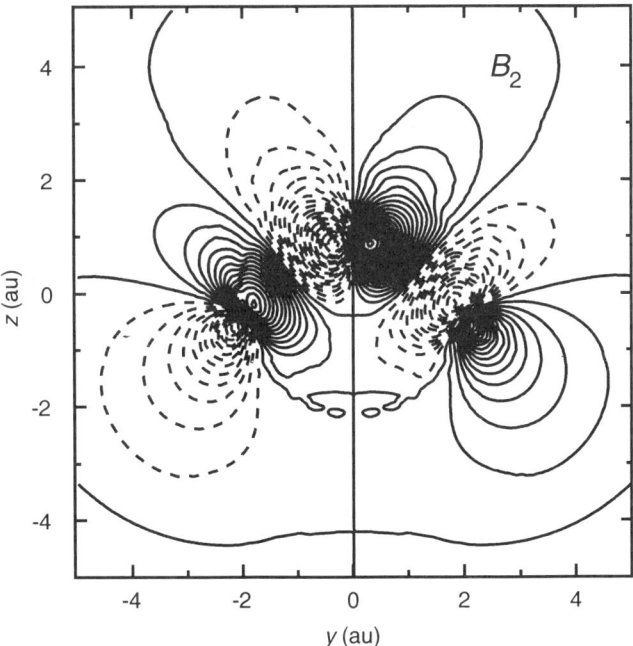

Figure 4.: Resonant state wavefunction for the $^2B_2$ resonance at 10.76 eV. The plane shown is the same as in Figure 2.

If we now return to the analysis of the scattering experiments which observed vibrational excitations of O_3, [22] we find that indeed the experimental resonance located around 4.2 eV is shown to excite both the ν_1 stretching vibration and the ν_2 bending vibration and therefore it is reasonable to say that the associated temporary negative ion belongs to the a_1 symmetry and causes vibronic coupling with both modes via its additional charge distribution within the molecular volume. This suggestion was also made in the analysis of the experimental data, [22] but it is interesting to see that our present calculations can attach some specific topological reasons for such physical behaviour.

The same experimental analysis [22] also found that a second, b_2, resonance appeared around 6.6 eV in the measurements but that the latter resonant process only excited the stretching mode and not the molecular bending mode. The present calculations for the corresponding piecewise diabatic potential of that I.R. are shown in figure 3, where we see that both the $l = 1$ and $l = 2$ components exhibit repulsive barriers, albeit at different energy positions. Hence, our calculations indicate the presence of a very-low energy, very narrow resonance at 0.35 eV, with a width of 80 meV and of a second resonance at higher energy (10.76 eV) with a broader width of 1.03 eV. Given the fact that the present model potential treats exchange interactions only in an approximate, localized form, it is not surprising to find that our computed resonances are usually shifted at higher energies with respect to experiments, especially when vibrationally inelastic channels are being examined (while our calculations are strictly within the FN approximation). Therefore, we assigned the experimental resonance at 4.1 eV to our calculated a_1 resonance at 6 eV and can now reasonably surmise that our computed b_2 resonance around 10 eV is describing the one found experimentally around 7 eV. The corresponding wavefunction of our computed metastable state for the dominant $l = 3$ component is seen in Figure 4, where the charge distribution in the molecular plane clearly indicates additional charge being located along the bonds but not in the intramolecular angular region. Hence, one can qualitatively say that the vibronic coupling with the stretching mode is likely to be caused by the perturbation of that force constant induced by the metastable electronic state which we see to populate primarily the spatial region along the two bonds. It is also interesting to note that, as suggested by the experiments [22], it is only the b_2 resonant state which couples with both the symmetric (ν_1) and antisymmetric (ν_3) stretching motions of the target but not with the bending motion (ν_2) as the latter excitation does not appear in the energy-loss experiments conducted at the higher resonance energy position [22]. Thus, our present calculations with the model diabatic potential clearly provide physical explanations for the excitation selectivity seen by the scattering experiments at the resonances: the differences in the spatial modifications of the molecular electronic density induced by the metastable anions of different symmetry are responsible for different vibronic couplings within the molecule hence cause the selective excitations of different vibrational modes.

As mentioned earlier, the Ozone molecule is also known to have a stable negative ionic structure, the ozonide anion, with an electron affinity (vertical) of 2.10 eV [16]. The present calculations for the b_1 negative ion are shown in Figure 5, where the model potential curves are reported for the partial waves up to $l = 3$. The dominant component of the resonant state which we find here is the $l = 2$ component, although the $l = 1$ p-wave also contributes to it. The corresponding wavefunction for the metastable state, which, from our calculations, turns out to be rather narrow ($\Gamma_R \sim 0.7eV$), is shown in Figure 6 and is seen to behave as a molecular π^*-orbital distributed fairly uniformly above and below the molecular plane shown in the figure as the vertical axis at x=0: it is antibonding along the oxygen bonds but it is bonding above and below the terminal O atoms. This is very much in agreement with the suggested experimental symmetry for the stable ozonide ion [16], and for the π^* nature of its extra electron. Thus, although the present model calculations locate it as a narrow resonance but shifted by about 4 eV on the high-energy side, they are very likely describing the same state.

Another interesting point to consider is the recent experimental suggestion of a fairly narrow resonance at the threshold of the dissociative attachment channel for the $O_2^- + O$ process [16, 23]. The present calculations allow us to also analyse this aspect of the metastable states of O_3^- and can suggest possible symmetries that may be responsible for that behaviour.

The present diabatic potential for the a_2 symmetry is shown in Figure 7, where we clearly see that the $l = 2$ potential curve is the only possible component that can contribute in the low-energy region and is the only one which shows very little coupling with the higher channels that could cause dynamical decay of the resonance by flux 'leaking' through the nonadiabatic coupling, as seen by us in previous polyatomic systems [5].

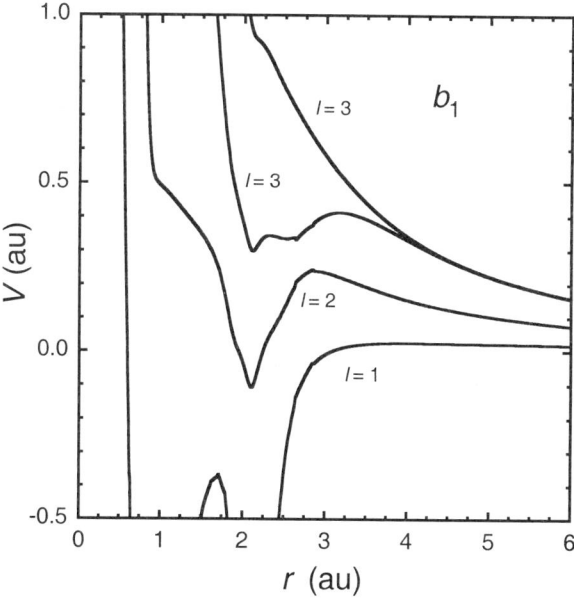

Figure 5.: Computed diabatic potential curves for the b_1 symmetry of the ozone molecule.

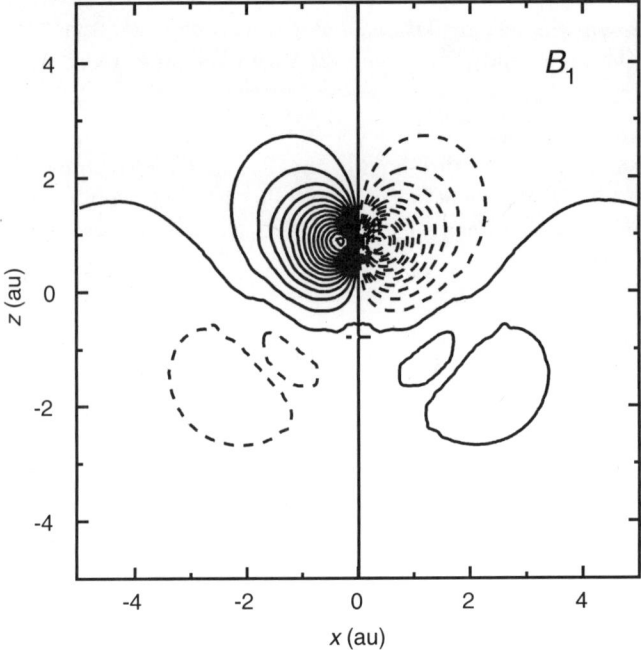

Figure 6.: Resonant state wavefunction for the $^2B_1$ resonance. The plane shown gives a sideview of the molecule, with the three oxygen atoms along the vertical axis at x=0 and with the central atom in the upper position.

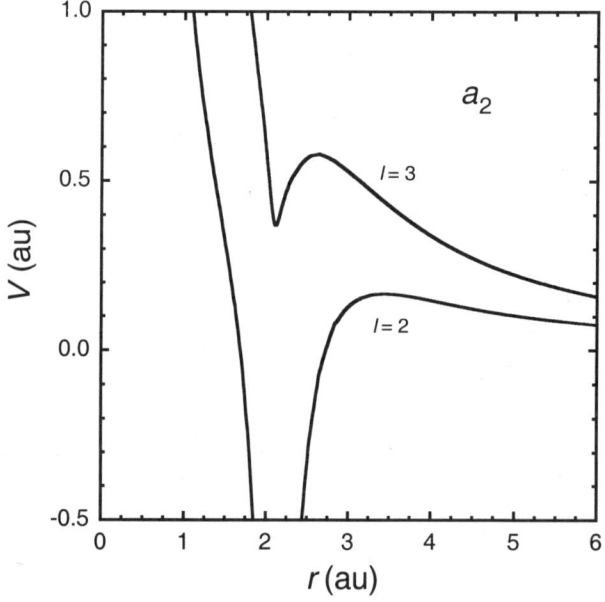

Figure 7.: Diabatic potential energy curves for the a_2 symmetry of the ozone molecule. Only the lowest two curves are shown here.

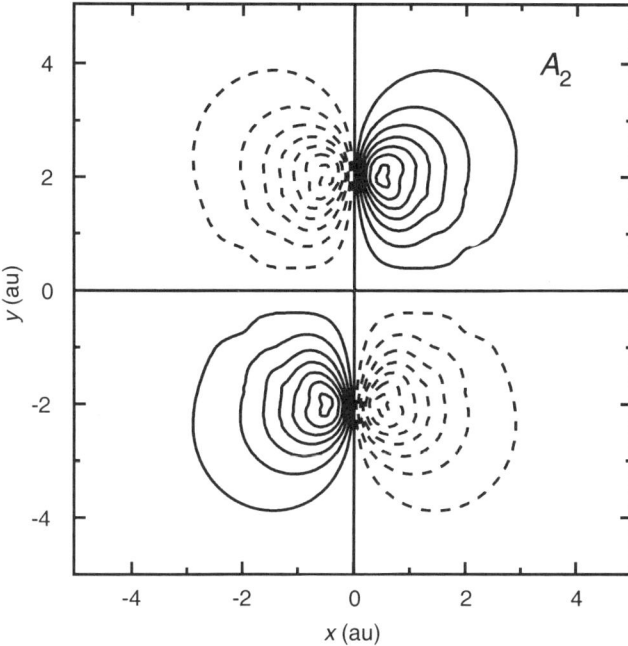

Figure 8.: Resonant state wavefunction for the $^2A_2$ resonance. The plane shown is the (xy) plane with the three oxygen atoms along the vertical axis at x=0. Here the central O atom lies at the crossing with the y=0 axis and the outer O atoms are above and below that crossing point.

The present calculations suggest that such potential indeed supports a metastable state at very low energy (0.34 eV) and with the smallest width which has been produced by our present calculations for Ozone ($\Gamma \sim 6 meV$!). The corresponding wavefunction is shown in Figure 8 and indicates that the resonant state, due nearly exclusively to trapping behind an $l = 2$ single barrier, has the shape of a molecular d*-orbital located away from the central Oxygen atom and with mostly a two-center distribution over the outer Oxygen partners. In contrast to the previous resonances causing the vibrational excitation discussed above (a_1 and b_2), which would be related to the σ_u^*. resonant state ($^4\Sigma_u^+$) of oxygen, we have here a metastable state which resembles more a $\pi*$ resonance of O_2 that could be formed during the detachment process.

A further possible candidate for such a low-lying resonance could be had by looking at the lower resonant state of b_2 symmetry which we are reporting in Figure 9. For this resonance also the $l = 2$ potential channel (see Figure 3) is present but also shows nonadiabatic coupling to the $l = 1$ component which was not present in the a_2 potential curves of Figures 7. Hence the lower b_2 resonance, although also dominated by the d-wave component, turns out to have a larger width of 80 meV because of this nonadiabatic dynamical coupling through which the metastable state can 'leak out' into the continuum. It however appears at the same very low energy of about 0.3 eV (see Table 1). The actual 3D shape of that metastable orbital is shown by the plot of Figure 9 and indicates that this latter resonance, contrary to the spatial topology of the a_2 resonance of Figure 8, is now mostly located on the molecular plane although it also has a major portion of its density on the outer O atoms. One could therefore suggest that, if both resonant states can exist at energies very close to the onset of the detachment process, the a_2 component is likely to live longer and less likely to undergo break-up by nonadiabatic dynamical coupling with other partial wave components. However, both of these resonant states appear to locate the extra electron mostly on the O_2^- fragment of the ozonide ion and therefore are likely to be reasonable candidates for the observed threshold resonance in the detachment channel [23].

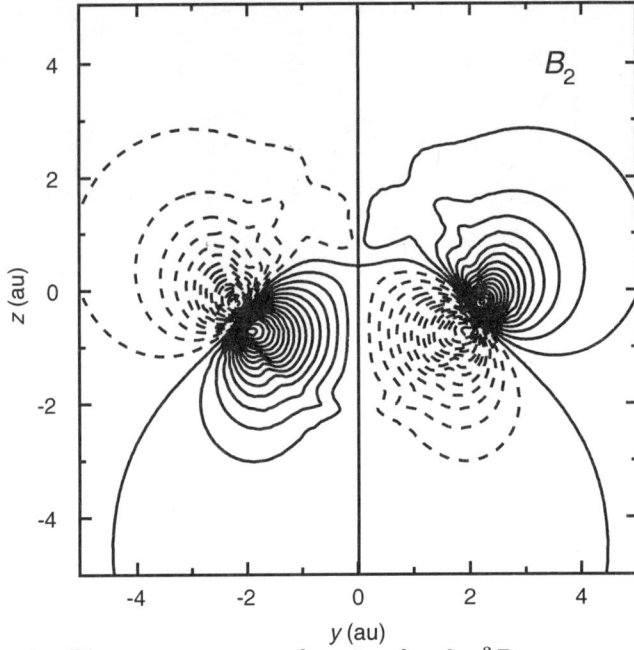

Figure 9.: Resonant state wavefunction for the $^2B_2$ resonance at 0.35 eV. The plane shown is the one of the molecule, with the central oxygen atom along the vertical axis at x=0 and located between 1.0 and 2.0 a_o along the z axis.

GENERAL CONCLUSIONS

In the present work we have presented the use of a model diabatic potential for polyatomic targets to carry out the analysis of their resonant behaviour in the low-energy electron-molecule scattering processes. The resonances which we have focused on are chiefly the one-electron resonances associated with the electronically elastic channels but which can still be coupled to the nuclear inelastic channels that are open at the considered energies.

The calculations were carried out within the FN approximation for the nuclear motion and, in order to construct the model potential in a local form, we have employed the Semiclassical Exchange Approximation described in our earlier work and outlined in the previous Section.

The multipolar components of the Piecewise Diabatic Potential were therefore computed for several contributing partial waves and for all the relevant Irreducible Representation of the title molecule. The calculations of the low energy metastable states, the analysis of the spatial features of their corresponding wavefunctions and the evaluation of their widths allow us to carry out a fairly detailed description of the single-particle resonances in ozone and of their bearings on the experimental findings. In particular, we were able to establish the following:

(i) the experimentally found resonance around 4.2 eV can be associated to a metastable anion of $^2A_1$ symmetry and to a dominant electron trapping by an $l = 2$ potential barrier which our calculations locate around 6 eV and with an FN width of 1.54 eV;

(ii) the second resonance at higher energies observed by the experiments [22] around 6.6 eV is also found by our calculations and is attributed to a metastable state of $^2B_2$ symmetry located around 10 eV. Here the dominant trapping appears to take place via a barrier with $l = 3$ angular momentum;

(iii) the experimentally observed peak at the threshold for the dissociative attachment process, a narrow peak at very low energy in the $O_2^- + O$ channel recently presented by the Märk and coworkers, [23] is also attributed by our calculations to a very narrow resonance (width of about 6 meV!) appearing at about 0.3 eV and in the $^2A_2$ symmetry. The narrow nature of this resonance can be related to the marked dominance of the $l = 2$ diabatic potential trapping with negligible coupling to other dissociative channels for different partial waves;

(iv) the experiments have further suggested that the a_1 resonant state causes vibronic coupling effects which give rise to excitation of both the stretching mode (ν_1) and the bending mode (ν_2). The inspection of our calculated metastable wavefunction shows that the (N+1) electron density is located both within the two bonds and along each of them, hence perturbing the associated force constants of both modes. In other words, our calculations appear to confirm that the electronuclear perturbation induced by the temporarily bound extra electron is capable, in this specific symmetry, to trigger both modes of vibrational excitation;

(v) the resonance located at the higher energy (around 6.6 eV) was seen in the experiments [22] to cause selective vibrational excitation of only the stretching mode (ν_1). The corresponding metastable wavefunction obtained from our calculations (see Figure 4) indeed shows that the extra electron distribution is mostly located along the two bonds, with little extra contribution in the angular region between those bonds. Hence, it becomes reasonable to explain, albeit qualitatively, the findings from the experiments as related to the existence of markedly reduced electronuclear perturbation by the metastable electron of b_2 symmetry of the bending force constant of the molecule;

In conclusion, the present application of a model potential approach to the study and to the spatial analysis of metastable states of multielectron, multiatomic molecules shows that it can provide a powerful tool for the pictorial description of the microscopic dynamics and of the electronuclear coupling effects which preside over the formation of such single-particle molecular resonances.

Acknowledgements

The present work is dedicated to Professor P.G. Burke, on the occasion of his retirement from active academic duties at Queen's University Belfast. We wish to pay here our most heartfelt respects to a great scientist, an inspiring guide and an indefatigable mentor of the crucial role of computational tools for our understanding of elementary scattering phenomena. We are also grateful and honoured by the long-standing friendship which we have with Phil and look forward to continue receiving his help and advice for many years to come.

FAG also thanks the Italian Nat. Research Council (CNR) and the Italian Ministry of Scientific Research (MURST) for financial support. RRL acknowledges partial support from the Robert A. Welch Foundation of Houston, Texas, under grant A-1020. Both RRL and FAG finally thank the NATO Organisation for the award of a Collaborative Research Grant (no. 950552).

REFERENCES

1. H. Erhardt and L.A. Morgan, Eds., 1994, Electron Collisions with Molecules, Clusters and Surfaces, Plenum Press, New York
2. T.D. Märk, and G.H. Dunn, Eds., 1985, Electron Impact Ionization, Springer Verlag, Berlin
3. e.g. see: B.H. Lengsfield, T.N. Rescigno and C.W. Mc Curdy, *Phys. Rev. A*, 44:4296 (1991)
4. F.A. Gianturco and D.G. Thompson, The theory of scattering of slow electrons from polyatomic molecules *Comm. At. Mol. Phys.* 16:307 (1985)
5. F.A. Gianturco, R.R. Lucchese and N. Sanna, Electron Scattering from sulphur hexafluoride, *J. Chem. Phys.*, 100: 6464 (1994)
6. F.A. Gianturco, R.R. Lucchese, One-electron resonances and computed cross sections in electron scattering from the benzene molecule, *J. Chem. Phys.*, 108:6144 (1998)
7. R.R. Lucchese and F.A. Gianturco, One particle resonances in electron-molecule scattering, *Int. Rev. Phys. Chem.*, 15: 429 (1996)
8. F.A. Gianturco, R.R. Lucchese, The elastic scattering of electrons from CO_2 molecules, *J. Phys. B: At. Mol. Opt. Phys.* 29:3955 (1996)
9. F.A. Gianturco and A.K. Jain, The theory of electron scattering from polyatomic molecules, *Phys. Rep.* 143:347 (1986)
10. F.A. Gianturco, A. Jain and L.C. Pantano, A parameter-free model potential for correlation forces in electron-methane scattering *J. Phys. B: At. Molt. Phys.*, 20:571 (1986)
11. F.A. Gianturco, R.R. Lucchese, N. Sanna and A. Talamo, A generalized single-center approach for treating electron scattering from polyatomic molecules: in: *Electron Collisions with Molecules, Clusters and Surfaces*, H. Erhardt and L.A. Morgan, Eds., Plenum Press, New York (1994)
12. T.W. Shyn and C.J. Sweeny, Angular distribution of electrons elastically scattered from ozone, *Phys. Rev. A*, 47:2919 (1993)
13. J.A. Davies, W.M. Johnsone, N.J. Mason, P. Biggs and R.P. Wayne, Vibrational excitation of ozone by electron impact, *J. Phys. B: At. Mol. Opt. Phys.*, 26:L767 (1993)
14. W.M. Johnsone, N.T. Mason, W.R. Newell, P. Biggs, G. Marston and R.P. Wayne, Differential oscillator strengths produced by electron impact energy-loss spectroscopy of ozone, *J. Phys. B: At. Mol. Opt. Phys.*, 25:3873 (1992)
15. K.A. Newson, S.M. Luc, S.D. Price and N.J. Mason, Electon -impact ionization of ozone, *Int. J. Mass. Spectr. Ion. Processes* 148:203 (1995)
16. I.C. Walker, J.M. Gingell, N.J. Mason and G. Marston, Dissociative electron attachment (DEA) in ozone 0-10 eV, *J. Phys. B: At. Mol. Opt. Phys.*, 29:4749 (1996)
17. A. Banichevich and S.D. Peyerimhoff, Theoretical study of ground and excited states of ozone in its symmetric nuclear arrangement, *Chem. Phys.*, 174:93 (1993)
18. Y. Okamoto and Y. Itikawa, Elastic scattering of electrons by ozone molecules, *Chem. Phys. Lett.*, 203:61 (1993)
19. B.K. Sarpal, K. Pfingst, B.M. Nestmann and S.D. Peyerimhoff, Resonances in the integral elastic cross sectio of ozone below 20 eV, *Chem. Phys. Lett.*, 230:231 (1994)
20. M.T. Lee, S.E. Michelin, T. Kroin and L.E. Machado, Elastic cross sections for electron-ozone collisions, *J. Phys. B: At. Mol. Opt. Phys.*, 31:1781 (1998)
21. F.A. Gianturco, P. Paioletti and N. Sanna, Angular distributions and rotational excitations for electron scattering from ozone molecules, *Phys Rev. A.*, 58 (1998)
22. M. Allan and D.B. Popovic, Vibronic coupling and selectivity of vibrational excitation in the negative ion resonances of ozone, *Chem. Phys. Lett.*, 268:50 (1997)
23. T.D. Märk, unpublished results (1998)
24. F.A. Gianturco and S. Scialla, Local approximations of exchange interactions in electon-molecule collisions *J. Phys. B: At. Mol. Opt. Phys.*, 20:3171 (1987).

ASPECTS OF AN *AB INITIO* APPROACH TO ELECTRON SCATTERING BY SMALL MOLECULES

T. N. Rescigno

Physics and Space Technology Directorate
Lawrence Livermore National Laboratory
Livermore, CA 94550 USA

INTRODUCTION

It has been twenty years since a group of about thirty scientists from both sides of the Atlantic met in Asilomar, California to discuss various theoretical and computational approaches that were being developed to study electron-molecule collisions. The monograph[1] that resulted from that workshop provides an interesting look at a field that was then in its infancy, at least from the point of view of computations. It was, not surprisingly, at this time that the first generation of vector supercomputers were coming on line.

Considerable progress has been made over the past twenty years in extending the scope of electron-molecule calculations to include polyatomic targets and processes such as electronic excitation. In 1978, theoretical calculations were largely confined to simple diatomics and, aside from a few perturbative Born or distorted wave calculations, there were no serious calculations of electronic excitation. Twenty years ago, a converged static-exchange calculation on a diatomic molecule was an undertaking that pushed the limits of available computing power. The field has progressed considerably since that time. Some of this development has involved the detailed study of processes at extremely low collision energies or at energies very close to vibrational and/or rotational thresholds[2]. Another area that has seen considerable progress is the development of accurate model potentials to study processes such as vibrational and rotational excitation in electronically elastic scattering of electrons by polyatomic targets[3]; Prof. Gianturco has reviewed much of that progress.

I would like to confine my remarks to *ab initio* methods for studying electron-molecule collisions. To avoid any semantic differences or misunderstanding, let me say that by *ab initio*, I am referring to methods that treat the electronic Schrödinger equation, not as that of one particle moving in a (non-spherical) potential field, but as a *many-electron* equation describing the motion of N+1 indistinguishable particles. This equation must be solved subject to asymptotic scattering boundary conditions. In considering the interactions between electrons, we may find it useful to think of a scattered electron subject to a static interaction, arising from the average Coulomb field of the other electrons, an exchange interaction, arising from the Pauli principle, and a polarization interaction, arising from the dynamic distortion of the target by the incident electron. This decomposition of the electron-molecule interaction may form the basis for subsequent approximations but such components are not directly visible in the many-electron Schrödinger equation. *Ab initio* electron-molecule scattering should be thought of as a problem in continuum electronic structure.

When discussing *ab initio* methods for electron-molecule scattering, it is useful to distinguish variational from direct methods. Direct methods are those that attempt to compute the scattering wave function by direct numerical solution of the Schrödinger equation. By contrast, variational methods use a finite basis of (N+1)-particle functions to construct a trial wave function with unknown linear coefficients; these coefficients are determined through a stationary principle for a quantity such as the T-matrix or scattering amplitude. It is interesting to compare the variety of *ab initio* techniques that were being proposed to handle electron-molecule collisions twenty years ago, with the relatively small numbers of methods that have been extensively developed since then. The direct methods, which were a natural extension of numerical close-coupling techniques used in electron-atom scattering calculations and rely on single-center expansions of the wave functions and interaction potentials, have proved difficult to implement in general multi-state expansions and are generally used only in the context of model potentials. Some of the early methods that were proposed – such as the T-matrix method[4] and the continuum multiple scattering method [5] – have gone the way of the wooly mammoth. Some elements of these early ideas have been incorporated into the more powerful variational methods that have survived, namely, the R-matrix[6], the Schwinger multichannel[7] and the complex Kohn[8] methods. Significantly, Phil Burke and his collaborators have been involved in the development of *ab initio* methods for electron-molecule scattering from the outset and are responsible for the modern development of one of the most successful methods currently in use.

It is somewhat sad to note that there are now fewer groups involved in the calculation of electron-molecule cross sections than there were twenty years ago and that only a handful are involved with *ab initio* calculations on polyatomic targets. In fact, the problem has turned out to be an immensely difficult computational undertaking. And just as the advent of vector supercomputers signaled the first wave of serious computational developments twenty years ago, it may well be the current generation of parallel computers based on large numbers of distributed memory microprocessors that provides the means for continued progress in the field. The need for electron-molecule collision data is perhaps now greater than ever. The increasing importance of low temperature plasma processing as a tool in the commercial manufacture of semiconductor microelectronic devices has created a critical need for electron-molecule collision data. For many of the technologically important feed gases, not to mention the reactive fragments and radicals produced in the plasmas, experimental data simply doesn't exist and increasing importance will be placed on theoretical calculations.

TARGET STATES AND BOUND-FREE CORRELATION

An important development in electron-molecule scattering theory was the simultaneous appearance in a 1985 issue of J. Physics B[9] of three *ab initio* studies on electron impact excitation of the lowest $\left(b^3 \Sigma_u^+\right)$ triplet state of H_2. These three studies, using the R-matrix, Schwinger, and linear-algebraic methods, respectively, gave cross sections that were in good mutual accord and, in contrast to previous theory, agreed well with available experimental data. I mention these calculations, not only because they were the first demonstration that consistently applied *ab initio* methods could provide accurate cross sections for electronic excitation, but also because they illustrate the need for a consistent treatment of bound-free correlation in electron-molecule scattering. Consider the form of a general trial wave function that is used in most variational methods:

$$\Psi_{\Gamma_o} = \sum_{\Gamma} A\chi_{\Gamma} F_{\Gamma\Gamma_o} + \sum_{\mu} d_{\mu}^{\Gamma_o} \Theta_{\mu} \qquad (1)$$

where the first sum is over target states χ_{Γ}, $F_{\Gamma\Gamma_o}$ is the channel function that describes a free electron incident in channel Γ_o and scattered into channel Γ and A antisymmetrizes the product $\chi_{\Gamma} F_{\Gamma\Gamma_o}$. The sum should, in principle, run over all energetically open channels but, in practice, is severely truncated. The second sum, which makes up the Hilbert-space component of the wave function, consists of (N+1)- electron configuration state functions Θ_{μ} constructed from a set orthonormal molecular orbitals. These terms can be used to include the effects of bound-free correlation such as target distortion and polarization that are required as a result of truncating the expansion over target states, but they can also serve another function. Practical considerations having to do with the evaluation of matrix elements between many-electron functions require the use of an orthogonal basis. The target functions χ_{Γ} and the terms Θ_{μ} are constructed from a common set of orthonormal target molecular orbitals. It is convenient to require the channel scattering functions $F_{\Gamma\Gamma_o}$ to be orthogonal to these target orbitals. However, this may represent an unphysical strong orthogonality constraint on the total wave function that must be relaxed by the inclusion of additional "penetration" terms Θ_{μ} to the Hilbert-space part of the wave function. The $H_2(b^3\Sigma_u^+)$ excitation example illustrates the case. All three 1985 calculations used a two-state trial function that included single configuration descriptions of the ground ($1\sigma_g^2$) and first excited ($1\sigma_g 1\sigma_u$, $^3\Sigma_u$) states. If the scattering functions are required to be orthogonal to the $1\sigma_g$ and $1\sigma_u$ orbitals, then a consistent treatment requires the inclusion of the terms $1\sigma_g^2 1\sigma_u$ and $1\sigma_u^2 1\sigma_g$ to the trial function. It was found that these terms, while negligibly affecting the elastic cross section, changed the excitation cross section by almost a factor of two. The importance of such terms had not been appreciated in earlier calculations[10].

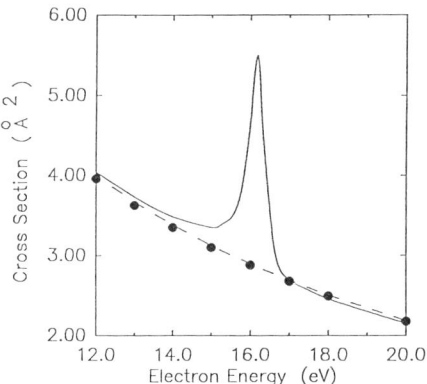

Figure 1. e^-+H_2 Σ_g^+ elastic cross sections with correlated ground state. Solid curve: Static-exchange plus penetration term; dashed curve: static-exchange without penetration term; solid circles: two-state calculation including penetration term.

Another simple example[11] illustrates the importance of bound-free target correlation and the importance of maintaining a consistent level of approximation in the N- and (N+1)-electron problems. Consider the simple case of elastic e^-+H_2 scattering in the static-exchange approximation. In this case, the first sum in Eq. (1) is truncated to a single term, but suppose we wish to use a correlated, two-term ground-state target function of the form $(c_1 1\sigma_g^2 + c_2 1\sigma_u^2)$, with $c_1 \approx .99$ and $c_2 \approx .1$. We would again think to include the two

69

"penetration" terms $1\sigma_g^2 1\sigma_u$ and $1\sigma_u^2 1\sigma_g$ for a consistent treatment. However, we find that inclusion of the $1\sigma_u^2 1\sigma_g$ term results in the appearance of a broad, spurious resonance near 16 eV, as seen in Fig. 1. No such feature is present in a calculation that uses a single term target state. When the calculation is performed without the $1\sigma_u^2 1\sigma_g$ term, the resonance disappears. The $1\sigma_u^2 1\sigma_g$ term can be thought of as a doubly excited negative ion term whose parent is the $(1\sigma_g 1\sigma_u)$, $b^3\Sigma_u$ state, which is physically open above 10 eV, but is not properly included in a single-channel calculation. A two-state calculation that includes the penetration terms also shows no spurious behavior.

The point I am trying to make with these simple examples is that trial functions of the form given by Eq. (1), which are generally referred to as close-coupling plus correlation, can give unphysical results unless some care is given to assure that there is a consistency between N-electron target states employed and the (N+1)-electron Hilbert space terms used to describe bound-free correlation. Traditional formulations of the close-coupling plus correlation problem[12] state that a trial wave function built with scattering functions constrained to be orthogonal to all the orbitals used to construct target states should include all penetration terms that arise from taking the direct product of a target orbital and the N-electron terms included in the included target states. What we have found is that this prescription can lead to disaster at intermediate energies if it results in the inclusion of bound-free penetration terms that correspond to target states not explicitly included in the close-coupling expansion. Using elaborate correlated target states exacerbates the problem. The proper solution to this problem, which happens to arise solely as a consequence of the Pauli principle and would not be present if we did have to antisymmetrize the incident electron with the target electrons, is to construct the penetration terms, not from the individual target state configurations, but from the direct product of target orbitals and the proper linear combination of configurations represented by the target state functions. This procedure is discussed at length in an earlier study[11]. In Fig. 2, we show excitation cross sections for the first two excited states of F_2 in a three-state complex Kohn variational calculation that employed correlated target states. The spurious resonances found in the close-coupling plus correlation calculations are found to be eliminated after the Hilbert space portion of the trial wave function is properly contracted. A simpler procedure, which we have found to yield almost equivalent results, is to only include those (N+1)-electron penetration terms that arise from the dominant configuration of each target state that is explicitly included.

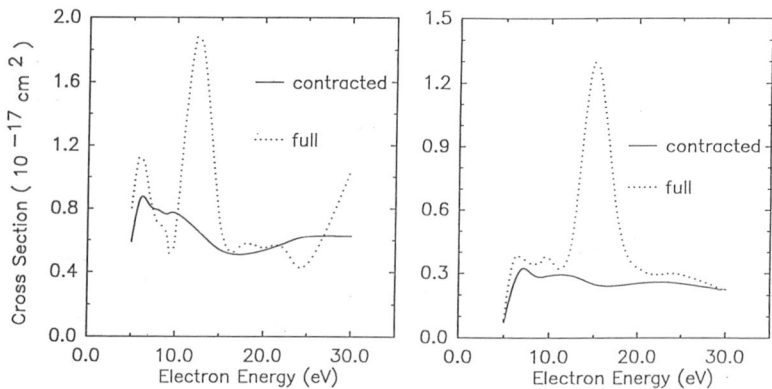

Figure 2. Total excitation cross sections for two states in e^-+F_2 scattering with and without contraction of penetration terms. Left panel: $X^1\Sigma_g^+ \rightarrow 1^3\Pi_g$, right panel: $X^1\Sigma_g^+ \rightarrow 1^1\Pi_g$.

INTERMEDIATE ENERGIES, PSEUDOSTATES AND ANALYTICITY

While the foregoing discussion addresses one of the problems encountered in carrying out close-coupling calculations at intermediate energies with correlated target states, namely that of *consistency*, it fails to address a larger issue, namely that of *convergence*, in the presence of an infinite number of open channels, including ionization continua. The use of pseudostates to accelerate the convergence of electron-atom close-coupling calculations in the low energy region below the ionization threshold was pioneered by Burke and coworkers[13] in the late 1960's. It is interesting to recall that, in 1973, Burke and Mitchell[14] published a paper on e⁻-H in the s-wave radial limit in which the pseudostates were treated as physical channels and convergence was examined above the ionization region as more pseudostates were added to the expansion. That calculation showed that there were broad resonances near the pseudostate thresholds, but that away from these regions, the cross sections appeared to converge to a well-defined limit. This work compared favorably with calculations that computed the T-matrix for complex values of the energy and then either extrapolated or analytically continued the results to physical energies above the ionization limit[15]. (See Fig. 3)

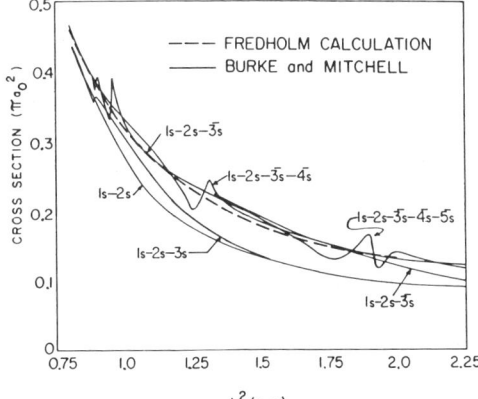

Figure 3. Singlet s-wave e-+H elastic cross section in the radial limit. Comparison of close coupling calculations (ref. 14) and calculations extrapolated from complex energies (ref. 15).

As computer power has increased over the years, so has the number of states that can be included in atomic close-coupling expansions. The culmination of this work is the convergent close-coupling method of Bray and Stelbovics[16], who have convincingly demonstrated that as the pseudostate expansion is systematically increased, the pseudostate resonance features gradually diminish and the cross sections converge to correct physical values. Intermediate R-matrix theory[17] and the R-matrix plus pseudostate[18] methods have achieved similar results, but with the latter methods, some averaging over pseudoresonance structure has generally been required. The uniform convergence found with the convergent close coupling method might well be the result of the particular Sturmian basis sets employed and their possible connection to an "equivalent quadrature" rule[19]. In any case, it remains to be seen if such techniques will be applicable to molecular problems. Here the low symmetry of the problem does not permit any separation of variables to be made, except in the special case of linear targets, and the consequent number of states needed for convergence is likely to be far greater than what is currently feasible.

We have been experimenting with a different approach, one more closely tied to early methods based on analytic continuation of the T-matrix from complex energies[20]. We first use the usual Feshbach formalism to partition the total space into P- and Q-space parts. P projects onto a set of physical target states that we wish to treat explicitly. With reference to Eq.(1), $P\Psi$ refers to the first sum to the right of the equality. $Q\Psi$ will consist, as before, of

(N+1)-electron square integrable terms, which can be thought of as products of N-electron target pseudostates and L^2 orbitals. The Schrödinger equation can be manipulated in standard fashion to derive an effective Hamiltonian that determines $P\Psi$. The effective Hamiltonian, H_{eff}, is the sum of PHP and an optical potential V_{opt}:

$$V_{opt} = PHQ \frac{1}{E - QHQ} QHP \qquad (2)$$

When P space includes all open channels, then the optical potential is real; however, when Q contains open channels, whether physical or pseudostates, V_{opt} is complex, reflecting loss of flux to open channels in Q, and the energy in Eq. (2) should be replaced by $E+i\varepsilon$, $\lim \varepsilon \to 0$. The corresponding S-matrix extracted from $P\Psi$ should not be unitary and the unitary defect of S can be used to compute the total inelastic cross section corresponding the channels contained in $Q\Psi$. Of course, in such cases, it is no longer proper to expand $Q\Psi$ solely in L^2 terms. With our prescription of expanding $Q\Psi$ solely in L^2 terms, V_{opt} will always be real and a variational calculation carried out with H_{eff} will give T-matrix elements for the channels delineated in P that oscillate erratically as the energy sweeps through values corresponding to the eigenvalues of QHQ.

Several years ago, Schneider and coworkers[21] suggested the idea of using complex rotation to obtain a convergent representation of Eq. (2) that would be valid above the ionization threshold and demonstrated the viability of the idea with a model calculation on e^- - H. Subsequently, we showed that complex (square-integrable) basis expansions could be used to extend the idea to many-electron atomic and molecular targets[22]. The use of complex basis functions, however, would require substantial modification to most electronic structure codes. Here, we outline something a bit simpler. If E_i and Φ_i are the eigenvalues and eigenfunctions of QHQ obtained in our basis of square-integrable terms, then we can write V_{opt} as

$$V_{opt} = PH|\Phi_i\rangle \frac{1}{E - E_i} \langle\Phi_i|HP \qquad (3)$$

If we add an imaginary part to E, ie let $E \to Z = E + iq$, then V_{opt} should converge to a well defined complex value as the number of basis states used to expand $Q\Psi$ is increased and the eigenvalues E_i become dense on the real axis. We propose to evaluate the (off-shell) T-matrix for several values of q and fit the calculated values to a matrix continued fraction in the variable Z, which can then be evaluated for real E to obtain a physical result. Note that this procedure differs from earlier work of Nuttall and coworkers[20] in which the entire T-matrix is evaluated for complex Z, with all channels represented by L^2 expansions, and then extrapolated to real energies. In our scheme, the P-space channels are properly accounted for with appropriate continuum basis functions; only the energy that appears in the denominator of V_{opt} is allowed to become complex.

To test this notion, we have carried out complex Kohn calculations on electron-Li_2 scattering in which $P\Psi$ contained only the $^1\Sigma_g^+$ ground state channel, represented by a single configuration $(1\sigma_g^2 1\sigma_u^2 2\sigma_g^2)$ wave function. The scattering wave function for this channel is expanded in a set of molecular orbitals constructed from Cartesian Gaussian functions, along with numerical continuum functions of appropriate symmetry. The Q-space part of the wave function, from which the optical potential is constructed, consisted of all 7-electron configurations that could be formed by keeping the $1\sigma_g$ and $1\sigma_u$ orbitals doubly occupied, singly occupying the $2\sigma_g$ orbital and allowing the remaining two electrons to occupy any remaining virtual orbital. For each physical energy, the T-matrix was evaluated for several complex optical potential energies, starting approximately 1 eV above the real axis, and fit to

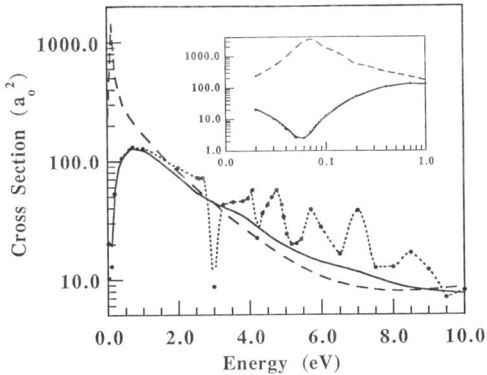

Figure 4. Elastic cross section for e-+Li$_2$ scattering in $^2\Sigma_u^+$ symmetry. Dashed line: static-exchange; solid line: optical potetial result extrapolated from complex energies; dotted line: optical potential result calculated with real energies.

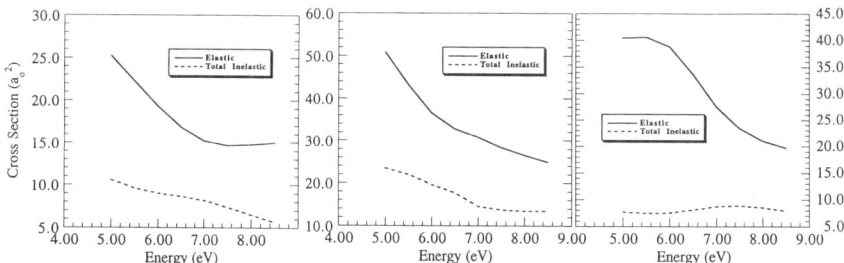

Figure 5. Elastic cross section for e-+Li$_2$ scattering in other symmetries computed from optical potential with extrapolation from complex energies. Left panel: $^2\Sigma_u^+$; center panel: $^2\Pi_u$; right panel: $^2\Pi_g$.

a matrix rational fraction which was in turn evaluated on shell. Fig. 4 summarizes our results for $^2\Sigma_u$ symmetry. We compare the static-exchange elastic cross section, which grossly overestimates the cross section below 1 eV, with optical potential calculations carried out on the real axis, which show the expected unphysical behavior above 1 eV, and with the results of extrapolation from complex Z, which are quite smooth. Fig. 5 shows results for other symmetries.

It would be interesting to experiment with including more than the ground state channel in $P\Psi$. In a molecule like H$_2$, for example, the ground state along with the first six excited states (which comprise all states that dissociate to n=1 and n=2 atoms), carry over 90% of the two-body flux. Thus the total inelasticity from such a calculation might provide a good measure of the total ionization cross section in a way that avoids explicit solution of the three-body Coulomb problem which, for a molecule, would be a formidable undertaking.

R-MATRIX THEORY WITHOUT A BOX

I have said little about the methods used to generate the results presented, except to mention that they were performed using the complex Kohn method. I also mentioned that virtually all *ab initio* (my definition of *ab initio*!) methods currently being used in electron-molecule scattering involve variational methods. So, in view of Phil Burke's long-standing history and association with the R-matrix method, I thought it fitting to close with some recent work we have done with the complex Kohn method to make it look more R-matrix like – in the spirit of showing that "all methods are the same method".

The R-matrix method has the attractive feature of being able to rapidly provide cross sections at any energy, after an initial setup that involves diagonalization of the Hamiltonian in an R-matrix basis that has no explicit energy dependence. This advantage, however, comes at a price – the R-matrix basis functions must be defined on a sphere of finite volume, thereby complicating the evaluation of matrix elements of the electronic interaction, $1/r_{12}$, particularly for multicenter targets. In the standard Kohn method, the expansion basis includes complex continuum functions that are explicitly connected to the total energy, so the Hamiltonian matrix (or at least parts of it) must be reformed and the variational equations resolved at each energy. This can become costly in situations where many points are needed to resolve a highly structured cross section.

To see exactly how the energy enters into the problem, recall that in the Kohn method, the T-matrix (for potential scattering, to keep the notation simple) can be expressed as

$$T_\ell = -\frac{2}{k}\left(\langle j_\ell|V|j_\ell\rangle + \sum_{q,q'}\langle j_\ell|V|q\rangle(E-H)^{-1}_{q,q'}\langle q'|V|j_\ell\rangle\right) \tag{4}$$

The explicit energy dependence arises from two places: the Ricatti-Bessel function, $j_\ell(kr)$, and the fact that the basis index q refers to a set of L^2 functions, augmented with an energy dependent outgoing wave that behaves asymptotically as $h_\ell^+(kr)$. (Note that $(E-H)_{q,q'}$ is a complex symmetric matrix, because the matrix elements with respect to the outgoing continuum function are defined without complex conjugation.) We will first address the second source of energy dependence.

We note that the expression

$$G_\ell^+(E) = \sum_{q,q'}|q\rangle(\langle q|E-H|q'\rangle)^{-1}\langle q'| \tag{5}$$

is merely a matrix representation of the full Green's function with the basis function $h_\ell^+(kr)$ included to insure the proper outgoing wave boundary condition. If we could find a representation in which the basis is not changed with each energy, the evaluation of the Green's function could be dramatically simplified. We have found[23] that we can accomplish this by choosing a basis of L^2 functions and augmenting it with *a set* of outgoing functions with different momentum values. Typically, for a range of about 25 eV in energy, we find that four or five continuum functions with evenly spaced values of k provide an accurate representation of the resolvent. The subtlety that requires special consideration is how to define the overlap and kinetic energy matrix elements between continuum functions with different k-values. This is not an issue when one function, $g_\ell \sim h_\ell^+(kr)$, is used with a value of k tied to the energy, $E=k^2/2$, since the quantity $(E-H) g_\ell$ vanishes for large r. When several continuum functions are used, the overlap and kinetic energy matrix elements must be defined by analytic continuation from the upper half k-plane. If the matrix elements are given by analytic formulas, those formulas provide the correct analytic continuation. When the continuum function are only known numerically, we can carry out the matrix element evaluation on the real axis to a large finite value of R and then complete the integration on a complex ray, $R+(r-R)e^{i\theta}$, using simple asymptotic forms for g_ℓ.[23] With this prescription, we can easily evaluate the Green's function at any energy after the Hamiltonian matrix is diagonalized. Although the Kohn expression can only be shown to provide a stationary value at the explicit k-values for which $E=k^2/2$, we have not noticed any significant errors at the intermediate k-values.

We now turn to the explicit appearance of the Ricatti-Bessel function, $j_\ell(kr)$, in the bound-free and free-free terms of the Kohn expression for the T-matrix[24]. To seek a different representation for $j_\ell(kr)$, we consider the function generated by operating with the partial-wave free-particle Green's function on some short-range, *energy independent*, function χ:

$$g_\ell = A_\ell G_\ell^{o+} \chi = -A_\ell (2/k) \left[h_\ell^+(kr) \int_0^r j_\ell(kr')\chi(r') + j_\ell(kr) \int_r^\infty h_\ell^+(kr')\chi(r') \right] \quad (6)$$

If we define the normalization constant as

$$A_\ell = -\frac{k}{2\int_0^\infty j_\ell(kr')\chi(r')} \quad (7)$$

then it is easy to show that

$$j_\ell(kr) = Im(g_\ell(r)) = \frac{A_\ell}{2}\left(G_\ell^{o+} - G_\ell^{o-} \right)\chi \quad (8)$$

Substituting Eq. (8) into Eq. (4) gives the desired result:

$$T_\ell = -\frac{(A_\ell)^2}{2k} \left(\langle \chi | \left(G_\ell^{o+} - G_\ell^{o-} \right)\left(V + VG_\ell^+ V \right)\left(G_\ell^{o+} - G_\ell^{o-} \right) | \chi \rangle \right) \quad (9)$$

If we now expand the free-particle Green's function in the same way we outlined for the full Green's function, then we have succeeded in expressing the T-matrix via an expression that has no explicit energy dependent matrix elements, aside from the trivial normalization constant, A_ℓ. The resulting expression has the key advantage associated with the R-matrix method in that it allows easy computation of the scattering amplitude at many energies from a single representation of the Hamiltonian – without the need for functions defined on a finite volume.

Although we have yet to implement this procedure for electron-molecule scattering, we have tested it on some atomic photoionization problems[23]. Fig. 6 shows the photoionization cross section for atomic Be in the 9-13 eV photon range below the threshold for producing

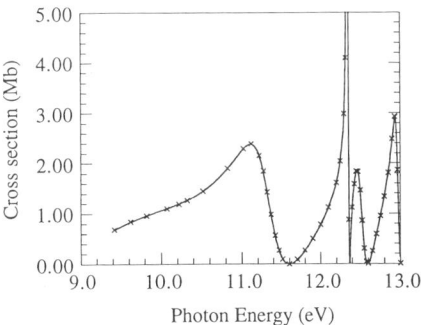

Figure 6. Photoionization cross section of Be calculated using the continuous energy version of the complex Kohn method.

the first excited state of Be$^+$ at 13.28 eV, a region dominated by autoionization. The cross sections over this entire range were obtained from a calculation in which only two continuum functions, corresponding to energies of 11 and 13 eV, were used in the expansion of the full

Green's function. Implementation of this method for electron-molecule scattering is currently underway.

ACKNOWLEDGEMENTS

I would like to acknowledge the contributions of Barry Schneider, Byron Lengsfield and Ann Orel to various aspects of the development of the Kohn method over the past decade. In particular, I want to acknowledge Bill McCurdy for his strong involvement in this work. This work was performed under the auspices of the US Department of Energy by the Lawrence Livermore National Laboratory under contract W-7405-Eng-48.

REFERENCES

1. *Electron-Molecule and Photon-Molecule Collisions*, eds. T. N. Rescigno, V. McKoy and B. I. Schneider, Plenum, New York, (1979).
2. M. A. Morrison, *Adv. At. Mol. Phys.* **24**, 51 (1988).
3. F. A. Gianturco, D. G. Thompson and A. Jain in *Computational Methods for Electron-Molecule Collisions*, eds. W. M. Huo and F. A. Gianturco, Plenum, New York 1995, p. 75.
4. See the article by A. W. Fliflet in ref. 1, p. 87.
5. See the article by J. L. Dehmer and D. Dill in ref. 1, p. 225.
6. P. G. Burke and K. A. Berrington, *Atomic and Molecular Processes: An R-Matrix Approach*, Institute of Physics, Bristol, 1993.
4. C. Winstead and V. McKoy in *Modern Electronic Structure Theory*, ed. D. Yarkony, World Scientific, Singapore (1995) p. 1375.
5. T. N. Rescigno, B. H. Lengsfield III and C. W. McCurdy in *Modern Electronic Structure Theory*, ed. D. Yarkony, World Scientific, Singapore (1995) p. 501.
6. K. L. Beluga, C. J. Noble and J. Tennyson, *J. Phys. B* **18**, L851 (1985); B. I. Schneider, *ibid* p. L857; M. A. Lima, T. L. Gibson, C. C. Lin and V. McKoy, *ibid* p. L865.
10. S. Chung and C. C. Lin, *Phys. Rev. A* **7**, 1874 (1978).
11. B. H. Lengsfield III and T. N. Rescigno, *Phys. Rev. A* **44**, 2913 (1991).
12. W. Eissner and M. J. Seaton, *J. Phys. B* **5**, 2187 (1972).
13. P. G. Burke, D. F. Gallaher and S. Geltman, *J. Phys. B* **2**, 1142 (1969).
14. P. G. Burke and J. F. B. Mitchell, *J. Phys. B* **6**, 320 (1973).
15. T. N. Rescigno and W. P. Reinhardt, *Phys. Rev. A* **10**, 158 (1974).
16. I. Bray and A. T. Stelbovics, *Phys. Rev. Letts.* **69**, 53 (1992); **70**, 746 (1993).
17. T. T. Scholz, H. R. J. Walters and P. G. Burke, *J. Phys. B* **23**, L467 (1990).
18. K. Bartschat and I. Bray, *Phys. Rev. A* **54**, R1002 (1996).
19. E. J. Heller, T. N. Rescigno and W. P. Reinhardt, *Phys. Rev. A* **6**, 2946 (1973).
20. F. A. McDonald and J. Nuttall, *Phys. Rev. Letts.* **23**, 361 (1969), *Phys. Rev. A* **4**, 1821 (1971).
21. A. K. Bhatia, B. I. Schneider and A. Temkin, *Phys. Rev. Letts.* **70**, 1936 (1993).
22. T. N. Rescigno and C. W. McCurdy, *Phys. Rev. Letts.* **73**, 3524 (1994).
23. T. N. Rescigno, A. E. Orel and C. W. McCurdy, *Phys. Rev. A* **55**, 342 (1997).
24. C. W. McCurdy, T. N. Rescigno, W. A. Isaacs and D. E. Manolopolous, *Phys. Rev. A* **57**, 3511 (1998).

ATOMS IN INTENSE LASER FIELDS

Charles J. Joachain

Physique Théorique
Université Libre de Bruxelles, C.P. 227
B-1050 Bruxelles, Belgium

1. INTRODUCTION

The study of laser-atom interactions at high intensities is a new and exciting aera of atomic physics, to which Phil Burke has made outstanding contributions. Rapid advances in this domain have become possible due to the development, using the "Chirped Pulse Amplification" (CPA) scheme[1], of lasers capable of yielding intensities of the order or exceeding the value $I_a = 3.5 \times 10^{16}$ W cm^{-2}, corresponding to the atomic unit of electric field strength $\mathcal{E}_a = 5.1 \times 10^9$ V cm^{-1}. Such laser fields are strong enough to compete with the Coulomb forces in controlling the electron dynamics in atomic systems. As a result, atoms and molecules in intense laser fields exhibit new properties which have been discovered via the study of multiphoton processes. These properties generate new behaviour of bulk matter in intense laser fields, with wide ranging potential applications such as the study of ultra-fast phenomena, the development of high frequency (XUV and X-ray) lasers, the investigation of the properties of plasmas and condensed matter under extreme conditions of temperature and pressure, and intense field control of atomic and molecular reactions. Over the last ten years, laser intensities have increased by four orders of magnitude[2], up to 10^{20} W cm^{-2} where relativistic effects in laser-atom interactions become important. On the theoretical side, the development of supercomputers has made it possible to perform calculations of unprecedented complexity, which have led to the prediction of novel properties of atomic systems in strong laser fields.

In this article I shall give an overview of the domain of high-intensity laser interactions with atoms and ions. Section 2 is devoted to a survey of the new phenomena discovered by studying atomic multiphoton processes in intense laser fields. The main non-perturbative theoretical methods currently used to analyze these processes are examined in Section 3. Possible future developments of this rapidly growing area of physics are considered in Section 4. Detailed reviews of the subject can be found in the book edited by Gavrila[3], and in the articles of Burnett et al.[4], Joachain[5], DiMauro and Agostini[6], Protopapas et al.[7] and Joachain et al.[8].

2. MULTIPHOTON PROCESSES IN ATOMS AND IONS

In this section, I shall discuss the three important processes occuring in atoms and ions: multiphoton ionization, harmonic generation and laser-assisted electron-atom collisions.

2.1. Multiphoton ionization and above threshold ionization

Let us begin by considering the multiphoton (single) ionization (MPI) reaction

$$n\hbar\omega + A^q \to A^{q+1} + e^- \qquad (1)$$

where q is the charge of the target atomic system A, expressed in atomic units (a.u.), $\hbar\omega$ is the photon energy and n is an integer.

This process was first observed by Voronov and Delone[9], who used a ruby laser to induce seven-photon ionization of xenon, and by Hall et al.[10], who recorded two-photon ionization from the negative ion I^-. In the following years, important results were obtained by several groups, in particular at Saclay where the dependence of the ionization yields on the intensity and the resonance-enhancement of MPI were studied.

A crucial step in the understanding of MPI was made when experiments detecting the energy-resolved photo-electrons were performed by Agostini et al.[11]. They discovered that the ejected electron in the reaction (1) could absorb photons in excess of the minimum required for ionization to occur. The study of this excess-photon ionization, known as "above threshold ionization" (ATI), has been one of the central themes of multiphoton physics in recent years.

A typical example of ATI photo-electron energy spectra, obtained by Petite et al.[12] is shown in Fig. 1. The spectra are seen to consist of several peaks, separated by the

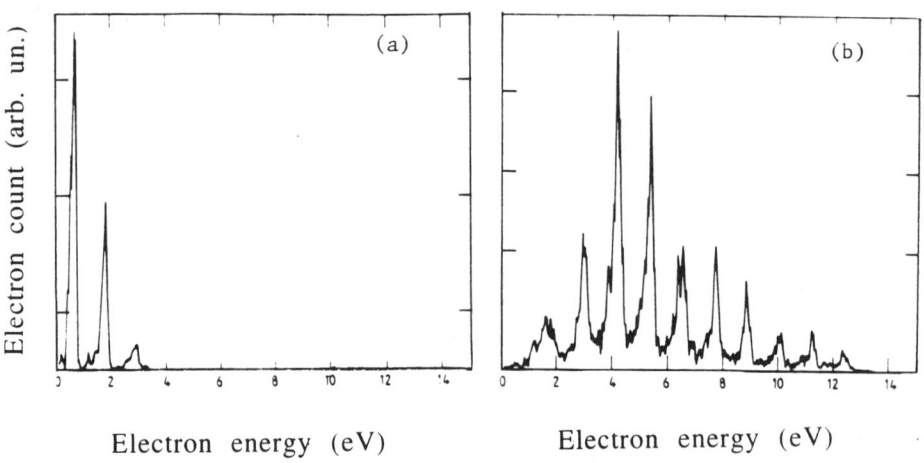

Figure 1. Electron energy spectra showing ATI of xenon at a wavelength $\lambda = 1064$ nm. (a) $I = 2 \times 10^{12}$ W cm^{-2}, (b) $I = 10^{13}$ W cm^{-2}. From Petite, Agostini and Muller[12].

photon energy $\hbar\omega$. As the intensity I increases [see Fig. 1(b)], peaks at higher energies appear, whose intensity dependence does not follow the lowest (non-vanishing) order perturbation theory (LOPT) prediction according to which the ionization rate for an n-photon process is proportional to I^n.

Another remarkable feature of the ATI spectra, also apparent in Fig. 1, is that as the intensity increases the low-energy peaks are reduced in magnitude. The reason for this peak suppression is that the energies of the atomic states are Stark-shifted in the presence of a laser field. For low laser frequencies, the AC Stark shifts of the lowest bound states are small in magnitude. On the other hand, the induced Stark shifts of the Rydberg and continuum states are essentially given by the electron ponderomotive energy U_p, which is the cycle-average kinetic energy of a quivering electron and is given (for non-relativistic velocities) by

$$U_p = \frac{e^2 \mathcal{E}_0^2}{4m\omega^2}, \qquad (2)$$

where m is the mass of the electron, e is the absolute value of its charge and $\mathcal{E}_0$ is the electric field strength. Since the energies of the Rydberg and continuum states are shifted upwards relative to the lower bound states by about U_p, there is a corresponding increase in the intensity- dependent ionization potential of the atom. If this increase is large enough, ionization by a given number $\bar{n}$ of photons becomes energetically forbidden (a phenomenon known as "channel closing") and more photons are needed to ionize the system. However, atoms interacting with smoothly varying pulses experience a range of intensities, so that the corresponding peak in the photo-electron spectrum will not disappear completely, as seen in Fig. 1(b).

For short (sub-picosecond) pulses, the ATI peaks exhibit a sub-structure[13] due to the fact that the intensity-dependent Stark shifts bring different states of the atom into multiphoton resonance during the laser pulse. This fine structure is not seen in long pulse experiments because in that case each electron regains its ponderomotive energy deficit from the field as it escapes the laser pulse adiabatically. Highly resolved spectra have been obtained recently by Hansch et al.[14]. Direct, simultaneous measurements of both the energy and the angular distribution of the electrons have been performed by Helm et al.[15].

For increasing laser field strengths approaching the Coulomb field which binds the electron and for low laser frequencies, the sharp ATI peaks of the photo-electron spectrum gradually blur into a continuous distribution[16,17]. In this regime, ionization can be analyzed by using a quasi-static model in which the bound electrons experience an effective potential formed by adding to the atomic potential the contribution due to the instantaneous laser electric field. This approach was used by Keldysh[18] to study tunneling ionization in the low-frequency limit, and pursued by Faisal[19], Reiss[20] and Amnosov et al.[21]. An important quantity in these studies is the Keldysh adiabaticity parameter γ, defined as the ratio of the laser and tunneling frequencies, which is given by

$$\gamma = \left(\frac{I_p}{2U_p}\right)^{1/2}, \qquad (3)$$

where I_p is the field-free atomic ionization potential. For small γ, tunneling dynamics dominates, the transition from multiphoton to tunneling ionization taking place in the region around $\gamma = 1$. Above a critical intensity I_c (which is equal to 1.4×10^{14} W cm^{-2} for atomic hydrogen in the ground state), the electron can classically "flow over the top" of the barrier (over-the-barrier, OTB, ionization), so that field ionization occurs and the atom ionizes in about one orbital period.

The semi-classical, "recollision model" developed recently by Corkum[22], Kulander et al.[23] and Lewenstein et al.[24] is based on the idea that strong field ionization dynamics at low frequencies proceeds via several steps. In the first (bound-free), step, an electron is liberated from its parent atom by tunneling or OTB ionization. In the second (free-free) step, the interaction with the laser field dominates, a fact which was noted earlier by using a simple classical picture of a quivering electron[25,26]. As the phase of the field reverses, the electron is accelerated back towards the atomic core. If the electron returns to the core, a third step takes place, in which scattering of the electron by the core then leads to single or multiple ionization, while radiative recombination leads to harmonic generation.

This semi-classical model has been very useful for explaining in terms of classical trajectories and return energies a number of features found in recent experiments, performed by using kilohertz-repetition rate, high-intensity lasers. These include the existence of a "plateau" in the ATI photo-electron energy spectra[27] (see Fig. 2) and sharp peaks in the angular distributions, called "rescattering rings". The prominent groups of ATI peaks within the plateau have been studied in atomic hydrogen by Paulus et al.[30] and in argon by Hertlein et al.[31] and Muller and Kooiman[32]. Measurements of ATI electron spectra in an elliptically polarized field have been carried out by Paulus et al.[33], who interpreted the observed ellipticity dependence of ionization rates into individual ATI peaks in terms of electron tunneling at different times during an optical cycle.

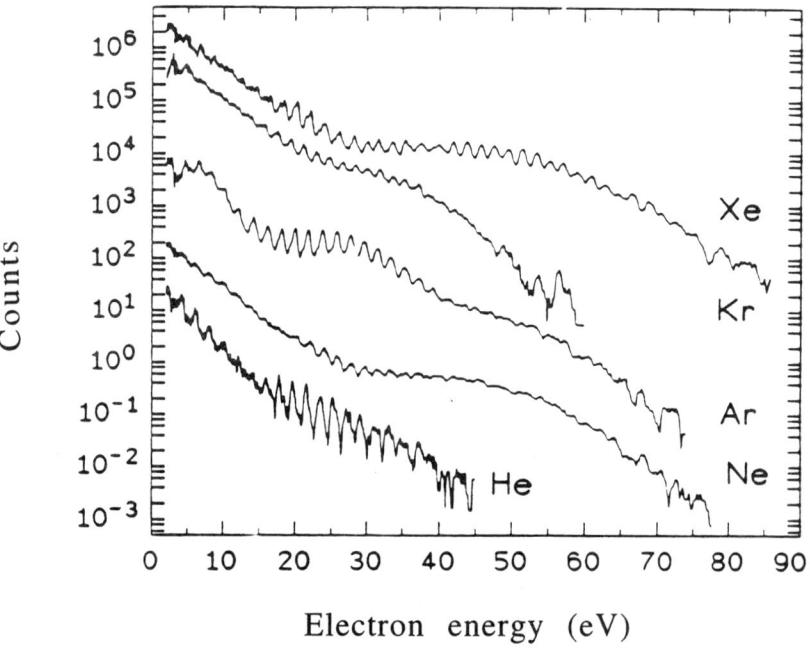

Figure 2. Photo-electron counts as a function of photo-electron energy, for various noble gases, at a laser wavelength $\lambda = 630$ nm and an intensity $I \simeq 2 \times 10^{14}$ W cm^{-2} (3×10^{14} W cm^{-2} for He). From Paulus et al.[27].

2.2. Harmonic generation

Atoms interacting with a strong laser field can emit radiation at higher-order multiples, or harmonics, of the angular frequency ω of the "pump" laser. This phenomenon, called harmonic generation, has attracted considerable interest in recent years, in particular as a promising source of coherent high-frequency radiation in the XUV and X-ray ranges. For an initial state of a given parity, the harmonics are emitted at angular frequencies ω_q which are odd multiples of the laser angular frequency, i.e. $\omega_q = q\omega$ with $q = 3, 5, \ldots$. The observation of the third harmonic was made in 1967 by New and Ward[34]. Later on, the availability of intense lasers has made it possible to produce high-order harmonics[35-45], and hence high-frequency photons, by focusing a laser beam into a gas jet. As an example, we see in Fig. 3 the emission spectra of various noble gases, obtained by l'Huillier and Balcou[42], who observed the harmonic $q= 133$ for neon at an intensity $I \simeq 1.5 \times 10^{15}$ W cm^{-2}.

In the recent experiments with ultra-short ($\simeq 25$ fs), high-intensity laser pulses[46-51], the atoms experience only a few laser cycles. The highest harmonic frequencies and harmonic orders ($q \simeq 297$) have been obtained under these conditions, reaching into the water window ($\lambda \simeq 2.7$ Å). Other important experimental developments have been spatially resolved measurements of the time-dependence[52] and direct measurements of the temporal coherence[53] of high-order harmonics.

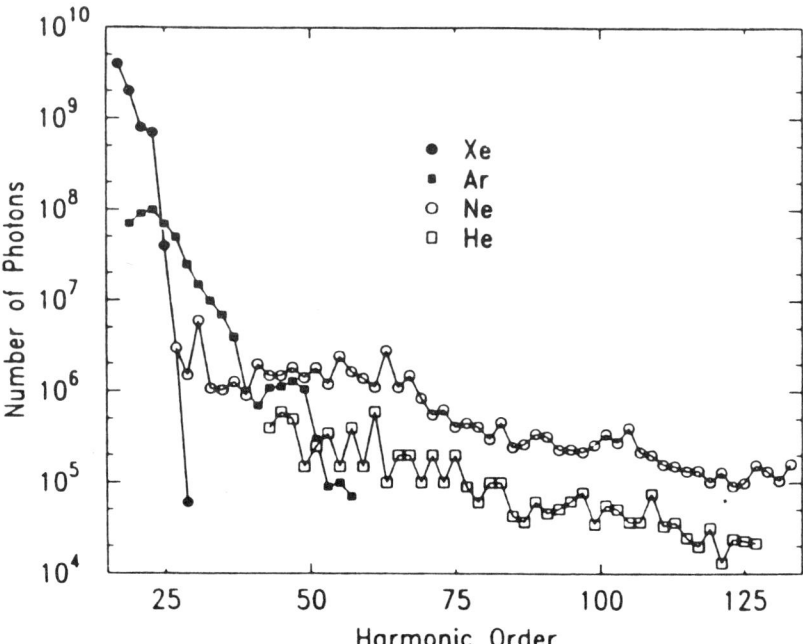

Figure 3. Harmonic emission spectra of various noble gases for a "pump" laser of wavelength $\lambda = 1053$ nm and intensity $I \simeq 1.5 \times 10^{15}$ W cm^{-2}. From L'Huillier and Balcou[42].

The theoretical treatment of harmonic generation by an intense laser pulse focused into a gaseous medium has two main aspects. First, the microscopic, singe atom response to the laser field must be analyzed. The single atom spectra must then be combined to obtain the macroscopic harmonic fields generated from the coherent emission of all the atoms in the laser focus. This is done by using the single-atom polarization fields as source terms in the Maxwelll equations. We shall only discuss here the microscopic aspect of the problem.

The power spectrum of the emitted radiation is proportional to the modulus squared of the Fourier transform of the dipole acceleration[54-55] $\mathbf{a}(t)$, where

$$\mathbf{a}(t) = \frac{d^2}{dt^2}\mathbf{d}(t) . \qquad (4)$$

In this equation $\mathbf{d}(t)$ denotes the laser-induced atomic dipole moment

$$\mathbf{d}(t) = <\Psi(t)| - e\mathbf{R}|\Psi(t)> , \qquad (5)$$

where $\Psi(t)$ is the atomic wave function in the presence of the laser field and

$$\mathbf{R} = \sum_{i=1}^{N}\mathbf{r}_i \qquad (6)$$

is the sum of the coordinates $\mathbf{r}_i$ of the N atomic electrons. When the atom is driven by a monochromatic field, the emitted harmonic radiation can be simply calculated from the induced dipole moment, which can be expanded as

$$\mathbf{d}(t) = \sum_q [\mathbf{d}(q\omega)e^{-iq\omega t} + c.c.] , \qquad (7)$$

where c.c. denotes the complex conjugate, and $\mathbf{d}(-q\omega) = [\mathbf{d}(q\omega)]^*$. The rate of emission of photons of angular frequency $q\omega$ is then proportional to $|\mathbf{d}(q\omega)|^2$.

At high intensities of the "pump" laser, the harmonic intensity distribution exhibits a rapid decrease over the first few harmonics, followed by a plateau of approximately constant intensity and then a cut-off, corresponding to an abrupt decrease of harmonic intensity. It is worth stressing that the existence of a plateau is a non-perturbative feature. It was discovered within the framework of time-dependent Schrödinger equation (TDSE) calculations[56] that the cut-off angular frequency ω_c of the harmonic spectrum is given approximately by the relation

$$\hbar\omega_c \simeq I_p + 3\, U_p . \qquad (8)$$

In the semi-classical "recollision model"[22-24], the maximum returning kinetic energy of a classical electron recolliding with the atomic core is given by 3.2 U_p, so that the highest energy which can be radiated is $I_p + 3.2\, U_p$, in good agreement with the TDSE calculations and with experiment.

An exciting new development is the possibility of using high-order harmonics to generate pulses of extremely short duration, in the range of hundreds of attoseconds (1 as = 10^{-18} s). There currently exist several proposed methods of attosecond pulse generation. The first one[57] involves the use of a very short "pump" laser pulse ($\simeq$ 20 fs) which should make it possible to generate single harmonic pulses of sub-femtosecond duration. The second method[58] is based on the high sensitivity of harmonic generation to the degree of ellipticity of the "pump" laser pulse. Indeed, using the "recollision model", it is easy to show that harmonics are essentially produced when the polarization of the laser field is linear. By creating a laser pulse whose polarization is linear only during a short time (close to a laser period), the harmonic emission can be limited to this interval, so that single sub-femtosecond pulses could be produced. Recent theoretical work[59-61] also predicts that high-order harmonics generated by an atom in a linearly polarized intense laser pulse form trains of ultra-short pulses corresponding to different trajectories of electrons that tunnel out of the atom and recombine with the parent ions. Under appropriate geometrical conditions, the macroscopic propagation in an atomic jet allows the selection of one of these trajectories, leading to a train of attosecond pulses, with one pulse per half-cycle.

2.3. Laser-assisted electron-atom collisions

An electron scattered by an atom (ion) in the presence of a laser field can absorb or emit radiation. Since these radiative collisions involve continuum states of the electron-atom (ion) system, they are often called "free-free transitions" (FFT). In weak fields, only one-photon processes have a large enough probability to be observed. However, as the field strength is increased, multiphoton processes become important. Examples of laser-assisted electron-atom collisions are "elastic" collisions:

$$e^- + A(i) + n\hbar\omega \rightarrow e^- + A(i) , \qquad (9)$$

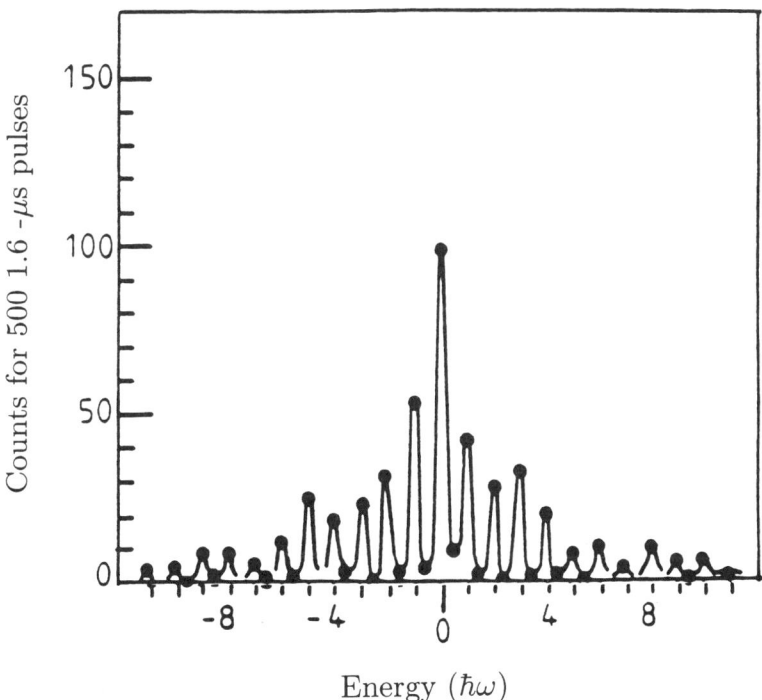

Figure 4. Energy spectrum of electrons scattered by argon atoms in the presence of a CO_2 laser of photon energy $\hbar\omega = 0.117$ eV. The open circles correspond to the experimental data. The full line is drawn to guide the eye. The abscissa gives the final electron energy in units of the photon energy, with the origin fixed at the initial electron energy $E_i = 15.8$ eV. The scattering angle $\theta = 155°$. The laser intensity $I = 10^8$ W cm^{-2}. From Weingartshofer et al.[64].

inelastic collisions:

$$e^- + A(i) + n\hbar\omega \to e^- + A(f) \tag{10}$$

and single ionization (e, 2e) collisions:

$$e^- + A(i) + n\hbar\omega \to A^+(f) + 2e^- \tag{11}$$

where $A(i)$ and $A(f)$ denote an atom A in the initial state i and the final state f, respectively, and $A^+(f)$ means the ion A^+ in the final state f. Positive values of n correspond to photon absorption (inverse bremsstrahlung), negative ones to photon emission (stimulated bremsstrahlung) and $n = 0$ to a collision process without net absorption or emission of photons, but in the presence of the laser field. A review of laser-assisted electron-atom collisions has been given recently by Ehlotzky et al.[62].

Direct information on laser-assisted electron-atom collisions is obtained by performing three-beam experiments, in which an atomic beam is crossed in coincidence by a laser beam and an electron beam, and the scattered electrons, having undergone FFT, are recorded. Several experiments of this kind have been done, in which the exchange of photons between the electron-atom system and the laser field has been observed in laser-assisted elastic[63-65] and inelastic[66-69] processes. As an illustration, we show in Fig. 4 the results of Weingartshofer et al.[64] for laser-assisted "elastic" electron-argon scattering. We remark that even at the modest intensity of 10^8 W cm^{-2} as many as eleven-photon emission and absorption transitions were observed. As seen in Fig. 4, the relative intensities of two successive FFT peaks are of the same order of magnitude, which indicates that perturbation theory cannot be used to analyze these results. Thus, as in the case of the other multiphoton processes discussed above, non-perturbative methods must be developed. It is to this subject that we now turn our attention.

3. THEORETICAL METHODS

In this section, following a brief summary of the basic equations of the theory, I shall give a survey of the main non-perturbative methods which have been used to study laser-atom interactions at high intensities. A more detailed account of the subject can be found in the review article of Joachain et al.[8].

3.1. Basic equations

In order to study the interaction of an atomic system with a laser field, we shall use a semi-classical approach in which the laser field is treated classically, while the atomic system is studied by using quantum mechanics. This approach is entirely justified for the intense fields considered here[5]. We neglect for the moment relativistic effects and treat the laser field in the dipole approximation as a spatially homogeneous electric field $\mathcal{E}(t)$, the corresponding vector potential being $\mathbf{A}(t)$, with $\mathcal{E}(t) = -d\mathbf{A}(t)/dt$. For example, if the field is linearly polarized, we have

$$\mathcal{E}(t) = \hat{\epsilon}\,\mathcal{E}_0 F(t) \cos(\omega t + \phi), \tag{12}$$

where $\hat{\epsilon}$ is the polarization vector, $\mathcal{E}_0$ is the electric field strength, $F(t)$ is the pulse shape function and ϕ is a phase.

Our starting point is the time-dependent Schrödinger equation

$$i\hbar \frac{\partial}{\partial t}\Psi(X,t) = H(t)\Psi(X,t) \tag{13}$$

where X denotes the ensemble of the atomic electron coordinates (i.e., their position coordinates $\mathbf{r}_i$ and spin variables). The Hamiltonian $H(t)$ of the system is given by

$$H(t) = H_{at} + H_{int}(t) \tag{14}$$

where

$$H_{at} = T + V \tag{15}$$

is the time-independent field-free atomic Hamiltonian. In the above equation T is the sum of the electron kinetic energy operators and V is the sum of the two-body Coulomb interactions between the constituents (nucleus and N electrons) of the atomic system. The laser-atom interaction term is

$$H_{int}(t) = \frac{e}{m}\mathbf{A}(t).\mathbf{P} + \frac{e^2 N}{2m} A^2(t) \tag{16}$$

where

$$\mathbf{P} = \sum_{i=1}^{N} \mathbf{p}_i \tag{17}$$

is the total momentum operator. The term in A^2 can be eliminated from the Schrödinger equation (13) by performing the gauge transformation

$$\Psi(X,t) = \exp\left[-\frac{i}{\hbar}\frac{e^2 N}{2m}\int^t A^2(t')dt'\right]\Psi_V(X,t), \tag{18}$$

which gives for $\Psi_V(X,t)$ the Schrödinger equation in the velocity gauge

$$i\hbar \frac{\partial}{\partial t}\Psi_V(X,t) = \left[H_{at} + \frac{e}{m}\mathbf{A}(t).\mathbf{P}\right]\Psi_V(X,t). \tag{19}$$

On the other hand, if we return to the Schrödinger equation (13) and perform the gauge transformation

$$\Psi(X,t) = \exp\left[-\frac{ie}{\hbar}\mathbf{A}(t).\mathbf{R}\right]\Psi_L(X,t) \tag{20}$$

where $\mathbf{R}$ is the sum of the coordinates $\mathbf{r}_i$ of the electrons [see eq. (6)], we obtain the Schrödinger equation in the length gauge

$$i\hbar \frac{\partial}{\partial t}\Psi_L(X,t) = [H_{at} + e\,\boldsymbol{\mathcal{E}}(t).\mathbf{R}]\Psi_L(X,t). \tag{21}$$

As we shall see below, it is sometimes convenient to study the interaction of an atomic system with a laser field in an accelerated frame, called the Kramers frame[70,71]. Starting from the Schrödinger equation (19) in the velocity gauge, we perform the unitary transformation

$$\Psi_V(X,t) = \exp\left[-\frac{i}{\hbar}\boldsymbol{\alpha}(t).\mathbf{P}\right]\Psi_A(X,t), \tag{22}$$

where

$$\alpha(t) = \frac{e}{m} \int^t \mathbf{A}(t')dt' \qquad (23)$$

is a vector corresponding to the displacement of a "classical" electron from its oscillation center in the electric field $\mathcal{E}(t)$. The Kramers transformation (22) therefore corresponds to a spatial translation, characterized by the vector $\boldsymbol{\alpha}(t)$, to a new frame moving with respect to the laboratory frame in the same way as a "classical" electron in the field $\mathcal{E}(t)$. In this accelerated Kramers frame the new Schrödinger equation for the wave function $\Psi_A(X,t)$ is

$$i\hbar \frac{\partial}{\partial t} \Psi_A(X,t) = \left[\sum_{i=1}^{N} \frac{p_i^2}{2m} + V(\mathbf{r}_1 + \boldsymbol{\alpha}(t), ... \mathbf{r}_N + \boldsymbol{\alpha}(t)) \right] \Psi_A(X,t) \qquad (24)$$

so that the interaction with the laser field is now incorporated via $\boldsymbol{\alpha}(t)$ into the potential V, which becomes time-dependent. We note that in the case of a linearly polarized monochromatic field

$$\mathcal{E}(t) = \hat{\boldsymbol{\epsilon}} \, \mathcal{E}_0 \cos(\omega t) , \qquad (25)$$

we have

$$\boldsymbol{\alpha}(t) = \hat{\boldsymbol{\epsilon}} \, \alpha_0 \cos(\omega t) , \qquad (26)$$

where

$$\alpha_0 = \frac{e\mathcal{E}_0}{m\omega^2} \qquad (27)$$

is called the "excursion" amplitude of the electron in the field.

3.2. Floquet theory

Let us now restrict our attention to monochromatic laser fields of angular frequency ω and of arbitrary polarization. The Hamiltonian $H(t)$ of the system is then periodic, $H(t+T) = H(t)$, where $T = 2\pi/\omega$. The Floquet method[72-74] can therefore be used to write the wave function $\Psi(X,t)$ in the form

$$\Psi(X,t) = e^{-iEt/\hbar} F(X,t) \qquad (28)$$

where the time-independent, complex quantity E is called the quasi-energy, and $F(X,t)$ is periodic in time, with period T, so that it can be expressed as the Fourier series

$$F(X,t) = \sum_{n=-\infty}^{+\infty} e^{-in\omega t} F_n(X) . \qquad (29)$$

The functions $F_n(X)$ are called the harmonic components of $F(X,t)$. Using eqs. (28) and (29), we obtain for $\Psi(X,t)$ the Floquet-Fourier expansion

$$\Psi(X,t) = e^{-iEt/\hbar} \sum_{n=-\infty}^{+\infty} e^{-in\omega t} F_n(X) . \qquad (30)$$

If we also make a Fourier analysis of the interaction Hamiltonian,

$$H_{int} = \sum_{n=-\infty}^{+\infty} e^{-in\omega t} (H_{int})_n \qquad (31)$$

and substitute both eqs. (30) and (31) into the Schrödinger equation (13), we obtain for the harmonic components $F_n(X)$ a system of time-independent coupled differential equations:

$$(E + n\hbar\omega - H_{at})F_n(X) = \sum_{k=-\infty}^{+\infty} (H_{int})_{n-k} F_k(X). \qquad (32)$$

with $n = 0, \pm 1, \pm 2....$ These equations, together with appropriate boundary conditions, form an eigenvalue problem, which we can write as

$$(\mathbf{H}_F - E)\,\mathbf{F} = 0 \qquad (33)$$

where the Floquet Hamiltonian $\mathbf{H}_F$ is an infinite matrix of operators. The quasi-energies can be expressed as

$$E = E_i + \Delta - i\frac{\Gamma}{2} \qquad (34)$$

where E_i is the energy of the initial unperturbed (field-free) state and Δ is the AC Stark shift of the state. The physical meaning of Γ can be deduced by noting that the integral over a finite volume of the electron density, averaged over one cycle, decreases like $\exp(-\Gamma t/\hbar)$. Hence the characteristic lifetime of an atom described by the Floquet state (28) is $\hbar/\Gamma$, which means that $\Gamma/\hbar$ is the total ionization rate of that state. We also note that, in the velocity or the length gauges, the interaction Hamiltonian can be written in the form

$$H_{int}(t) = H_+ e^{-i\omega t} + H_- e^{i\omega t} \qquad (35)$$

where H_+ and H_- are time-independent operators. The coupled equations (32) then take the simpler form

$$(E + n\hbar\omega - H_{at})F_n(X) = H_+ F_{n-1}(X) + H_- F_{n+1}(X) \qquad (36)$$

and the Floquet Hamiltonian $\mathbf{H}_F$ is an infinite tridiagonal matrix of operators.

The Floquet theory has been used extensively to study multiphoton processes in atomic systems. In particular, elaborate calculations have been performed for one-electron atoms. Denoting by $\mathbf{r}$ the electron position vector, and following Maquet et al.[75], the system of coupled equations (36) can be solved by expanding each harmonic component $F_n(\mathbf{r})$ on a discrete basis set consisting of spherical harmonics and complex Sturmian radial functions. This Sturmian-Floquet method has been applied extensively by Potvliege, Shakeshaft, Dörr et al.[76] to study multiphoton ionization and harmonic generation in atomic hydrogen. As an example, we show in Fig. 5 the photo-electron yield for the lowest ATI peak, calculated by Dörr et al.[77] for the multiphoton ionization of H(1s) by a 608 nm laser pulse of peak intensity 6.6×10^{13} W cm^{-2}, compared with the experimental data of Rottke et al.[78]. The subpeaks are due to Rydberg states moving in and out of resonance when the intensity of the pulse rises and falls[13], as explained in Section 2.1.

Gavrila et al.[79-82] have used the Floquet theory to study the behaviour of atoms interacting with high intensity monochromatic laser fields whose frequency is much larger than the threshold frequency for one-photon ionization. Their approach is formulated in the Kramers frame discussed in Section 3.1. For the case of an atom with one active electron, the Schrödinger equation (24) satisfied by the wave function Ψ_A is

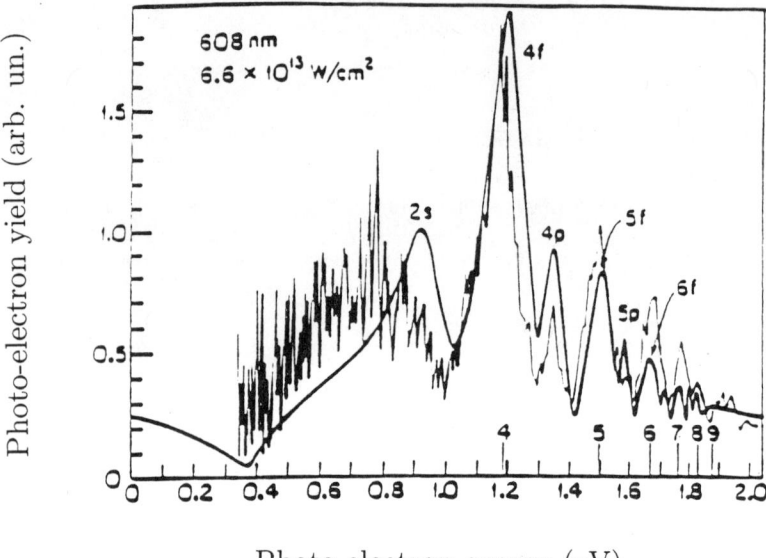

Figure 5. Yield of photo-electrons, into the lowest open channel, versus photo-electron energy, for ionization of H(1s) by a 608 nm pulse whose peak intensity is 6.6 ×10^{13} W cm^{-2} and whose duration is 0.5 ps. The bold curve is the result of Sturmian-Floquet calculations of Dörr et al.[77]. The thin curve represents the experimental data of Rottke et al.[78]. Some of the subpeaks are labelled by the dominant configuration of the resonant Floquet state. From Potvliege and Shakeshaft[76].

$$i\hbar\frac{\partial}{\partial t}\Psi_A(\mathbf{r},t) = \left[\frac{p^2}{2m} + V(\mathbf{r}+\boldsymbol{\alpha}(t))\right]\Psi_A(\mathbf{r},t) , \quad (37)$$

where $\boldsymbol{\alpha}(t)$ is the displacement vector, given by eq. (26) for the case of linear polarization. Since the equation (37) has periodic coefficients, one can seek solutions having the Floquet-Fourier form (30). Making also a Fourier analysis of the potential

$$V(\mathbf{r}+\boldsymbol{\alpha}(t)) = \sum_{n=-\infty}^{+\infty} e^{-in\omega t} V_n(\alpha_0,\mathbf{r}) \quad (38)$$

one obtains for the harmonic components $F_n(\mathbf{r})$ of $\Psi_A(\mathbf{r},t)$ the system of coupled equations

$$\left[E + n\hbar\omega - \left(\frac{p^2}{2m} + V_0(\alpha_0,\mathbf{r})\right)\right]F_n(\mathbf{r}) = \sum_{\substack{k=-\infty \\ (k\neq n)}}^{+\infty} V_{n-k}(\alpha_0,\mathbf{r})F_k(\mathbf{r}) . \quad (39)$$

Gavrila and co-workers have shown that in the high-intensity, high-frequency limit, the atomic structure in the laser field is essentially governed by the potential $V_0(\alpha_0,\mathbf{r})$, which is the static (time-averaged), "dressed" potential associated with the interaction potential V in the Kramers frame. Detailed calculations of the energies and eigenstates, performed for the hydrogen atom in the Kramers frame, show that the atom undergoes "dichotomy", i.e. the electronic cloud in the static potential $V_0(\alpha, \mathbf{r})$ splits into two disjoints parts as the excursion amplitude α_0 increases. For example, in the case of the ground state of atomic hydrogen in a linearly polarized laser field, the wave function exhibits two pronounced maxima around the end points $\pm\alpha_0\hat{\epsilon}$ of the "classical" electron excursion when $\alpha_0 = 30$ a.u. Dichotomy sets in around $\alpha_0 = 50$ a.u. and is complete by $\alpha_0 = 70$ a.u.

An important prediction of the high-frequency Floquet theory is that at sufficiently high intensity, and when the frequency of the laser field is substantially larger than the threshold frequency for one-photon ionization, the ionization rate decreases when the intensity increases. This phenomenon, called adiabatic stabilization, is due to the fact that the quiver motion of the bound electron(s) in the field becomes large, thus inhibiting ionization. Adiabatic stabilization was initially investigated by Gersten and Mittleman[83] and studied in detail for atomic hydrogen, not only by using the approximate high-frequency Floquet theory[79-82], but also the Sturmian-Floquet theory[84] and the R-matrix-Floquet theory[85-87] which I shall discuss in Section 3.3. It should be noted that in contrast to the high-frequency Floquet calculations, both the Sturmian-Floquet and the R-matrix-Floquet methods provide "exact" Floquet results for atomic hydrogen.

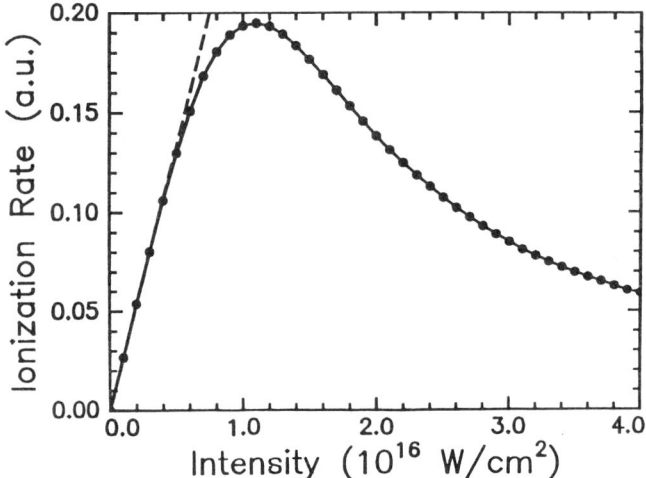

Figure 6. Total ionization rate (in a.u.) versus intensity for H(1s) in a laser field of angular frequency $\omega = 0.65$ a.u. The solid line corresponds to the Sturmian-Floquet[84] or R-matrix-Floquet[88] calculations, which are in excellent agreement. The LOPT result is given by the broken line.

As an example, we show in Fig. 6 the total rate Γ (in a.u.) for ionization of H(1s), as obtained from the Sturmian-Floquet[84] and the R-matrix-Floquet methods for an angular frequency $\omega = 0.65$ a.u. which is larger than the threshold value ($\omega = 0.5$ a.u.) for one-photon ionization. The Floquet results are seen to increase linearly at low intensities, as predicted by lowest order (in the present case, first order) perturbation theory (LOPT). They have a peak near 10^{16} W cm^{-2} and then decrease with increasing intensity, thus exhibiting the stabilization behaviour.

In the Floquet analysis discussed above, a monochromatic laser field of constant intensity is assumed to be present at all times. In reality, an atom cannot be instantaneously released into a super-intense field, but must be subjected to a laser pulse of finite duration. This implies that during the laser pulse turn on, where the intensity is lower, substantial ionization will occur. In other words, before the atoms experience a super-intense field where they can stabilize, they must pass through a "death valley" where their lifetime is very short[89]. Therefore, if adiabatic stabilization is to be observed experimentally, the laser pulse risetime must not be so long that saturation (complete ionization) occurs during the turn-on. However, it must be long enough so that the atom adiabatically remains in the ground state in the Kramers frame. By comparing Floquet calculations with time-dependent calculations, it has been shown that these criteria can both be fulfilled[90-92]. As the laser pulses become shorter, excited states of the Kramers static potential $V_0(\alpha_0, \mathbf{r})$ become populated. As a result, dichotomy will become less apparent[93-94], but stabilization still persists.

Due to the adiabaticity condition and the high-frequency, high-intensity requirements, the experimental verification of adiabatic stabilization is therefore a very difficult task. Fortunately, the stabilization conditions can be met at lower intensities and frequencies if the atom is prepared in an initial state which is not the ground state, but a "circular" Rydberg state with large n, l and $|m|$ quantum numbers, such that the lifetime in the "death valley" regime is sufficiently long[82,95]. An experiment of this kind, using two laser pulses, has been performed by de Boer et al.[96,97] to study the stabilization behaviour of the hydrogen-like, "circular" 5 g state of neon. The experimental data show a suppression of ionization as the intensity increases, in accordance with the prediction of adiabatic stabilization, and are in fair agreement with theoretical calculations[98].

Another interesting prediction of Floquet calculations is the occurrence of light-induced states (LIS). The emergence of LIS has been seen in one-dimensional model atom calculations[99,100] and has been analyzed in terms of shadow poles[84,101], which arise from solutions of the Floquet equations corresponding to unphysical boundary conditions. These poles can, however, lead to physically observable consequences following the crossing of a multiphoton threshold, when a shadow pole becomes a physical pole. This can give rise to the appearance of "new discrete states" for certain laser intensities. Although LIS arising from shadow poles have been predicted by a number of theoretical calculations[84, 99-102], experimental verification has yet to come forth.

The Floquet method can also be used to study laser-assisted electron-atom collisions. Starting with the Floquet-Fourier form (30) of the wave function, and expanding the harmonic components in terms of partial waves, Floquet close-coupling (FCC) equations can be derived which are satisfied by the corresponding radial functions. Using the FCC equations, Dimou and Faisal[103] have studied laser- induced resonances in electron-proton scattering. Within the context of the high-frequency Floquet theory, Gavrila and Kaminski[104] have considered laser-assisted potential scattering. Recently, a momentum space formulation of the problem has been given by Kylstra and Joachain[105], who derived an integral equation - the Floquet-Lippmann-Schwinger (FLS) equation -

for the scattering T-matrix elements. They solved the FLS equation numerically for laser-assisted low-energy electron scattering by various potentials and analyzed the validity of the low frequency approximation of Kroll and Watson[106]. Finally, we mention that the Floquet theory has also been used to construct "dressed" atomic target states in a non-perturbative way in order to study laser-assisted electron-atom collisions[107-109]. This approach is particularly useful to analyze resonant cases, where the laser frequency matches a transition frequency in the atom.

3.3. R-matrix-Floquet theory

The R-matrix-Floquet (RMF) theory is a non-perturbative approach which has been proposed by Burke, Francken and Joachain[85,86] to analyze atomic multiphoton processes in intense laser fields. The RMF theory treats multiphoton ionization, harmonic generation and laser-assisted electron-atom collisions in a unified way. It is completely ab-initio and is applicable to an arbitrary atom or ion, allowing an accurate description of electron correlation effects.

Let us consider an atomic system, composed of a nucleus of atomic number Z and N electrons, in a laser field which is treated classically as a spatially homogeneous electric field $\mathcal{E}(t)$. Although more general cases (such as two-colour fields) have been studied, we shall assume for the moment that the laser field is monochromatic and linearly polarized, so that the electric field $\mathcal{E}(t)$ is given by eq.(25). Neglecting relativistic effects, the atomic system in the presence of this laser field is then described by the time-dependent Schrödinger equation [see eqs.(13) - (16)].

$$i\hbar \frac{\partial}{\partial t}\Psi(X,t) = [H_{at} + \frac{e}{m}\mathbf{A}(t).\mathbf{P} + \frac{e^2 N}{2m}A^2(t)]\Psi(X,t) \qquad (40)$$

where the vector potential is $\mathbf{A}(t) = \hat{\epsilon} A_0 \sin(\omega t)$ with $A_0 = -\mathcal{E}_0/\omega$. We shall be interested in the following three processes involving at most one unbound electron: multiphoton single ionization of atoms and ions, harmonic generation and laser-assisted electron-atom (ion) elastic and inelastic collisions.

According to the R-matrix method[110-112], configuration space is subdivided into two regions. The internal region is defined by the condition that the radial coordinates r_i of all N electrons are such that $r_i \leq a$ ($i = 1, 2, ...N$), where the sphere of radius a envelops the charge distribution of the target atom states retained in the calculation. In this region, exchange effects involving all N electrons are important. The external region is defined so that one of the electrons (say electron N) has a radial coordinate $r_N \geq a$, while the remaining $N-1$ electrons are confined within the sphere of radius a. Hence in this region exchange effects between the "external" electron and the remaining $N-1$ electrons can be neglected.

Having divided configuration space into an internal and an external region, we must solve the time-dependent Schrödinger equation (40) in these two regions separately. This is done by using the Floquet method which, as we have seen in Section 3.2, reduces the problem to solving an infinite set of coupled time-independent equations for the harmonic components $F_n(X)$ of the wave function $\Psi(X,t)$. The solutions in the internal and external regions are then matched on the boundary at $r = a$.

In the internal region it is convenient to use the length gauge since in this gauge the laser-atom coupling tends to zero at the origin. We remark that in this region the Floquet Hamiltonian $\mathbf{H}_F$ is not Hermitian, due to surface terms at $r = a$ arising from the kinetic energy operator in H_{at}. These terms can be eliminated by introducing a

Bloch operator[113] $\mathbf{L}_B$, so that $\mathbf{H}_F + \mathbf{L}_B$ is Hermitian in the internal region. Following the R-matrix procedure[111,112], an elaborate basis set is then constructed, in which the operator $\mathbf{H}_F + \mathbf{L}_B$ is diagonalized. Using the spectral representation of this operator, one obtains on the boundary the important relation[85,86]

$$\mathbf{u}(a) = {}^L\mathbf{R}(E)\left[r\frac{d\mathbf{u}}{dr}\right]_{r=a} \tag{41}$$

where $\mathbf{u}(r)$ denotes the set of reduced radial wave functions (i.e. radial wave functions multiplied by r) and ${}^L\mathbf{R}(E)$ is the R-matrix in the length gauge. The logarithmic derivatives of the reduced radial wave functions on the boundary $r = a$, which provide the initial conditions for solving the problem in the external region are then given by eq. (41).

In the external region, we have only one electron ($r_N \geq a$), whose motion is studied by using the velocity gauge. Here a simple close-coupling expansion can be used for the harmonic components, since exchange effects between this "external" electron and the remaining $N - 1$ electrons are negligible. The resulting set of coupled differential equations is then solved[87], subject to boundary conditions at $r = a$ and $r \to \infty$. At $r = a$, the matching of the internal and external region solutions provides ${}^V\mathbf{R}(E)$, the R-matrix in the velocity gauge. The coupled equations are solved from $r = a$ to a large value $r = a'$ of the radial coordinate by propagating the R-matrix ${}^V\mathbf{R}(E)$. The solutions obtained in this way are matched at $r = a'$ with the solutions satisfying given boundary conditions for $r \to \infty$, calculated by using asymptotic expansions[87].

The boundary conditions for $r \to \infty$ are formulated in the Kramers frame[70,71], because in this frame the channels decouple asymptotically. These boundary conditions differ according to the process considered. For the case of multiphoton ionization and harmonic generation, there are only outgoing waves corresponding to Siegert[114] boundary conditions. It is then found that solutions will only occur for certain complex values of the energy [see eq.(34)]. From the knowledge of the eigenvectors, one may obtain all the other physical quantities, such as the branching ratios into the channels, the angular distribution of the ejected electrons, etc. In the case of laser-assisted electron-atom (ion) collisions, one must impose S-matrix (or T - or K-matrix) asymptotic boundary conditions[115,116]. The scattering amplitudes and cross sections are then given in terms of the elements of the S, T or K-matrix.

The RMF theory summarized above has been applied to a number of multiphoton processes. We begin by considering multiphoton ionization of atoms in the presence of a linearly polarized monochromatic laser field [see eq.(25)]. We have already shown in Fig. 6 the total RMF ionization rate[88] as a function of the intensity for H(1s) in a laser field of high angular frequency ($\omega = 0.65$ a.u.), in connection with the discussion of adiabatic stabilization. Similar results have also been obtained[88] for H(2s) in a laser field of angular frequency $\omega = 0.184$ a.u., corresponding to a KrF laser. The RMF theory has also been used to calculate the total ionization rate, the branching ratios into the dominant ionization channels and the angular distribution of the ejected electrons for H(1s) in a laser field of angular frequency $\omega = 0.184$ a.u., such that at low intensities three-photon absorption is required for ionization.

We now turn to two-electron systems, for which the RMF theory has provided non-perturbative multiphoton ionization rates including electron correlation effects[117-119]. As an example, we display in Fig. 7 the RMF results for the multiphoton detachment of H^-, obtained at an angular frequency $\omega = 0.0149$ a.u. such that at least two photons are necessary to detach an electron. The total detachment rate is shown, as well as the

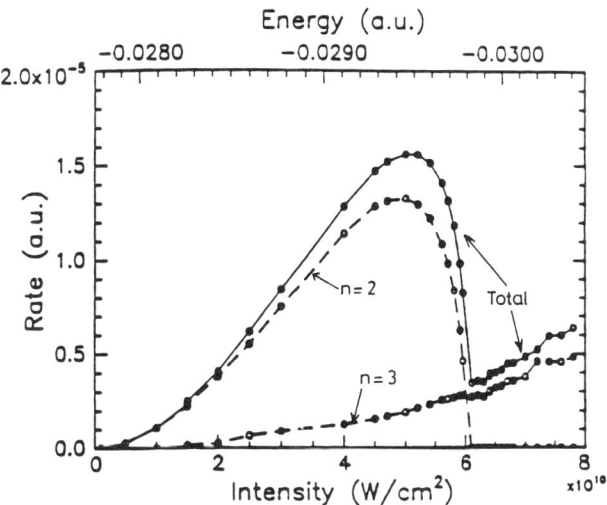

Figure 7. Total and partial (into the $n = 2$ and $n = 3$ photon channels, broken curves) RMF detachment rates of H$^-$ in a laser field of angular frequency $\omega = 0.0149$ a.u., versus intensity. The open circles correspond to the calculated results. From Dörr et al.[118]

partial rates into the two- and three-photon detachment channels. At low intensities the total detachment rate increases perturbatively with the second power of the intensity, since the dominant detachment channel is the two-photon channel. At these low intensities the three-photon partial rate is very small, being proportional to the third power of the intensity. When the intensity reaches the value 6×10^{10} W cm^{-2}, the two-photon channel closes, due to the dynamic Stark shift, and only three- and higher photon detachment processes are possible. Above this intensity the difference between the total detachment rate and the partial rate into the three-photon detachment channel is due to higher order processes. The RMF theory has also been applied[120,121] to study the multiphoton detachment of several other negative ions.

The RMF theory has allowed to study ab-initio a wide variety of interesting resonance effects in multiphoton ionization[117-125]. For example, in the case of two-photon ionization, due to the dipole selection rules, resonances can in general occur between the ground state and the members of two different Rydberg series. Now, when two Rydberg states, each belonging to a different Rydberg series and lying close in energy are resonantly coupled to the ground state by a single photon transition, interference between the two resonant pathways will occur. This interference, which depends on the laser intensity, will in turn modify the ionization rate of the ground state. This effect has been demonstrated in neon, using the RMF theory to study resonant two-photon ionization via the 5s and 4d Rydberg states[122].

A spectacular effect which has been predicted by the RMF theory is the occurrence of laser-induced degenerate states (LIDS) involving autoionizing states in complex atoms[123]. To understand this phenomenon, we first recall that autoionizing states of atoms produce characteristic resonance structures in the photo-electron yield, not only in (one-photon) photo-ionization, but also in multiphoton ionization. At low intensities, these structures can be reproduced by using perturbation theory to treat the interaction of the atom with the radiation field. By contrast, for the case of atoms in intense fields, a perturbative description of the ground state-autoionizing state coupling fails when the intensity is large enough so that the laser-induced width of the ground state becomes comparable to the width of the autoionizing state near resonance[126,127]. A detailed study[123] shows that there exist critical points in the complex energy plane (to each of which corresponds a critical intensity and angular frequency) such that the complex quasi-energies of the two states coupled by the laser field are exactly degenerate, i.e. where LIDS occur. The existence of LIDS is a general phenomenon, which has been observed in RMF calculations for multiphoton transitions[123-125] and understood by constructing models[123-125] that retain the essential ingredients of the full RMF calculations. We remark that LIDS can be viewed as an extension of the work of Berry[128], where the adiabatic passage around degeneracies in a parameter space was described, and which has attracted considerable interest, particularly with respect to the associated geometric phase.

The RMF theory has also been applied to study harmonic generation[129-131] and laser-assisted collisions[132-134]. In the latter case, the first application of the RMF method has been a study[132] of electron-proton scattering in a laser field of the type described by eq.(25). The laser field induced resonances[103] due to the temporary capture of the projectile electron into atomic hydrogen bound states, and structures corresponding to different sublevels can appear. The RMF theory has also been applied to electron-argon scattering in a laser field, where the argon target was represented as a model potential[133], and to laser-assisted electron-atomic hydrogen scattering[134].

Recently, the RMF method has been generalized[135,136] to bichromatic laser fields such as

$$\mathcal{E}(t) = \hat{\epsilon} \left[\mathcal{E}_1 \cos(\omega_1 t) + \mathcal{E}_2 \cos(\omega_2 t + \phi) \right] \qquad (42)$$

where $\hat{\epsilon}$ is the (common) unit polarization vector, $\mathcal{E}_1$ and $\mathcal{E}_2$ are the amplitudes of the electric fields oscillating with the angular frequencies ω_1 and ω_2, respectively, and ϕ is a phase. The extension of the method for atoms in two laser fields with incommensurable frequencies has been used to analyze light-induced continuum structures (LICS) in helium[135,137], as well as doubly and triply resonant multiphoton processes involving autoionizing resonances in magnesium[136]. Within the context of these multiply resonant processes, coherent control of the ionization can be exercised in the sense that by changing the laser parameters, the degree of interaction between the resonant processes can be varied. In addition, the RMF calculations performed for the case of magnesium[136] predict the occurrence of LIDS between autoionizing levels at laser intensities and frequencies which are accessible to experimental studies.

To conclude this section on the RMF theory, we remark that recent reviews of this method and its applications have been given by Joachain[138], Dörr[139] and Joachain et al.[140].

3.4. Numerical solution of the time-dependent Schrödinger equation

The non-perturbative Floquet and R-matrix Floquet methods which I have considered so far are based on the assumption that the Hamiltonian of the atomic system in the laser field is periodic in time. Although this is not true for a realistic laser pulse, it is still possible to incorporate pulse shape effects into the theoretical Floquet or R-matrix Floquet calculations for laser pulses which are relatively short, down to fractions of picoseconds (10^{-12} s). However, for ultra-short pulses, in the femtosecond (10^{-15} s) range one must in general obtain information about the multiphoton processes by direct numerical integration of the time-dependent equations of motion. In the non-relativistic regime, this amounts to solving the time-dependent Schrödinger equation (TDSE). This approach has the advantage that no restrictions need to be made about the type of laser pulse and that solutions can, in principle, be obtained for all regimes of intensity and frequency. It has the disadvantage that it is computationally very intensive.

A straightforward way of reducing the computational load is to study one-dimensional model atoms. Since one-dimensional codes are relatively fast, it is possible to conduct "numerical experiments" by investigating a large range of parameters. However, one dimensional calculations present a number of disadvantages due to the over-simplification of the problem. In particular, the effects related to the transverse direction, such as those due to circularly polarized radiation, are neglected. In addition, the singularity at the origin associated with the Coulomb potential is overemphasized in one dimension and requires the use of smoothed model potentials. Nevertheless, a number of interesting results have been obtained in one-dimensional studies of the TDSE. These have been reviewed by Eberly et al.[141] and Protopapas et al.[7]. Of particular interest are the recent one-dimensional, time-dependent R-matrix calculations of Phil and Val Burke[142] which open the door to the solution of the TDSE by using the R-matrix method.

Advances in computer technology over the past ten years have made possible the numerical integration of the TDSE for atoms or ions with one single active electron (SAE) in laser fields. The first calculations of this kind were carried out by Kulander[143]. These single electron calculations are "exact" for hydrogenic systems. However, for atoms or ions with more than one electron, dynamic electron correlations are neglected in the SAE model.

As an illustration of the "exact" TDSE calculations performed for atomic hydrogen, we show in Fig. 8 the ionized fraction of atoms as a function of time, for H(1s) in high frequency ($\omega = 2$ a.u.), very intense laser fields which are turned on very rapidly (with a two-cycle ramp), as obtained by Latinne, Joachain and Dörr[144]. The Floquet results at the corresponding field strengths are also displayed, and are seen to be in good agreement with the values obtained by solving numerically the TDSE. We also see from Fig. 8 that both the Floquet and the time-dependent calculations exhibit the stabilization behaviour, the ionized fraction corresponding to $\mathcal{E}_0 = 10$ a.u. being inferior to that for $\mathcal{E}_0 = 5$ a.u. The extra beat structure visible for the curve corresponding to $\mathcal{E}_0 = 10$ a.u. in Fig. 8 is due to interference between the populations of the 1s and higher lying Floquet states (mainly the 2s state) which are populated during the turn-on of the field. Detailed comparisons between Floquet and "exact" TDSE calculations for atomic hydrogen in very strong, ultra-short, high frequency laser pulses of various shapes have been made by Dörr, Latinne and Joachain[145], who have also studied the population transfer from the ground state to excited states, for one-and two-colour linearly polarized fields. Corrections to the dipole approximation have been

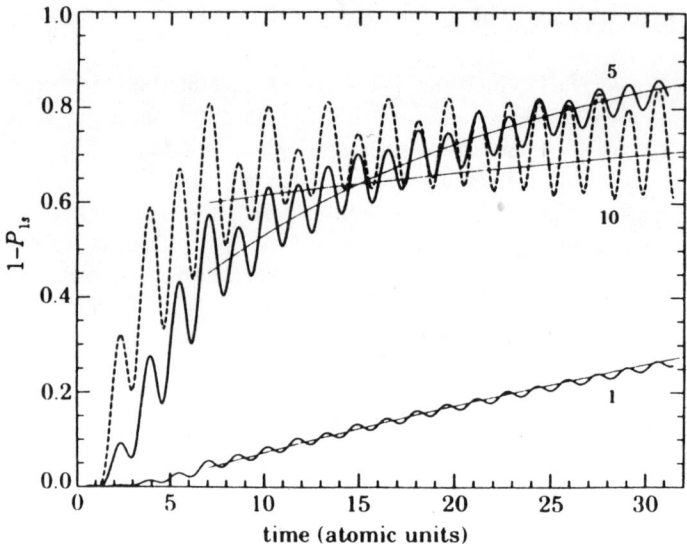

Figure 8. Ionized fraction versus time (in a.u.) for H(1s) in linearly polarized laser fields with peak electric field strengths $\mathcal{E}_0 = 1$, 5 and 10 a.u., indicated next to the curves. The thin lines give the corresponding Floquet results. From Latinne, Joachain and Dörr[144].

investigated by Latinne, Joachain and Dörr[144], who found that for atomic hydrogen in a high frequency laser field these corrections modify the ionization probability only slightly, up to the maximum intensity considered (2.5×10^{19} W cm^{-2}). The coupling of the magnetic field to the spin of the electron was also examined[144] by solving the Pauli equation. No significant differences regarding the ionization probability were found.

A second example of TDSE calculations performed for atomic hydrogen is provided by the high-order ATI spectra obtained by Cormier and Lambropoulos[146]. These confirmed the existence of a plateau in the spectra, observed in experiments[27], which can be understood qualitatively by using the semi-classical "recollision model" mentioned in Section 2.1.

A final example of TDSE calculations for atomic hydrogen is the work of Taïeb et al.[147], who obtained two-colour photo-electron spectra for the case of hydrogen atoms exposed to an intense radiation pulse containing the fundamental frequency of a laser and a weaker harmonic whose frequency is high enough so that the atom can be ionized by a single photon. Interferences arise between multiphoton "above threshold ionization" (ATI) and laser-assisted single photon ionization (LASPI). These interference effects can lead to a partial coherent control of the photo-ionization process[147,148].

In the case of complex atoms, nearly all the TDSE calculations have been performed by using the single active electron (SAE) approximation[149]. This is due to the fact that "exact" TDSE calculations for atoms or ions with two or more electrons in laser fields are prohibitively large, even on the most recent computers. Among the few

Nearly all the quantum mechanical calculations of laser-atom interactions in the relativistic domain have been restricted until now to lower dimensional treatments. In particular, Protopapas, Keitel and Knight[164] have solved the time-dependent relativistic Schrödinger equation in the Kramers frame, while Kylstra, Ermolaev and Joachain[165] have used a B-spline expansion in momentum space to solve numerically the time-dependent Dirac equation. In both cases, the calculations were performed for a model, one-dimensional atom in the high frequency, high-intensity (stabilization) regime. Magnetic field and retardation effects are clearly not included in one-dimensional model calculations, since their description requires an additional spatial dimension. On the other hand, relativistic effects due to the dressing of the electron mass by the laser field[166] and the influence of the negative energy states can be investigated in such calculations. Kylstra et al.[165] found that for a peak electric field strength $\mathcal{E}_0 = 175$ a.u. and an angular frequency $\omega = 1$ a.u. (such that $q = 0.4$) relativistic effects become apparent. Even under these extreme conditions, however, the Dirac wave function remains localized in a superposition of field-free bound states and low-energy continuum states. We show in Fig. 9 the results of Kylstra et al.[165] for the Dirac and Schrödinger probability densities (the latter one being obtained from the numerical solution of the time-dependent non-relativistic Schrödinger equation) at the end of the 9 th laser cycle, when the electric field is maximum. The peak in the Dirac probability density corresponds to the relativistic "classical" excursion amplitude, $x_0 = 124$ a.u. Likewise, the peak in the Schrödinger probability density occurs at $x_0 = 175$ a.u., the non-relativistic classical excursion amplitude. At the end of the pulse, the ionization probabilities are, respectively, 0.52 for the Dirac wave function and 0.58 for the Schrödinger wave func-

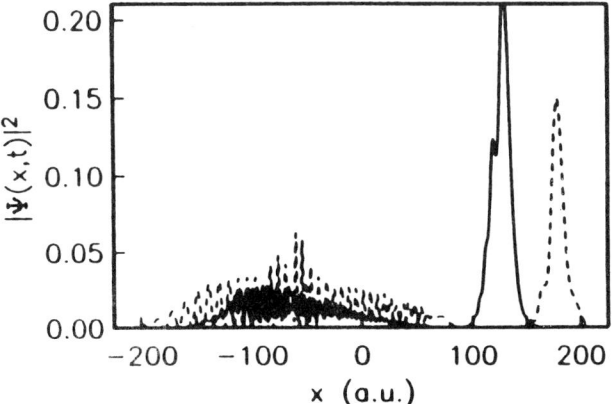

Figure 9. The Dirac (solid line) and Schrödinger (dotted line) probability densities, at the end of the 9 th cycle, for a laser pulse with a four cycle $\sin^2$ turn-n, an angular frequency $\omega = 1$ a.u. and a peak electric field strength $\mathcal{E}_0 = 175$ a.u. From Kylstra, Ermolaev and Joachain[165].

TDSE computations performed until now beyond the SAE approximation are those of Zhang and Lambropoulos, who have calculated multiphoton photo-electron spectra for helium[150] and magnesium[151] using B-splines to construct basis functions, and those of Taylor et al.[152-154], who have studied multiphoton processes in helium by direct numerical integration of the TDSE on a Cray T3D. Results on ionization and double excitation of helium by short intense laser pulses, obtained by direct numerical integration of the TDSE using the method of complex scaling, have also been reported by Scrinzi and Piraux[155].

An important test for theories of multiphoton processes including correlation effects is to calculate accurately double ionization yields for two-electron systems in intense laser fields. Promising results have been obtained by Watson et al.[156] using a "semi-independent" electron approach, which requires the solution of two single-active-electron problems, the second incorporating the results from the first and therefore subject to interelectronic correlation. The calculations of Watson et al.[156] reproduce the large enhancement ("knee") of the double ionization yield (with respect to the predictions of the SAE model) observed in the experiment of Walker et al.[157]. This "knee" structure has also been reproduced qualitatively in calculations performed for one-dimensional two-electron atoms with interelectronic correlations[156,158] and in the many-body S-matrix calculations of Faisal and Becker[159,160]. A review of the current state of the theory of two-electron atoms in strong laser field has been written by Lambropoulos et al.[161].

For atoms or ions with more than two active electrons the TDSE can only be solved on restricted bases, built on judiciously chosen orbitals for each atomic system under consideration. Several other methods, however, have been used. In particular, time-dependent density functional theory allows the time-dependent problem to be formulated in terms of single-particle orbitals moving in an approximate effective potential. A review of this method has been given by Ullrich and Gross[162].

3.5 Relativistic effects in laser-atom interactions

In this section, we shall consider the relativistic effects which arise when atoms interact with very strong laser fields. Such effects are expected to become important when the "quiver" velocity of the electron is of the order of the velocity of light, i.e. when its ponderomotive energy U_p approaches its rest mass energy mc^2. Using eq.(2), this means that the quantity

$$q = \frac{U_p}{mc^2} = \frac{e^2 \mathcal{E}_0^2}{4m^2\omega^2 c^2} \tag{43}$$

must then be of the order of unity. If the electric field strength $\mathcal{E}_0$ and the angular frequency ω are expressed in atomic units (a.u.), we have $q = 1.33 \times 10^{-5}(\mathcal{E}_0/\omega)^2$. Thus, if $\omega = 0.043$ a.u. (corresponding to a Nd-YAG laser), we see that $q = 1$ when $\mathcal{E}_0 = 11.8$ a.u., i.e. when the intensity $I = 4.9 \times 10^{18}$ W cm^{-2}. On the other hand, for a laser of higher angular frequency $\omega = 1$ a.u., we have $q = 1$ when $\mathcal{E}_0 = 274$ a.u., corresponding to the very large intensity $I = 2.6 \times 10^{21}$ W cm^{-2}.

Three-dimensional Monte-Carlo classical simulations of ionization and harmonic generation in the relativistic regime have been performed by Keitel and Knight[163] for hydrogen atoms in very intense laser fields. They found that the magnetic field component of the laser field may induce a significant motion of the electron in the propagation direction of the field. This leads to additional ionization, a possible breakdown of stabilization and reduced harmonic generation.

tion, indicating that the Dirac wave function is slightly more stable against ionization. Kylstra et al. calculated from the Dirac wave function the energy distribution of the ionized electrons at the end of the pulse. They found that the majority of the electrons is emitted with very low energies, a result also obtained by Faisal and Radozycki[167], who considered a separable potential model of a bound Klein-Gordon particle interacting with an intense laser field. The time-dependent Klein-Gordon equation has also been solved numerically by Taïeb et al.[168] for a model one-dimensional atom in order to look for the signature of relativistic effects at ultra-high intensities. They found that a clear signature of these effects is the occurrence of modifications in ATI spectra resulting from a two-color scheme involving simultaneously a high frequency and a low frequency laser field. Recently, Rathe et al.[169] have investigated the relativistic wave packet dynamics of a single electron initially bound by a smoothed Coulomb potential in two dimensions, using the Dirac equation, and Ermolaev[170] has studied atomic states of one-electron atoms in super-intense laser fields, using a high-frequency approximation proposed by Krstić and Mittleman[171], which generalizes the high-frequency theory described in Section 3.2 to the relativistic domain. The calculations of Ermolaev show the existence of a wide intermediate range of field strengths where, in the lowest order of the high-frequency approximation, the binding is stabilized by the laser field.

4. CONCLUSIONS AND FUTURE DEVELOPMENTS

The availability of very intense laser pulses whose electric field strength approaches or even exceeds the Coulomb binding field within atoms has made possible, in recent years, the discovery of new phenomena in laser-atom interactions. These include the "above threshold ionization" process in multiphoton ionization, the stabilization of atoms at super-high intensities, the emission by atoms of very high order harmonics of the exciting laser light and multiphoton processes in laser-assisted collisions. In this article, I have given a survey of these phenomena and of the main theoretical methods which have been used to analyze them. A similar review could be written about molecules in intense laser fields, where new effects such as Coulomb explosions, bond softening and coherent control of dissociation have been discovered[172-174]. The interaction of clusters of atoms with strong laser fields constitutes a new area of multiphoton physics, where enhanced yields of high harmonics and the generation of very energetic ionization fragments have been observed[175,176]. Solid targets and plasmas interacting with intense laser pulses also exhibit a wide range of interesting phenomena and potential applications such as electron acceleration to relativistic energies and recombination X-ray lasers[177]. Finally, the petawatt (10^{15} W) laser now being developed at Livermore[178] will yield intensities of the order of 10^{21} W cm^{-2}. At such intensities, the electric field strength $\mathcal{E}_0 \simeq 10^{12}$ V cm^{-1} is about two hundred times larger than the atomic unit ($\mathcal{E}_a = 5.1 \times 10^9$ V cm^{-1}), the radiation pressure $P = I/c$ reaches the enormous value of 300 Gbar and the quiver motion of the ejected electrons (for an angular frequency $\omega = 0.043$ a.u. of a Nd-YAG laser) is fully relativistic. Much work will be required to understand the phenomena occurring in this regime, which is relevant for studies of the Fast Ignitor concept of inertial confinement fusion as well as for astrophysical applications[2].

Acknowledgments

It is a pleasure to thank Dr. M. Dörr, Dr. A.M. Ermolaev and Dr. N.J. Kylstra for many fruitful discussions.

REFERENCES

1. D. Strickland and G.A. Mourou, Opt. Commun. 56: 219 (1985).
2. G.A. Mourou, C.P.J. Barty and M.D. Perry, Physics Today 51: 22 (1998).
3. M. Gavrila, ed., *Atoms in Intense Laser Fields*, Adv. At. Mol. Opt. Phys. Suppl. 1 (1992).
4. K. Burnett, V.C. Reed and P.L. Knight, J. Phys. B 26: 561 (1993).
5. C.J. Joachain, Theory of laser-atom interactions, in: *Laser Interactions with Atoms, Solids and Plasmas*, R.M. More, ed., Plenum Press, New York (1994), p. 39.
6. L.F. DiMauro and P. Agostini, Adv. At. Mol. Opt. Phys. 35: 79 (1995).
7. M. Protopapas, C.H. Keitel and P.L. Knight, Rep. Progr. Phys. 60: 389 (1997).
8. C.J. Joachain, M. Dörr and N.J. Kylstra, Adv. At. Mol. Opt. Phys. (to be published).
9. G.S. Voronov and N.B. Delone, JETP Letters 1: 66 (1965).
10. J.L. Hall, E.J. Robinson and L.M. Branscomb, Phys. Rev. Lett. 14: 1013 (1965).
11. P. Agostini, F. Fabre, G. Mainfray, G. Petite and N.K. Rahman, Phys. Rev. Lett. 42: 1127 (1979).
12. G. Petite, P. Agostini and H.G. Muller, J. Phys. B 21: 4097 (1988).
13. R.R. Freeman, P.H. Bucksbaum, H. Milchberg, S. Darack, D. Schumacher and M.E. Geusic, Phys. Rev. Lett. 59: 1092 (1987).
14. P. Hansch, M.A. Walker and L.D. van Woerkom, Phys. Rev. A 55: R 2535 (1997); Phys. Rev. A 57: R 709 (1998).
15. H. Helm and M.J. Dyer, Phys. Rev. A 49: 2726 (1994); V. Schyja, T. Lang and H. Helm, Phys. Rev. A 57: 3692 (1998).
16. S. Augst, D. Strickland, D.D. Meyerhofer, S.L. Chin and J.H. Eberly, Phys. Rev. Lett. 63: 2212 (1989).
17. E. Mevel, P. Breger, R. Trainham, G. Petite, P. Agostini, A. Migus, J.P. Chambaret and A. Antonetti, Phys. Rev. Lett. 70: 406 (1993).
18. L.V. Keldysh, Sov. Phys.-JETP 20: 1307 (1965).
19. F.H.M. Faisal, J. Phys. B 6: L 89 (1973).
20. H.R. Reiss, Phys. Rev. A 22: 1786 (1980).
21. M.V. Ammosov, N.B. Delone and V.P. Krainov, Sov. Phys.-JETP 64: 1191 (1986).
22. P.B. Corkum, Phys. Rev. Lett. 71: 1994 (1993).
23. K.C. Kulander, K.J. Schafer and J.L. Krause, Dynamics of short-pulse excitation, ionization and harmonic conversion, in *Super-Intense Laser-Atom Physics*, B. Piraux, A. L'Huillier and K. Rzazewski, eds, Plenum Press, New York (1993), p. 95.
24. M. Lewenstein, Ph. Balcou, M. Yu. Ivanov, A. L'Huillier and P.B. Corkum, Phys. Rev. A 49, 2117 (1994).
25. M.Y. Kuchiev, JETP. Lett. 45: 404 (1987).
26. H.B. van Linden van den Heuvell and H.G. Mu'ler, in *Multiphoton Processes*, S.J. Smith and P.L. Knight, eds, Cambridge Univ. Press, London (1988).
27. G.G. Paulus, W. Nicklich, H. Xu, P. Lambropoulos and H. Walther, Phys. Rev. Lett. 72: 2851 (1994).
28. B. Yang, K.J. Schafer, B. Walker, K.C. Kulander, P. Agostini and L.F. DiMauro, Phys. Rev. Lett. 71: 3770 (1993).
29. G.G. Paulus, W. Nicklich and H. Walther, Europhys. Lett. 27: 267 (1994).
30. G.G. Paulus, W. Nicklich, F. Zacher, P. Lambropoulos and H. Walther, J. Phys. B 29: L 249 (1996).
31. M.P. Hertlein, P.H. Bucksbaum and H.G. Muller, J. Phys. B 30: L 197 (1997).
32. H.G. Muller and F.C. Kooiman, Phys. Rev. Lett. 81: 1207 (1998).
33. G.G. Paulus, F. Zacher, H. Walther, A. Lohr, W. Becker and M. Kleber, Phys. Rev. Lett. 80: 484 (1998).
34. G.H.C. New and J.F. Ward, Phys. Rev. Lett. 19: 556 (1967).
35. A. McPherson, G. Gibson, H. Jara, U. Johann, T.S. Luk, I.A. McIntyre, K. Boyer and C.K. Rhodes, J. Opt. Soc. Am. B 4: 595 (1987).
36. J. Wildenauer, J. Applied Phys. 62: 41 (1987).
37. R. Rosman, G. Gibson, K. Boyer, H. Jara, T.S. Luk, I.A. McIntyre, A. McPherson, J.C. Salem and C.K. Rhodes, J. Opt. Soc. Am. B 5: 1237 (1988).
38. M. Ferray, A. L'Huillier, X.F. Li, L.A. Lompré, G. Mainfray and C. Manus, J. Phys. B 21: L 31 (1988).
39. L.A. Lompré, A. L'Huillier, P. Monot, M. Ferray, G. Mainfray and C. Manus, J. Opt. Soc. Am. B 7: 754 (1990).

40. A. L'Huillier, L.A. Lompré, G. Mainfray and C. Manus, Adv. At. Mol. Opt. Phys. Suppl. 1: 139 (1992).
41. J.J. Macklin, J.D. Kmetec and C.L. Gordon III, Phys. Rev. Lett. 70: 766 (1993).
42. A. L'Huillier and Ph. Balcou, Phys. Rev. Lett. 70: 774 (1993).
43. K. Miyazaki and H. Takada, Phys. Rev. A 52: 3007 (1995).
44. Y. Nagata, K. Midorikawa, M. Obara and K. Toyoda, Opt. Lett. 21: 15 (1996).
45. S.G. Preston, A. Sanpera, M. Zepf, W.J. Blyth, C.G. Smith, J.S. Wark, M.H. Key, K. Burnett, N. Nakai, D. Neely and A.A. Offenberger, Phys. Rev. A 53: R31 (1996).
46. J. Zhou, J. Peatross, M.M. Murnane and H.C. Kapteyn, Phys. Rev. Lett. 76: 752 (1996).
47. I.P. Christov, J. Zhou, J. Peatross, A. Rundquist, M.M. Murnane and H.C. Kapteyn, Phys. Rev. Lett. 77: 1743 (1996).
48. C. Spielmann, N.H. Burnett, S. Sartania, R. Koppitsch, M. Schnürer, C. Kan, M. Lenzner, P. Wobrauschek and F. Krausz, Science 278: 661 (1997).
49. Z. Chang, A. Rundquist, H. Wang, M.M. Murnane and H.C. Kapteyn, Phys. Rev. Lett. 79: 2967 (1997).
50. M. Schnürer, C. Spielmann, P. Wobrauschek, C. Streli, N.H. Burnett, C. Kan, F. Ferencz, R. Koppitsch, Z. Cheng, T. Brabec and F. Krausz, Phys. Rev. Lett. 80: 3236 (1998).
51. A. Rundquist, C.G. Durfee, Z.H. Chang, C. Herne, S. Backus, M.M. Murnane and H.C. Kapteyn, Science 280: 1412 (1998).
52. J.W.G. Tisch, D.D. Meyerhofer, T. Ditmire, N. Hey, M.B. Mason and M.H.R. Hutchinson, Phys. Rev. Lett. 80: 1204 (1998).
53. C. Bellini, C. Lyngä, A. Tozzi, M.B. Gaarde, T.W. Hänsch, A. L'Huillier and C.G. Wahlström, Phys. Rev. Lett. 81: 297 (1998).
54. K. Burnett, V.C. Reed, J. Cooper and P.L. Knight, Phys. Rev. A 45: 3347 (1992).
55. J.L. Krause, K.J. Schafer and K.C. Kulander, Phys. Rev. A 45: 4998 (1992).
56. J.L. Krause, K.J. Schafer and K.C. Kulander, Phys. Rev. Lett. 68: 3535 (1992).
57. K.J. Schafer and K.C. Kulander, Phys. Rev. Lett. 78: 638 (1997).
58. P.B. Corkum, N.H. Burnett and M. Yu. Ivanov, Opt. Lett. 19: 1870 (1994).
59. P. Antoine, A. L'Huillier and M. Lewenstein, Phys. Rev. Lett. 77: 1234 (1996).
60. P. Antoine, B. Piraux, D.B. Milosevic and M. Gajda, Phys. Rev. A 54: R 1761 (1996).
61. P. Salières, A. L'Huillier, P. Antoine and M. Lewenstein, Adv. At. Mol. Opt. (to be published).
62. F. Ehlotzky, A. Jaron and J.Z. Kaminski, Phys. Rep. 297: 63 (1998).
63. A. Weingartshofer, J.K. Holmes, G. Caudle, E.M. Clarke and H. Kruger, Phys. Rev. Lett. 39: 269 (1977).
64. A. Weingartshofer, J.K. Holmes, J. Sabbagh and S.L. Chin, J. Phys. B 16: 1805 (1983).
65. B. Wallbank and J.K. Holmes, J. Phys. B 27: 1221 (1994).
66. N.J. Mason and W.R. Newell, J. Phys. B 20: L 323 (1987).
67. B. Wallbank, J.K. Holmes, L. Le Blanc and A. Weingarshofer, Z. Phys. 10: 467 (1988).
68. B. Wallbank, J.K. Holmes and A. Weingartshofer, J. Phys. B 23: 2997 (1990).
69. S. Luan, R. Hippler and H.O. Lutz, J. Phys. B 24: 3241 (1991).
70. H.A. Kramers, *Collected Scientific Papers* (North Holland, Amsterdam, 1956), p. 272.
71. W.C. Henneberger, Phys. Rev. Lett. 21: 838 (1968).
72. G. Floquet, Ann. Ec. Norm. (2) 13: 47 (1883).
73. J.H. Shirley, Phys. Rev. B 138: 979 (1965).
74. S.I. Chu, Adv. At. Mol. Phys. 21: 197 (1985).
75. A. Maquet, S.I. Chu and W.P. Reinhardt, Phys. Rev. A 27: 2946 (1983).
76. See for example the review article of R.M. Potvliege and R. Shakeshaft, Adv. At. Mol. Opt. Phys. Suppl. 1: 373 (1992).
77. M. Dörr, R.M. Potvliege and R. Shakeshaft, Phys. Rev. Lett. 64: 2003 (1990).
78. H. Rottke, B. Wolff, M. Brickwedde, D. Feldmann and K.H. Welge, Phys. Rev. Lett. 64: 404 (1990).
79. M. Gavrila, Adv. At. Mol. Opt. Phys. Suppl. 1, 435 (1992).
80. M. Pont, N. Walet, M. Gavrila and C.W. McCurdy, Phys. Rev. Lett. 61, 939 (1988).
81. M. Pont and M. Gavrila, Phys. Rev. Lett. 65, 2362 (1990).
82. R.J. Vos and M. Gavrila, Phys. Rev. Lett. 68, 170 (1992).
83. J.I. Gersten and M.H. Mittleman, Phys. B 9: 2561 (1976).
84. M. Dörr, R.M. Potvliege, D. Proulx and R. Shakeshaft, Phys. Rev. A 43: 3729 (1991).
85. P.G. Burke, P. Francken and C.J. Joachain, Europhys. Lett. 13: 617 (1990).
86. P.G. Burke, P. Francken and C.J. Joachain, J. Phys. B 24: 761 (1991).
87. M. Dörr, M. Terao-Dunseath, J. Purvis, C.J. Noble, P.G. Burke and C.J. Joachain, J. Phys. B 25: 2809 (1992).

88. M. Dörr, P.G. Burke, C.J. Joachain, C.J. Noble, J. Purvis and M. Terao-Dunseath, J. Phys. B 26: L 275 (1993).
89. P. Lambropoulos, Phys. Rev. Lett. 55: 2141 (1985).
90. M. Pont and R. Shakeshaft, Phys. Rev. A 44: R 4110 (1991).
91. J. Zakrzewski and D. Delande, J. Phys. B 28: L 667 (1995).
92. B. Piraux and R.M. Potvliege, Phys. Rev. A 57: 5009 (1998).
93. K.C. Kulander, K.J. Schafer and J.L. Krause, Phys. Rev. Lett. 66: 2601 (1991).
94. V.C. Reed, P.L. Knight and K. Burnett, Phys. Rev. Lett. 67: 1415 (1991).
95. R.M. Potvliege and P.H.G. Smith, Phys. Rev. A 48, R 46 (1993).
96. M.P. de Boer, J.H. Hoogenraad, R.B. Vrijen, L.D. Noordam and H.G. Muller, Phys. Rev. Lett. 71, 3263 (1993).
97. M.P. de Boer, J.H. Hoogenraad, R.B. Vrijen, R.C. Canstantinescu, L.D. Noordam and H.G. Muller, Phys. Rev. A 50: 4085 (1994).
98. B. Piraux and R.M. Potvliege, Phys. Rev. A 57: 5009 (1998).
99. J.N. Bardsley, A. Szöke and M. Comella, J. Phys. B 21: 3899 (1988).
100. R. Bhatt, B. Piraux and K. Burnett, Phys. Rev. A 37: 98 (1988).
101. M. Dörr and R.M. Potvliege, Phys. Rev. A 41: 1472 (1990).
102. J.C. Wells, I. Simbotin and M. Gavrila, Phys. Rev. Lett. 80: 3479 (1998).
103. L. Dimou and F.H.M. Faisal, Phys. Rev. Lett. 59: 872 (1987).
104. M. Gavrila and J.Z. Kaminski, Phys. Rev. Lett. 52: 613 (1984).
105. N.J. Kylstra and C.J. Joachain, Phys. Rev. A 58: R26 (1998).
106. N.M. Kroll and K.M. Watson, Phys. Rev. A8: 804 (1973).
107. P. Francken and C.J. Joachain, Phys. Rev. A 41: 3770 (1990).
108. P. Francken and C.J. Joachain, J. Opt. Soc. Am. B 7: 554 (1990).
109. M. Dörr, C.J. Joachain, R.M. Potvliege and S. Vucic, Phys. Rev. A 49: 4852 (1994).
110. E.P. Wigner, Phys. Rev. 70: 15 (1946), E.P. Wigner and L. Eisenbud, Phys. Rev. 72: 29 (1947).
111. P.G. Burke and W.D. Robb, Adv. At. Mol. Phys. 11: 143 (1975).
112. P.G. Burke and K.A. Berrington, eds., *Atomic and Molecular Processs: an R-matrix Approach*, Institute of Physics Publ., Bristol (1993).
113. C. Bloch, Nucl. Phys. 4: 503 (1957).
114. A.J.F. Siegert, Phys. Rev. A 56: 750 (1939).
115. C.J. Joachain, *Quantum Collision Theory*, North-Holland, Amsterdam, 3rd ed. (1983).
116. P.G. Burke and C.J. Joachain, *Theory of Electron-Atom Collisions*, Plenum Press, New York (1995).
117. J. Purvis, M. Dörr, M. Terao-Dunseath, C.J. Joachain, P.G. Burke and C.J. Noble, Phys. Rev. Lett. 71: 3943 (1993).
118. M. Dörr, J. Purvis, M. Terao-Dunseath, P.G. Burke, C.J. Joachain and C.J. Noble, J. Phys. B 28: 4481 (1995).
119. A.S. Fearnside, J. Phys. B 31: 275 (1998).
120. H.W. van der Hart, J. Phys. B 29: 3059 (1996).
121. D.H. Glass, P.G. Burke, C.J. Noble and G.B. Wöste, J. Phys. B 31: L667 (1998).
122. N.J. Kylstra, M. Dörr, C.J. Joachain and P.G. Burke, J. Phys. B 28: L685 (1995).
123. O. Latinne, N.J. Kylstra, M. Dörr, J. Purvis, M. Terao-Dunseath, C.J. Joachain, P.G. Burke and C.J. Noble, Phys. Rev. Lett. 74: 46 (1995).
124. A. Cyr, O. Latinne and P.G. Burke, J. Phys. B 30: 659 (1997).
125. N.J. Kylstra, Resonance Effects in Multiphoton Ionization, in *Photon and Electron Collisions with Atoms and Molecules*, P.G. Burke and C.J. Joachain, eds., Plenum Press, New York (1997), p.205.
126. P. Lambropoulos and P. Zoller, Phys. Rev. A 24: 379 (1981).
127. K. Rzazewski and J.H. Eberly, Phys. Rev. Lett. 47: 408 (1981).
128. M.V. Berry, Proc. Roy. Soc. London A 392: 45 (1984).
129. R. Gebarowski, P.G. Burke, K.T. Taylor, M. Dörr, M. Bensaid and C.J. Joachain, J. Phys. B 30: 1837 (1997).
130. R. Gebarowski, K.T. Taylor and P.G. Burke, J. Phys. B 30: 2505 (1997).
131. M. Bensaid, M. Dörr and C.J. Joachain, J. Phys. B (to be published).
132. M. Dörr, M. Terao-Dunseath, P.G. Burke, C.J. Joachain, C.J. Noble and J. Purvis, J. Phys. B 28: 3545 (1995).
133. C.T. Chen and F. Robicheaux, J. Phys. B 29: 345 (1996).
134. D. Charlo, M. Terao-Dunseath, K.M. Dunseath and J.M. Launay, J. Phys. B 31: L 539 (1998).
135. H.W. van der Hart, J. Phys. B 29: 2217 (1996).

136. N.J. Kylstra, H.W. van der Hart, P.G. Burke and C.J. Joachain, J. Phys. B 31: 3089 (1998).
137. N.J. Kylstra, E. Paspalakis and P.L. Knight, J. Phys. B 31: L 719 (1998).
138. C.J. Joachain, R-Matrix-Floquet Theory of Multiphoton Processes, in *Multiphoton Processes 1996*, P. Lambropoulos and H. Walther, eds., Institute of Physics, Bristol (1997), p. 46.
139. M. Dörr, R-Matrix-Floquet Theory of Multiphoton Processes, in *Photon and Electron Collisions with Atoms and Molecules*, P.G. Burke and C.J. Joachain, eds., Plenum Press, New York (1997), p.191.
140. C.J. Joachain, M. Dörr and N.J. Kylstra, Comments At. Mol. Phys. 33: 247 (1997).
141. J.H. Eberly, R. Grobe, C.K. Law and Q. Su, Adv. At. Mol. Opt. Phys. Suppl. 1: 301 (1992).
142. P.G. Burke and V.M. Burke, J. Phys. B 30: L 383 (1997).
143. K.C. Kulander, Phys. Rev. A 36: 2726 (1987); Phys. Rev. A 38: 778 (1988).
144. O. Latinne, C.J. Joachain and M. Dörr, Europhys. Lett. 26: 333 (1994).
145. M. Dörr, O. Latinne and C.J. Joachain, Phys. Rev. A 52: 4289 (1995); Phys. Rev. A 55: 3697 (1997).
146. E. Cormier and P. Lambropoulos, J. Phys. B 30: 77 (1997).
147. R. Taïeb, V. Véniard and A. Maquet, J. Opt. Soc. Am. B 13: 363 (1996).
148. V. Véniard, R. Taïeb and A. Maquet, Phys. Rev. Lett. 74: 4161 (1995).
149. K.C. Kulander, K.J. Schafer and J.C. Krause, Adv. At. Mol. Phys. Suppl. 1: 247 (1992).
150. J. Zhang and P. Lambropoulos, J. Phys. B 28: L 101 (1995).
151. J. Zhang and P. Lambropoulos, Phys. Rev. Lett. 77: 2186 (1996).
152. K.T. Taylor, J.S. Parker, D. Dundas, E. Smyth and S. Vivirito, Time-Dependent Multiphoton Processes in Helium using a Cray T3D, in *Multiphoton Processes 1996*, P. Lambropoulos and H. Walther, eds., Institute of Physics, Bristol (1997), p.56.
153. J. Parker, K.T. Taylor, C.W. Clark and S. Blodgett-Ford, J. Phys. B 29: L 33 (1996).
154. K.T. Taylor, J.S. Parker D. Dundas, E. Smyth and S. Vivirito, Multiphoton Processes in a Two-Electron Atom Using a Cray T3D, in *Photon and Electron Collisions with Atoms and Molecules*, P.G. Burke and C.J. Joachain, eds., Plenum Press, New York (1997), p.223.
155. A. Scrinzi and B. Piraux, Phys. Rev. A 56: R 13 (1997).
156. J.B. Watson, A. Sanpera, K. Burnett, D.G. Lappas and P.L. Knight, Phys. Rev. Lett. 78: 3770 (1997).
157. B. Walker, B. Sheehy, L.F. DiMauro, P. Agostini, K.J. Schafer and K.C. Kulander, Phys. Rev. Lett. 73: 1227 (1994).
158. D. Bauer, Phys. Rev. A 56: 3028 (1997).
159. F.H.M. Faisal and A. Becker, Effect of Rescattering on ATI and e-e Correlation on Double Ionization in Intense Laser Fields, in *Multiphoton Processes 1996*, P. Lambropoulos and H. Walther, eds., Institute of Physics, Bristol (1997), p. 118.
160. A. Becker and F.H.M. Faisal, Laser Phys. 7: 684 (1997).
161. P. Lambropoulos, P. Maragakis and J. Zhang, Phys. Rep. (to be published).
162. C.A. Ullrich and E.K.U. Gross, Comm. At. Mol. Phys. 33: 211 (1997).
163. C.H. Keitel and P.L. Knight, Phys. Rev. A 51: 1420 (1995).
164. M. Protopapas, C.H. Keitel and P.L. Knight, J. Phys. B 29: L 591 (1996).
165. N.J. Kylstra, A.M. Ermolaev and C.J. Joachain, J. Phys. B 30: L 449 (1997).
166. L.S. Brown and T.W.B. Kibble, Phys. Rev. 133 A: 705 (1964).
167. F.H.M. Faisal and T. Radozycki, Phys. Rev. A 47: 4464 (1993), Phys. Rev. A 48, 554 (1993).
168. R. Taïeb, V. Véniard and A. Maquet, Phys. Rev. Lett. 81: 2882 (1998).
169. U.W. Rathe, C.H. Keitel, M. Protopapas and P.L. Knight, J. Phys. B 30: L 531 (1997).
170. A.M. Ermolaev, J. Phys. B 31: L 65 (1998).
171. P.S. Krstić and M.H. Mittleman, Phys. Rev. A 42: 4037 (1990).
172. K. Codling and L.J. Frasinski, Contemp. Phys. 35: 243 (1994).
173. A.D. Bandrauk, ed., *Molecules in Laser Fields*, M. Dekker, New York (1993).
174. A. Giusti-Suzor, F.H. Mies, L.F. DiMauro, E. Charron and B. Yan, J. Phys. B 28: 309 (1995).
175. T. Ditmire, T.D. Donnelly, R.W. Falcone and M.D. Perry, Phys. Rev. Lett. 75: 3122 (1995).
176. J.W.G. Tisch, T. Ditmire, D.J. Fraser, N. Hay, M.B. Mason, E. Springate, J.P. Marangos and M.H.R. Hutchinson, High-Harmonic Generation from Xenon Atom Clusters, in *Multiphoton Processes 1996*, P. Lambropoulos and H. Walther, eds., Institute of Physics, Bristol (1997), p. 177.
177. M.D. Perry and G. Mourou, Science 264: 917 (1994).
178. M.D. Perry, B.C. Stuart, G. Tietbohl, J. Miller, J.A. Britten, R. Boyd, M. Everett, S. Herman, H. Nguyen, H.T. Powell and B.W. Shore, CLEO '96, 1996 Technical Digest Series, vol 9, Opt. Soc. Am., Washington, DC (1996), p. 307.

MANIPULATING SMALL MOLECULES WITH INTENSE LASER FIELDS

Jan H. Posthumus, Keith Codling and Leszek J. Frasinski

JJ Thomson Laboratory
University of Reading
PO Box 220 Whiteknights
Reading RG6 6AF UK

INTRODUCTION

When molecules interact with intense lasers, a number of phenomena are observed that are absent at the lower intensities. For example one can observe above threshold ionisation (Agostini et al 1979), above threshold dissociation (Giusti-Suzor et al 1990) and high harmonic generation (McPherson et al 1987). More importantly, the photoelectron, photoion and photon emission spectra can be modified quite substantially by changes in the intensity, pulse length, wavelength or state of polarisation of the laser radiation. In this article we concentrate almost exclusively on the process of dissociative ionisation and in particular on how the behaviour of the resulting ions, as determined by their time-of-flight (TOF) spectra, can be modified by manipulating one or more of the above laser parameters.

The behaviour of small molecules in intense laser fields has been studied for just over a decade; by intense we mean focused intensities in excess of 10^{15} W cm^{-2} (or laser E-fields of about 10V Å^{-1}). Such E-fields exceed those that bind outer electrons in atoms and molecules and so it is not surprising that such lasers can produce very high stages of ionisation. In the multiphoton picture this would inevitably involve many (≥ 100) low energy photons (a Ti : sapphire laser has a wavelength of about 750 nm or an energy of 1.65 eV). One might, therefore, expect to be able to describe the ionisation process in terms of a classical field ionisation model and indeed, in the case of atoms, calculated appearance intensities for specific charge states agree well with experiment.

In their first experiments on the multielectron dissociative ionisation (MEDI) of N_2, Frasinski et al (1987) suggested that the ion TOF spectra might be explained in terms of a similar field ionisation approach. In molecules, of course, the process is more complex because ionisation and dissociation occur on roughly the same time scales, depending on the molecule concerned and the rise time of the laser. Moreover, a diatomic molecule has an axis of symmetry and it was argued that, because of the molecule's elongated shape, the potential difference created by the laser E-field is larger along its axis than at right angles to it. Consequently the molecule would be more easily field ionised when its axis was along the E-field direction. The first experiments on N_2 showed a strong peaking of fragment ions along the laser E-field, consistent with this simple field ionisation approach.

Subsequently Codling et al (1987, 1989) introduced the idea of electron localisation and pointed out that there must be a particular combination of laser field and inter-ion separation that maximises the ionisation probability, since further dissociation raises the inner potential barrier and thereby inhibits ionisation.

In the early '90s it was noted that the kinetic energy release, E_c, of *all* (Q_1, Q_2) channels were, to a good approximation, a constant fraction, C_m, of the Coulomb energy at $R = R_e$, the neutral molecule equilibrium internuclear distance. That is

$$E_c (eV) = C_m \frac{14.4 Q_1 Q_2}{R_e} = \frac{14.4 Q_1 Q_2}{R_c} \quad (1)$$

The simplest explanation of this behaviour was that the transient molecular ions were stabilised by the laser field at a critical distance R_c ($= R_e/C_m$). This idea of laser-induced stabilisation was supported by other experiments, which suggested that the fragment energies were *independent* of laser pulse rise time, see Posthumus et al (1997).

Although vibrational trapping (a kind of laser-induced stabilisation) had been introduced by Zavriyev et al (1993) to explain aspects of their H_2 TOF spectra, it was hard to conceive of any mechanism that could stabilise the higher channels, such as the (5,5) channel of I_2. Posthumus et al (1996), encouraged by publications by Seideman et al (1995) and Zuo and Bandrauk (1995), refined and extended the field ionisation model, to see if there could be a more convincing explanation of the above behaviour. This field ionisation, Coulomb explosion model is discussed briefly in the next section.

At the same time it was clear that more sophisticated experiments were required to probe the MEDI process. Most of these experiments, which were typically of the pulse-probe or double-pulse variety, could be described generally as 'manipulating the dissociative ionisation process' in one way or another. Therefore the following sections discuss laser-induced dissociation, laser-induced charge-asymmetry, laser-induced stabilisation, laser-induced alignment and coherent control of dissociation, a technique that could lead to laser-induced orientation. The more esoteric suggestions, such as laser-induced trapping will be mentioned briefly in the conclusions.

THE FIELD IONISATION, COULOMB EXPLOSION MODEL

Let us consider a diatomic molecule aligned with the laser E-field and focus our attention on the outer electron, combining the remaining electrons and nuclei into two point-like atomic ions of charge Q_1 and Q_2. Figure 1 illustrates the process of finding the appearance intensities for the (1, 1) channel of I_2. The double-well potential in which the outer electron moves is given by:

$$U = - \frac{Q/2}{|x + R/2|} - \frac{Q/2}{|x - R/2|} - \epsilon x \quad (2)$$

when $Q = Q_1 + Q_2$ is the sum of the atomic core charges, x, the axial position, R, the internuclear separation and ϵ the laser E-field.

The energy level, E_L, of the outer electron in this symmetric double well can be approximated by the expression:

$$E_L = \frac{(-E_1 - Q_2/R) + (-E_2 - Q_1/R)}{2} \qquad (3)$$

where E_1 and E_2 are the ionisation potentials of the atomic ions.

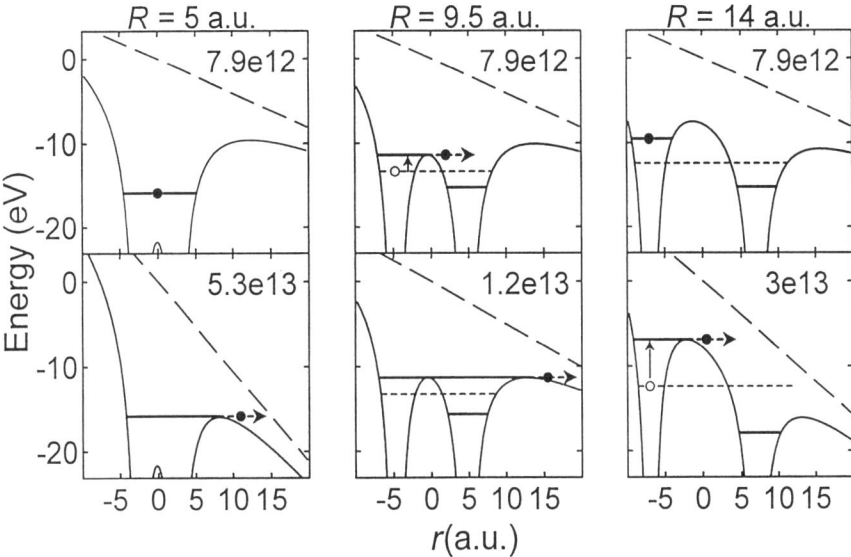

FIGURE 1. Double well potentials for I_2^+ at three internuclear distances and a number of laser intensities.

The upper figures show the situation at a laser intensity of 7.9×10^{12} W cm^{-2} and three values of internuclear separation. When R = 5 a.u. the energy level is well above the central barrier and an intensity of 5.3×10^{13} Wcm^{-2} is required for over-the-barrier ionisation. For a separation of R = 7.5 a.u. (not shown) the inner barrier rises to localise the electron in the left-hand well. Between 7.5 and 9.8 a.u. the electron is Stark shifted by the laser E-field and at R = 9.5 a.u. (figure 1, bottom centre), an intensity of only 1.2×10^{13} W cm^{-2} is required for over-the-barrier ionisation. This effect is responsible for the rapid fall in appearance intensity seen in figure 2.

Above R = 10 a.u. the central barrier, which now rises above the outer barrier, becomes the real obstacle to ionisation and, since the level of the inner barrier rises steadily with increasing R, so does the appearance intensity. For example at R = 14 a.u. (bottom right) an intensity of 3×10^{13} W cm^{-2} is required for ionisation. Indeed, beyond 20 a.u. the dissociating ions behave almost as free atomic ions. In summary, one sees a well-defined minimum in the appearance intensity (a maximum in the ionisation rate) at about 10 a.u. for the (1,1) channel of I_2.

More interesting, perhaps, is the fact that this minimum is at almost the *same* distance for all (Q_1, Q_2) channels, see figure 2. This is because the enhanced ionisation phenomenon is closely related to the process of electron localisation, which occurs at roughly the same internuclear distance for all channels, see Posthumus et al (1995). This distance at which localisation occurs (the kink in the curves of figure 2) was initially associated with the critical distance, but R_c is now used to denote the minimum.

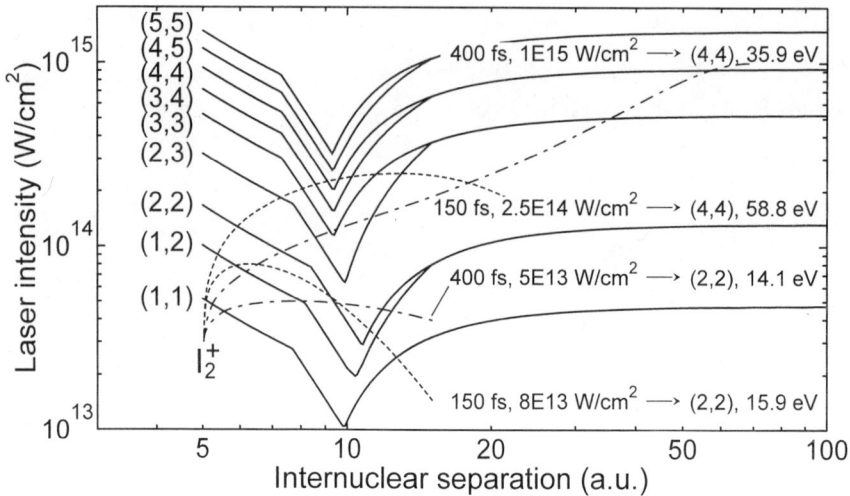

FIGURE 2. Threshold intensities of the (Q_1, Q_2) channels of I_2 (solid curves) and trajectories (dashed curves).

The fate of a particular molecule depends both on its location within the focal volume and the laser pulse length. The two dashed curves in figure 2 trace the temporal evolution of the laser intensity versus internuclear separation for a 150 fs pulse; the dash-dot curves show two trajectories for a 400 fs pulse. The process of over-the-barrier ionisation to the next stage of ionisation occurs each time a specific trajectory first crosses an appearance intensity curve. One sees that for a 150 fs pulse and peak intensity of 8×10^{13} W cm^{-2} (the lower dashed curve) the (2,2) channel is created at R = 9.5 a.u. giving a dissociation energy of 15.9 eV. With a 400 fs pulse of intensity 5×10^{13} W cm^{-2}, the trajectory results in an energy of 14.1 eV for the (2,2) channel. That is, the energy release is virtually *independent* of pulse length (or rise time).

The (4, 4) channel is produced with a dissociation energy of 58.8 eV (68% of Coulomb energy at R_c) for a pulse length of 150 fs and peak intensity of 2.5×10^{14} W cm^{-2}, but for a pulse length of 400 fs and an increased laser intensity, the dissociation energy is only 35.9 eV. That is, one can expect to observe the process of post dissociative ionisation (PDI).

MANIPULATING THE DISSOCIATIVE IONISATION PROCESS

Having given a brief description of the field ionisation model, we will now look at various aspects of the MEDI process and see where it may be able to throw light on past and more recent experiments. Before doing so, we need to describe one particular aspect of the experimentally observed ion TOF spectra.

Using linearly polarised light, the fragment ions are ejected preferentially along the E-field. If the E-field lies along the axis of the drift tube of the TOF spectrometer, then the ions are ejected either towards (forward ions) or away from (backward ions) the microchannel plate detector. If voltages +U and −U are applied across the interaction region, both the forward and backward ions travel towards the detector, with the forward ions arriving first. The TOF, t, is given by the expression:

$$t = A\left(\frac{m}{qU}\right)^{1/2} + B\left(\frac{mv_\parallel}{qU}\right) \tag{4}$$

where $mv_\parallel$ is the initial momentum component parallel to the drift tube axis and A and B are instrumental constants. In this spatial focusing regime, the first term gives the TOF of zero energy ions and the difference in TOF of the forward and backward ions gives their kinetic energies.

Laser-induced dissociation

When a diatomic molecule is subjected to an intense laser field, the process of above threshold dissociation (ATD) occurs. The signature of ATD is a series of ion peaks separated, in the case of a homonuclear molecule, by half the photon energy. To explain the ion TOF spectra in detail one must consider that the various bound and unbound states of the molecular ion are 'dressed' by the laser field, with the result that multiphoton couplings soften the molecular bond, resulting in fragments with low kinetic energies, see Bucksbaum et al (1990).

The field ionisation model has little to say about the early stages of the MEDI process. Bond softening is implicitly assumed to occur and the experimentally determined energies of the dissociating ions are used as input data in tracking the subsequent dynamics. Comparisons between experiment and theory are somewhat problematic because calculations on bond softening assume that molecular ions are in their ground vibrational state, whereas experimentally the vibrational population is determined by the frequency and pulse length of the laser and the various Franck-Condon factors between the ground and ionic states. Moreover, to probe bond-softening one needs to use the more definitive double-pulse experiments; these are at present impossible to perform with light molecules, since the dissociation time scales are extremely short.

In the context of the present article, it is safe to say that one can manipulate the ratio of ionisation-to-dissociation by simply adjusting the laser pulse rise time; the shorter the rise time, the more likely it is that one will observe the parent fragment ion. This is clearly the case in I_2, where the molecular ions I_2^{2+} and I_2^{3+} are enhanced for short rise times. In the context of more complex molecules, the use of femtosecond pulses means that one can observe parent molecular ions with considerable strength, since the relative importance of laser-induced dissociation is reduced, see Singhal et al (1996). Thus femtosecond laser mass spectrometry (FLMS) is likely to become important for the detection of drugs, explosives etc.

Laser-induced charge asymmetry

There has been considerable discussion in the past with regard to the charge-symmetry of the fragmentation process. Earlier disagreements stemmed from an inability to unambiguously correlate the ion fragments using conventional TOF techniques. Such ambiguities were eliminated by the introduction of covariance mapping (Frasinski et al 1989). In the context of the field ionisation model and a laser field that is reversing its polarity roughly every femtosecond, it would seem reasonable that a molecular ion such as $[N_2^{4+}]$ would fragment symmetrically, since a lower E-field is required to produce the symmetric outcome. However, when Hill et al (1992) studied N_2 at 193 nm using pulses of 10-15 ps duration, they explained their low energy N^+ and N^{3+} ions in terms of asymmetric fragmentation. They suggested that the symmetric fragmentation seen typically in MEDI experiments was the result of PDI. Dietrich et al (1993a) agreed and suggested that their observation of the asymmetric (1, 3) channel in I_2 was consistent with their pulse length of 80 fs; they were observing the nascent process. In fact Hatherly et al (1994), using 150 fs pulses and covariance mapping, saw very little evidence of this channel.

The first convincing evidence for charge-asymmetric fragmentation was the observation by Strickland et al (1992) of the (0, 2) channel of I_2 using 30 and 80 fs pulses at 630 nm. As argued above, one might have expected the (1, 1) channel to dominate, but I_2^+ ions become strongly polarised in a laser E-field that is almost strong enough to induce field ionisation and, for a small range of internuclear separations and laser intensities, one can expect to observe the (0, 2) channel. That is, transfer of electron charge from the up-field potential well to the down-field well can occur. At this point two electrons must oscillate back and forth between the two wells.

Figure 3 (a) shows examples of trajectories for the (0, 2) channel for 130 and 220 fs; they are based on the measured dissociation energy for this channel. Both show dissociation along the (0, 1) potential curve until they cross the [I_2^{2+}] appearance intensity curve (the solid line). At this point, transfer to the (0, 2) channel can occur because the intensity equals or exceeds 4×10^{13} W cm^{-2}, see Posthumus et al (1997). In order to continue to observe the (0, 2) channel, one must ensure that the laser intensity does not exceed the appearance intensity of the (0, 2) to (1, 2) transition (the dash-dot curve) at larger values of R. If the pulse width exceeds 220 fs, the (1, 2) is created and the (0, 2) channel disappears.

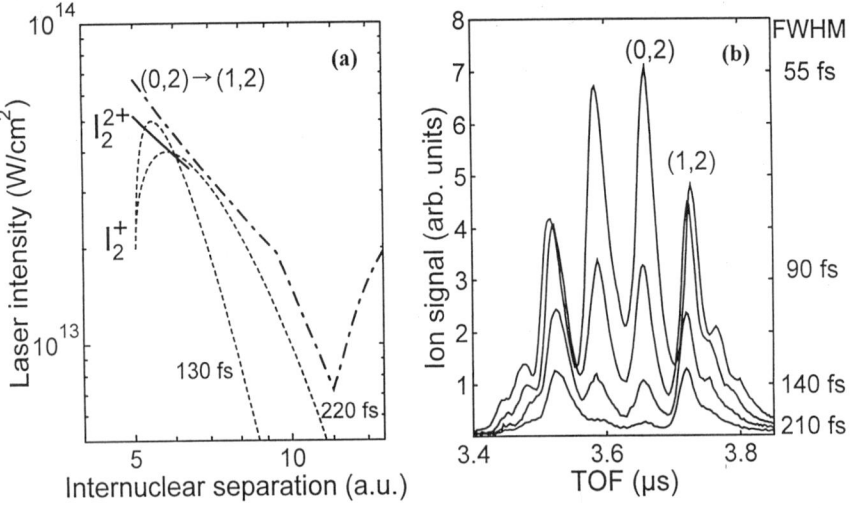

FIGURE 3. (a) The dash-dot curve is the threshold intensity curve for the (0, 2) to (1, 2) transition. The dashed curves show the trajectories for 130 and 150 fs; (b) The (0, 2) channel decreases steadily from 55 to 210 fs.

Figure 3 (b) is a section of the I_2 TOF spectrum centred at 3.63 μs, showing the I^{2+} fragment ions. It was obtained at a wavelength of 750 nm and pulse lengths ranging from 55 to 210 fs. One sees the usual forward and backward peaks associated with the (0, 2), (1, 2), (2, 2) and (3, 2) channels. The I^{2+} ions associated with the (0, 2) channel are seen to disappear as the pulse length exceeds 210 fs. Presumably one would be able to see similar charge-asymmetric channels for the lighter molecules such as CO and N_2 at considerably shorter pulse lengths.

Laser-induced stabilisation

As pointed out in the introduction, experiments in the early '90s gave results that could be interpreted in terms of laser-induced stabilisation. When I_2 was studied with pulses of 90 fs, the dissociation energies were a certain fraction of the Coulomb value ie $C_m \approx 0.75$ (equation 1); with 150 fs pulses $C_m \approx 0.70$ and with 200 fs pulses $C_m \approx 0.65$. Since these values were obtained at different laboratories and subject to errors of 5-10%, it was assumed that these results constituted further evidence for laser-induced stabilisation, that is $C_m = 0.70$ (±0.05). However, interferometric control of laser pulse rise time has been used (Giles et al 1994) to show categorically that C_m varies with pulse rise time and we now believe that the above variations may have been real.

An interferometer of the Michelson type was used to produce two pulses with a time delay of typically 50 fs between the two. It was possible to switch rapidly from pulses of effective width 55 fs to ones of 100 fs simply by changing from destructive to constructive interference; the other laser parameters, such as peak intensity, were virtually unchanged. The ion TOF spectra were seen to change, in that the time difference between the forward and backward peaks decreased with increasing laser pulse rise time, in agreement with the field ionisation model and in contradiction to the ideas of laser-induced stabilisation.

These results were combined with the earlier ones and, considering the simplicity of the field ionisation model, the agreement between experiment and theory was surprisingly good, see Posthumus et al (1996a). Experiments on N_2 using pulse lengths of 55 and 400 fs and employing covariance mapping also showed that longer pulses yielded ion fragments with lower kinetic energies.

In summary, the results of such experiments are consistent with the idea of rapid sequential ionisation as the dissociating molecule arrives at R_c, the critical internuclear separation; the concept of laser-induced stabilisation is not required. We have studied the MEDI of H_2 and have been unable to reproduce the vibrational structure observed in the bond-hardening experiment of Zavriyev et al (1993). We suggest that the inherent stability of H_2 may be responsible for observing a strong (1, 1) channel rather than vibrational trapping.

Since the field ionisation model and indeed the quantum approach predict considerable enhancement of ionisation at R_c, it was important to perform experiments to check whether this enhancement actually occurs. Constant et al (1996) employed 80 fs pulses at 625 nm to create the slowly dissociating (0, 2) channel in I_2. A second pulse, which caused the (0, 2) to (1, 2) transition, was delayed with respect to the first. Posthumus et al (1997) used a laser of wavelength 750 um and 55 fs duration and the results of their experiment are shown in figure 4 (a). For delays ranging from 100 to 480 fs one observes an extra peak (arrowed) due to creation of the (1, 2) channel.

The curves in figure 4 (b) explain why the peak moves to lower energies (lower TOFs for the backward ions) as the delay is increased. Calculations indicated that the second pulse should be most effective when the delay is about 250 fs, that is when the I and I^{2+} fragments have arrived at R_c; experimentally the greatest enhancement was found to occur at a delay of about 200 fs, in reasonably good agreement.

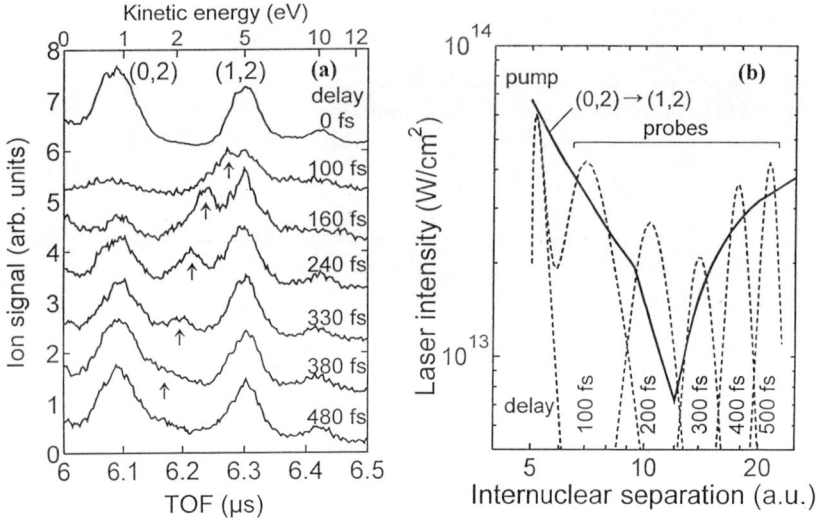

FIGURE 4. (a) I^{2+} fragments in the I_2 TOF spectra, the arrow denoting the (0, 2) to (1,2) transition; (b) Pump-probe curves showing why the TOF of the arrowed 'backward' peak decreases with increasing delay.

Laser-induced alignment

In their first publication, Frasinski et al (1987) tentatively suggested that the strong peaking of fragment ions along the laser E-field might be explained in purely geometrical terms; molecules that happen to be lying along the laser E-field are more easily ionised. It was then pointed out by Strickland et al (1992) that the peaked angular distribution of I_2 fragment ions obtained using sub-100 fs pulses could not be explained in this way but required dynamic, laser-induced alignment, associated with the molecule's polarizability. Subsequently double-pulse experiment performed by Normand et al (1992) on CO using 30 ps pulses and Dietrich et al (1993b) on I_2, using sub-100 fs pulses appeared to confirm that laser-induced alignment does occur in both. In fact the interpretation of these experiments is somewhat questionable, since the phenomenon of enhanced ionisation at R_c was not appreciated at the time.

Posthumus et al (1998a) have recently performed experiments on I_2, N_2 and H_2 and in the case of I_2 in particular, have attempted to determine the extent to which angle-dependent enhanced ionisation, a purely geometric effect, could explain the highly anisotropic angular distributions. Threshold intensities were calculated for a full range of angles, θ, between the laser E-field and the molecular axis. The Stark shift is now represented by the dot product ½**E.R**. Figure 5 shows the ionisation threshold intensity versus internuclear distance for a range of θ values, for the $I^+ + I^+$ channel. At the critical distance, the threshold varies by a factor of 5 for orthogonal polarisations.

When a molecule dissociates, it is the laser intensity that exists at R_c that determines the final charge state. One sees, for example, that when a molecule lies at an angle of 75° the required intensity is 3×10^{13} W cm^{-2}, a factor of 3 greater than at 0°. Because of the form of the variation in intensity with location in the laser focus, the focal volume for molecules that can be ionised when oriented at 75° is considerably less than for those that are fully aligned.

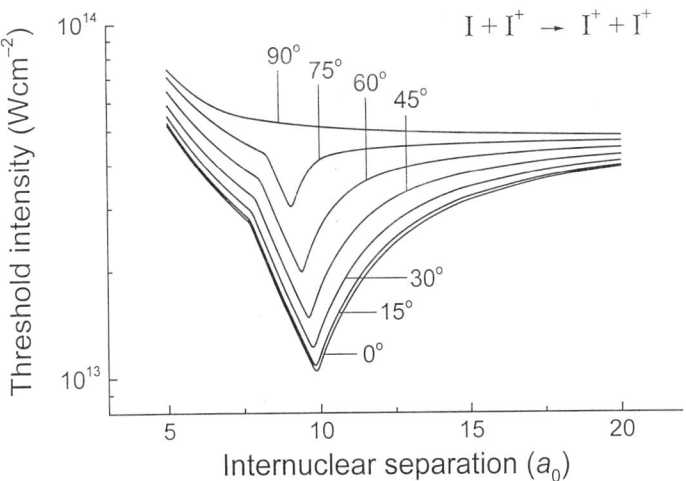

FIGURE 5. Threshold intensities as a function of ion separation for a series of angles between laser E-field and molecular axis.

Assuming random orientation of the I_2 molecules and no laser-induced rotation, the angular distributions depend critically upon the shell volumes as defined by the threshold intensities. Figure 6 (a) shows the relative sizes of the shells as a function of angle for the (1, 2) channel of I_2, for a series of laser intensities. The experimental data on the angular distribution of this channel shown in figure 6 (b) were obtained at 790 nm using a 50 fs pulse duration. The overall agreement between the two confirms that the angular anisotropy in I_2 is fundamentally the result of geometric rather than dynamic alignment.

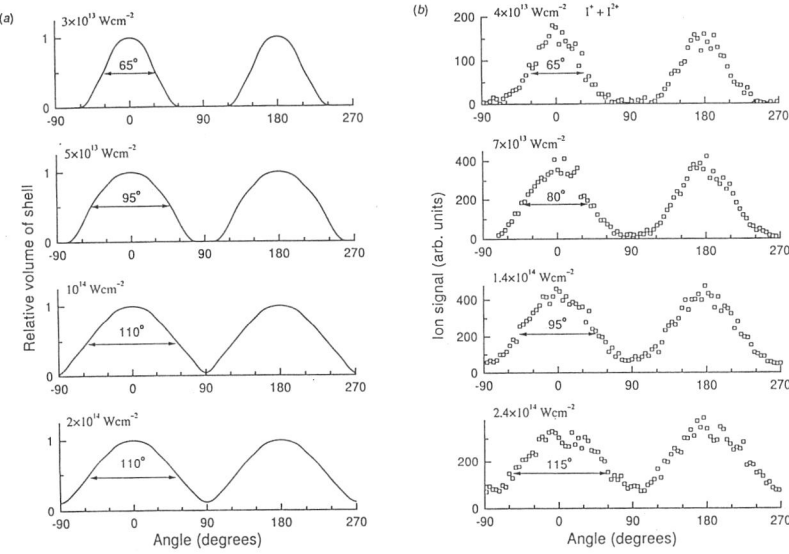

FIGURE 6. (a) Relative volume of the $I^+ + I^{2+}$ shell as a function of angle between laser E-field and molecular axis; (b) Experimental angular distributions.

Posthumus et al (1998a) have also measured the angular distribution of H_2, a much lighter molecule that is susceptible to re-orientation. Furthermore, due to its higher ionisation potential, this molecule is more resistant to field ionisation and can therefore be subjected to higher laser E-fields and torques before it dissociates. The angular distribution is considerably sharper than that of I_2. At an intensity of 1.5×10^{14} W cm^{-2} the angular distribution of H^+ fragments associated with the (1, 1) channel can be fitted to a $\cos^{22}\theta$ distribution; no H^+ ions are ejected at 90° to the E-field.

This considerable difference in behaviour of the two molecules, H_2 and I_2, has been confirmed by recent double-pulse experiments of Posthumus et al (1998b). The experimental arrangement is shown schematically in figure 7. A Ti : sapphire laser provides pulses of 50 fs duration at a wavelength of 790 nm. the laser beam is fed through a Mach-Zehnder-like optical arrangement so that two laser beams, A and B, are brought to focus on the beam of H_2 molecules. The laser pulse A precedes pulse B by about 1 ps.

The H^+ ions are extracted from the interaction region by an electric field, pass through a field free region and are detected by microchannel plates in conjunction with a digital oscilloscope, as usual. Beam A is linearly polarised with its E-field orthogonal to the axis of the TOF analyser and therefore H^+ fragments produced by these pulses are ejected in a direction perpendicular to the drift tube axis and miss the detector unless their initial velocities are very small, see figure 7. Beam B is polarised with its E-field along the axis and therefore H^+ ions produced by these pulses consist of the usual forward and backward pairs.

In the present experiment, beam B is used to probe *spatially* the effect of beam A. In order to improve the spatial resolution, the probe beam is arranged to have a sharper focus but both have roughly the same peak intensity of 5×10^{14} W cm^{-2}, see figure 7. The pump focus is scanned systematically through the probe focus by step-wise rotation of mirror M.

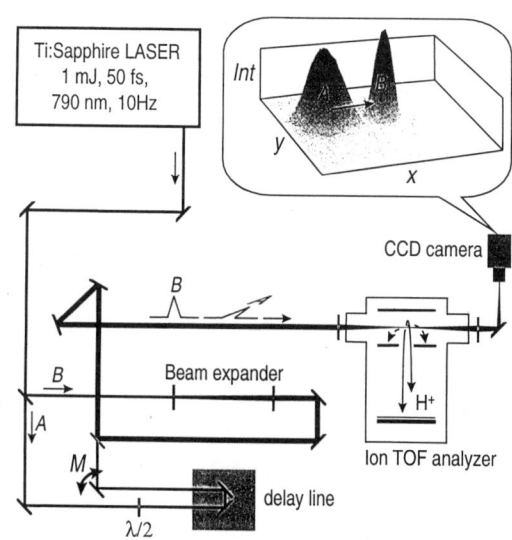

FIGURE 7. Pump-probe experiment. The inset shows a 3-D impression of the two foci as observed by the CCD camera.

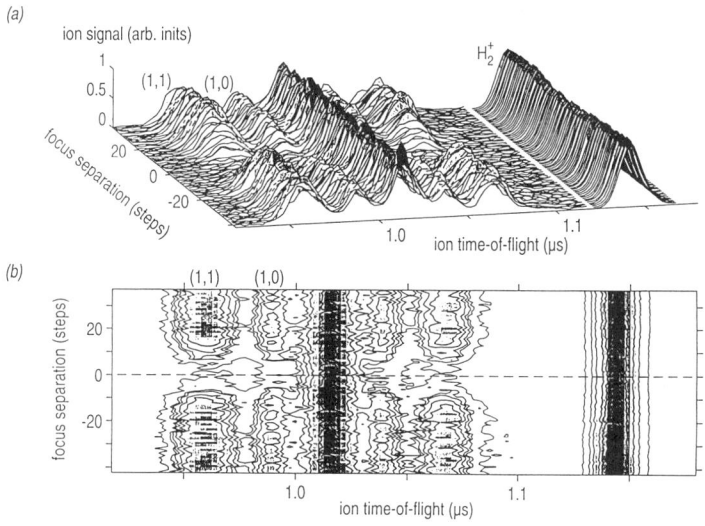

FIGURE 8. (a) H_2 TOF spectra as a function of spatial overlap between pump and probe; (b) contour maps of the same data.

At each step an ion TOF spectrum is taken. Typical results are shown in figures 8 (a) and (b). The TOF spectra show five H^+ peaks, two forward and two backward peaks produced by beam B and one central peak of low energy fragments, produced by beam A; the forward peaks are labelled (1, 1) and (1, 0).

Note that the fragments produced by beam B disappear completely when the beams overlap significantly. This shows that the first pulse fragments *all* H_2 molecules in the orthogonal direction. The molecules must therefore have been rotated parallel to the E-field, since they were initially randomly oriented.

When the same double pulse experiment was performed with I_2, quite different results were obtained. The forward and backward ions did not disappear when the foci were overlapped, see Posthumus et al (1998b).

Coherent control of dissociation dynamics

A few years ago Sheehy et al (1995) reported on phase control in the two-colour photo dissociation of HD. They used the fundamental (1053 nm) and second harmonic (527 nm) of a 50 ps Nd : YLF laser in order to control the spatial asymmetry of the dissociation channels $H + D^+$ and $H^+ + D$. Thompson et al (1997) performed a similar experiment on H_2 with laser pulses that were three orders of magnitude shorter and an order of magnitude more intense. Despite these large differences and the fact that shorter wavelengths were used, 750 and 375 nm, they obtained the same result as the earlier experiment; the ions were emitted counter-intuitively, that is in a direction opposite to that in which the maximum of the combined two-colour laser E-field pointed, see figure 9.

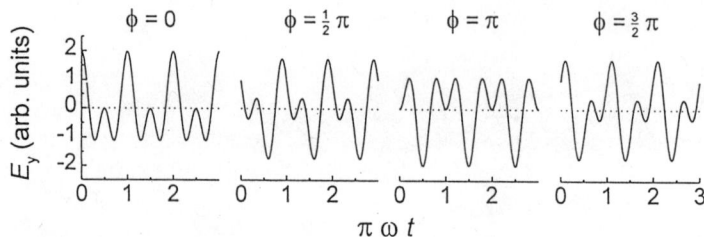

FIGURE 9. The two-colour laser E-field cos(ωt) + cos(2ωt + φ), for equal intensities and four different phases.

The experiment utilised, once again, an interferometer of the Michelson type to create a phase difference between fundamental and second harmonic and filters were used to ensure that the two had similar intensities. In order to restrict the production of the (0, 1) channel to the centre of the laser focus, both beams were attenuated to give peak intensities of about 10^{14} W cm^{-2}. Figure 10 (a) shows an enhancement of the forward protons. A change of π in the relative phase causes the backward proton emission to dominate, see figure 10 (b).

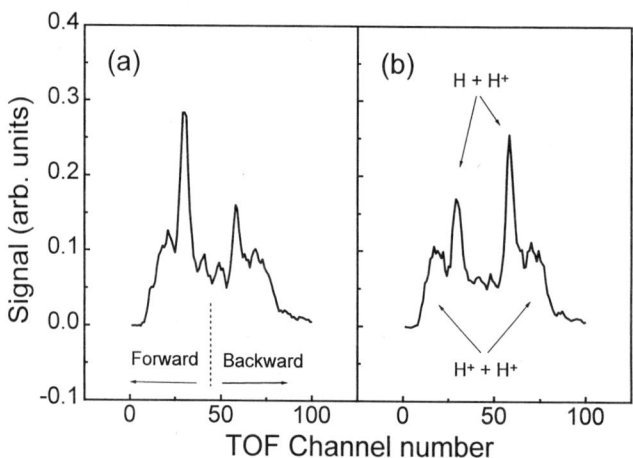

FIGURE 10. The proton emission changes from (a) predominantly forward to (b) predominantly backward when φ is changed by π.

In the short pulse experiment, it was impossible to define the direction of the peak in the superposed fields. To achieve this end, an electron drift tube was located opposite the ion drift tube, see Thompson et al (1997). An extraction field directed both types of particle to their respective detectors. Assuming that the electrons behaved intuitively, the ions were found to be emitted counter-intuitively, as in the earlier experiment. This behaviour was explained by Posthumus et al (1996b) in the context of the field ionisation model.

Charron et al (1995) suggested that coherent control in the high intensity regime could be utilised to orient, rather than align a diatomic molecule. They showed that, by using an appropriately long wavelength, it ought to be possible to induce different asymmetries in the H$^+$ and D$^+$ photofragmentation of HD and to control the spatial separation of these isotopes. As far as we are aware, this has not yet been achieved experimentally.

CONCLUSIONS

We have seen that many of the experiments described above can be discussed in terms of the manipulation of the dissociation dynamics. One can, for example, change the energy spectrum of a particular channel by simply increasing the laser intensity, when the mechanism of bond-softening can occur. Alternatively a double-pulse experiment can excite from a bond-softened state to a Coulomb repulsive state. In fact, by using laser pulses of short duration for the second pulse, one can hope to use the technique of Coulomb explosion imaging (Ellert et al 1998) to monitor the time evolution of a specific dissociation process. Moreover, using Coulomb explosion imaging and position-sensitive detectors one could expect to be able to monitor a change in the geometry of a triatomic molecule on ionisation, for example.

We have seen that one can align light molecules such as H_2 and N_2 in an intense laser field and this may be useful, since the orientation of molecules can play a crucial role in physical and chemical interactions. Moreover, the Stark shift creates a potential minimum for a ground state molecule at a position where the laser intensity is a maximum. This may lead to a method of focusing and trapping molecules, see Stapelfeldt et al (1997). Finally, one major aim of chemists is to manipulate the outcome of a chemical reaction using phase control and intense lasers.

ACKNOWLEDGEMENTS

We are pleased to acknowledge the Engineering and Physical Sciences Research Council (UK) for their financial support. Special thanks go to Drs P.F. Taday and A.J. Langley of Rutherford Appleton Laboratory for their expert assistance with experiments described here.

REFERENCES

Agostini P., Fabré F., Mainfray G., et al, 1979, Phys.Rev.Lett 42:1127
Bucksbaum P.H., Zavriyev A., Muller H.G., and Schumacher D.W., 1990, Phys.Rev.Lett 64:1883
Charron E., Giusti-Suzor A., and Mies F.H., 1995, Phys.Rev.Lett. 75:2815
Codling K., Frasinski L.J., Hatherly P.A., and Barr J.R.M., 1987, J.Phys.B. 20:L525
Codling K., Frasinski L.J., and Hatherly P.A., 1989, J.Phys.B 22:L321
Constant E., Stapelfeldt H., and Corkum P.B., 1996, Phys.Rev.Lett.76:4140
Dietrich P., Strickland D.T., and Corkum P.B., 1993a, J.Phys.B 26:2323
Dietrich P., Strickland D.T., Laberge M., and Corkum P.B., 1993b, Phys.Rev. A 47:2305
Ellert C.H., Stapelfeldt H., Constant E., et al, 1998, Phil.Trans.R.Soc.Lond. 356:329
Frasinski L.J., Codling K., Hatherly P., et al, 1987, Phys.Rev.Lett 58:2424
Frasinski L.J., Codling K., and Hatherly P.A., 1989, Science 246:973
Giles A.J., Posthumus J.H., Thompson M.R., et al, 1994, Opt.Commun. 118:537
Guisti-Suzor A., He X., and Atabek O., 1990, Phys.Rev.Lett. 64:515
Hatherly P.A., Stankiewicz M., Codling K., et al, 1994, J.Phys.B 27:2993
Hill W.T., Zhu J., Hatten D.L., et al, 1992, Phys.Rev.Lett. 69:2646
McPherson A., Gibson G., Jara H., et al, 1987, J.Opt.Soc.Am.B 4:595
Normand D., Lompré L.A., and Cornaggia C., 1992, J.Phys.B 25:L497
Posthumus J.H., Frasinski L.J., Giles A.J., and Codling K., 1995, J.Phys.B 28:L349
Posthumus J.H., Giles A.J., Thompson M.R., et al, 1996a, J.Phys.B 29:L525
Posthumus J.H., Thompson M.R., Giles A.J., and Codling K., 1996b, Phys.Rev.A 54:955
Posthumus J.H., Codling K., Frasinski L.H., and Thompson M.R., 1997, Laser Phys. 7:813

Posthumus J.H., Plumridge J., Thomas M.K., et al, 1998a, J.Phys B 31:L553
Posthumus J.H., Plumridge J., Frasinski L.J., et al, 1998b, J.Phys.B submitted
Thompson M.R., Thomas M.K., Taday P.F., et al, 1997, J.Phys B 30:5755
Seideman T., Ivanov M.Yu., and Corkum P.B., 1995, Phys.Rev.Lett. 75:2819
Sheehy B., Walker B., and Di Mauro, L.F., 1995, Phys.Rev.Lett. 74: 4799
Singhal R.P., Ledingham K.W.D., Kosmidis C., et al, Chem.Phys.Lett. 253:81
Stapelfeldt H., Sakai H., Constant E., and Corkum P.B., 1997, Phys.Rev.Lett. 79:2787
Strickland D.T., Beaudoin Y., Dietrich P., and Corkum P.B., 1992, Phys.Rev.Lett.68:2755
Zavriyev A., Bucksbaum P.H., Squier J., and Saline F., 1993, Phys.Rev.Lett. 70:1077
Zuo T., and Bandrauk A.D., 1995, Phys.Rev. A 52:R2511

MATRIX METHODS

Iain S. Duff

Department for Computation and Information
CCLRC-Rutherford Appleton Laboratory
Didcot, Chilton
Oxfordshire OX11 0QX, UK

INTRODUCTION

The intention of this paper is to describe current matrix methods for large-scale problems to an audience of computational physicists and chemists. We will discuss both the solution of the linear equations

$$\mathbf{Ax} = \mathbf{b}, \tag{1}$$

and the solution of the eigensystem

$$\mathbf{Ax} = \lambda \mathbf{Bx}, \tag{2}$$

where the matrices $\mathbf{A}$ and $\mathbf{B}$ are large and sparse. The problem (2) is called the generalized eigenproblem. The particular, commonly occurring, case where $\mathbf{B} = \mathbf{I}$, viz.

$$\mathbf{Ax} = \lambda \mathbf{x}, \tag{3}$$

is called the eigenproblem, or *standard* eigenproblem. The solution of the eigenproblem for all values of λ and $\mathbf{x}$ corresponds to finding a similarity transformation for diagonalizing the matrix and is often called *matrix diagonalization*. In many cases, however, the full diagonal is not required and only a few eigenvalues λ are needed.

Although it is possible to use the solution of the eigenproblem to facilitate solutions of the linear system (1), and I have known people to use this route, I should stress that the problem (3) is usually much more complicated to solve than the problem (1), not least because (3) is nonlinear in the unknowns λ and $\mathbf{x}$. Thus, if the solution to (1) is all that is required (even for several right-hand sides $\mathbf{b}$), then this should be tackled directly.

Sparse systems arise in very many application areas. We list just a few such areas in Table 1.

This table, reproduced from Duff *et al* (1989), shows the number of matrices from each discipline present in the Harwell-Boeing Sparse Matrix Collection. This standard set of test problems is currently being upgraded to a new Collection called

Table 1. A list of some application areas for sparse matrices

acoustic scattering	4	demography	3	network flow	1
air traffic control	1	economics	11	numerical analysis	4
astrophysics	2	electric power	18	oceanography	4
biochemical	2	electrical engineering	1	petroleum engineering	19
chemical engineering	16	finite elements	50	reactor modeling	3
chemical kinetics	14	fluid flow	6	statistics	1
circuit physics	1	laser optics	1	structural engineering	95
computer simulation	7	linear programming	16	survey data	11

the Rutherford-Boeing Sparse Matrix Collection (Duff et al 1997) that will include far larger systems and matrices from an even wider range of disciplines. This new Collection will be available from netlib (http://www.netlib.org) and the Matrix Market (http://math.nist.gov/MatrixMarket).

The definition of a large sparse matrix is a matter for some debate. Suffice it to say that we regard a matrix as large if it cannot be factorized efficiently using a code for general linear systems from a standard package for dense systems, such as LAPACK (Anderson et al 1995). The order of a matrix that is considered large is thus a function of time depending on the development of both dense and sparse codes and advances in computer architecture. Partly for amusement, we show in Table 2 the order of general unstructured matrices which sparse methods have been used to solve as a function of the date. I think this alone serves to illustrate some of the advances in sparse solution methods over the last 25 years.

Table 2. Order of general sparse matrices solved by direct methods as a function of date

date	order
1970	200
1975	1000
1980	10000
1985	100000
1990	250000
1995	1000000

The matrix is sparse if the presence of zeros within the matrix enables us to exploit this fact and obtain an economical solution.

There are two main classes of techniques for solving (1), iterative methods and direct methods. In a direct method, we use a factored representation and solve the system using these factors in a predetermined amount of memory and time, usually to a high degree of accuracy. In iterative methods, we construct a sequence of approximations to the solution, often the "best" approximation in subspaces of increasing dimension. The work is generally low per iteration but the number of iterations is usually not known *a priori* and may be high, particularly if an accurate solution is required. We consider general aspects of the solution of large sparse linear equations in the next section. We discuss direct methods of solution and iterative techniques in the following two sections, and then compare and combine these approaches. We then briefly review matrix diagonalization indicating the relationship of techniques used in this case with those used in iterative methods for solving linear equations. We finally make a few comments on software availability and concluding remarks in our last two sections.

THE SOLUTION OF LINEAR EQUATIONS

It is a notational convenience to denote by $\mathbf{A}^{-1}$ the inverse of the matrix $\mathbf{A}$ so that the solution to (1) is given by $\mathbf{A}^{-1}\mathbf{b}$. However, there is almost no occasion when it is appropriate to compute the inverse in order to solve a set of linear equations. Even if explicit entries of the inverse are required, for example for sensitivity analysis, there are usually far more computationally efficient ways of doing this than to compute the inverse. For example, the ith column of the inverse can be obtained by solving a set of equations with $\mathbf{e}_i$, the ith column of the identity matrix, as the right-hand side vector and, if specified entries are required, for example the diagonal of $\mathbf{A}^{-1}$, then advantage can be taken of sparsity to compute this efficiently (Erisman and Tinney 1975).

Accuracy, Stability, and Conditioning

As a numerical analyst, I am of course concerned about the accuracy of the solution, a concern which I hope is shared by the applications scientist or engineer. Before we continue our discussion on accuracy, it might be useful to first distinguish two concepts which are often confused, namely conditioning and stability. Conditioning is a property of the problem being solved. If the problem is badly conditioned, then small perturbations to the given data could give large changes to the solution, even if it is computed exactly. Stability is a property of the algorithm that is used to effect the solution. An algorithm is *backward* stable if the solution it computes in finite precision arithmetic is the exact solution of a slightly perturbed problem.

A good measure of accuracy would be to measure the difference between computed and exact solutions in some norm, say the l_2 norm, but that is rather difficult since, if you already know the exact solution, there seems little point in going to the trouble of solving the system. A measure which is more easy to compute is the residual $\mathbf{b} - \mathbf{A}\tilde{\mathbf{x}}$, where $\tilde{\mathbf{x}}$ is the computed solution. The residual is a measure of how well your computed solution satisfies the equation. As is common when we do not know (or cannot control) the scaling of the system, we use a relative measure of the residual (dividing by $(\|\mathbf{A}\| \|\tilde{\mathbf{x}}\|)$ or some such quantity*). This is directly related to the perturbation to the original data that would be needed to ensure that we have computed an exact solution to the perturbed system and is called the backward error, a concept which was pioneered by Jim Wilkinson (for example, Wilkinson (1961)) and which revolutionized the thinking of numerical analysts. Now, the backward error is related to the actual (or forward) error through the relationship

$$\text{Forward error} \leq \text{Condition number} \times \text{Backward error} \qquad (4)$$

where, as we remarked earlier, the *Condition number* is a property of the problem (not the solution technique). There are many different condition numbers (Higham 1996)

---

*When we use norm signs, $\|..\|$ without a suffix, then the actual norm used is not of great importance, although one would normally use consistent norms within a single analysis or computation.

and one of the most common is given by

$$\text{Condition number} = ||\mathbf{A}|| \, ||\mathbf{A}^{-1}||, \qquad (5)$$

often denoted by $\kappa(\mathbf{A})$.

A major problem for large sparse systems is that the condition number (even if the original system is scaled) can be far greater than the reciprocal of machine precision. Thus the bound (4) indicates that, even if we solve with a backward error of machine precision, our solution may contain no correct digits. It is then a mute point how one decides whether the problem has been solved or not. Usually it is apparent from the underlying problem, so often the applications scientist can judge this better than the numerical analyst. Before you lose all faith in numerical analysis, I should say that this is more alarming than it might at first appear. A simple scaling[†] often helps and sometimes a more appropriate condition number, for example a component-wise one, might give a more realistic bound. However, I should stress that a small (scaled) residual does mean that we have not introduced instability in the solution process. In a sense, we are doing as well as we can even if the solution is far from what was expected.

Effect of Symmetry

When an applications scientist or engineer is deciding which algorithm or software to choose, one of the first questions is to ask if the matrix $\mathbf{A}$ is symmetric (or is Hermitian in the complex case). This makes a crucial difference whether direct or iterative techniques are being used for solution. For direct methods, not only are work and storage nearly halved but, particularly in the commonly occurring positive definite case, more efficient sparse data structures and sparsity based orderings can be used. For iterative methods, not only are the algorithms and software more reliable and robust, but there is often some theory to guarantee convergence. For matrix diagonalization, the normality of symmetric matrices (see section on Matrix Diagonalization) means that robust and accurate methods of determining eigenvalues and eigenvectors exist.

DIRECT METHODS

Direct methods use a factorization of the coefficient matrix to facilitate the solution. The most common factorization for unsymmetric systems is an LU factorization where the matrix $\mathbf{A}$ (or rather a permutation of it) is expressed as the product of a lower triangular matrix $\mathbf{L}$ and an upper triangular matrix $\mathbf{U}$. Thus

$$\mathbf{PAQ} = \mathbf{LU}, \qquad (6)$$

where $\mathbf{P}$ and $\mathbf{Q}$ are permutation matrices. This factorization can then be used to solve the system (1) through the two steps:

$$\mathbf{Ly} = \mathbf{Pb}, \qquad (7)$$

and

$$\mathbf{Uz} = \mathbf{y}, \qquad (8)$$

whence the solution $\mathbf{x}$ is just a permutation of the vector $\mathbf{z}$, viz.

$$\mathbf{x} = \mathbf{Qz}. \qquad (9)$$

[†]By *scaling* we mean choosing diagonal matrices $\mathbf{D}_1$ and $\mathbf{D}_2$ so that the nonzero entries of the scaled matrix $\mathbf{D}_1 \mathbf{A} \mathbf{D}_2$ have similar magnitude.

This use of *LU* factorization to solve systems of equations is usually termed *Gaussian elimination* and indeed the terms are often used synonymously. Another way of viewing Gaussian elimination is as a multistage algorithm which processes the equations in some order. At each stage, a variable is chosen in the equation and is eliminated from all subsequent equations by subtracting an appropriate multiple of that equation from all subsequent ones. The coefficient of the chosen variable is called the *pivot* in Gaussian elimination and the multiple of the pivot row or equation is called the *multiplier*. Clearly, there must be some reordering performed (called *pivoting*) if a pivot is zero but equally pivoting will normally be necessary if the pivot is very small (in fact if the multipliers are large) relative to other entries since then original information could be lost (from adding very large numbers to relatively very small numbers in finite-precision arithmetic) and we could solve a problem quite different from that originally intended. In the sparse case, pivoting is also required to preserve sparsity in the factors. For example, if the matrix **A** is an arrowhead matrix[‡], then selecting entry (n, n) as pivot will give dense triangular factors while choosing pivots from the diagonal in any order with entry (n, n) chosen last will give no *fill-in* (that is, there will be no entries in positions that were not entries in the original matrix). Of course, such a choice could be bad for the numerical criterion just mentioned above. The reconciliation of these possibly conflicting goals of pivoting has been a topic for research. We touch on this briefly below.

If the matrix **A** is symmetric positive definite, it is normal to use the factorization

$$\mathbf{PAP}^T = \mathbf{LL}^T. \qquad (10)$$

The factorization (10) is called a Cholesky factorization. For more general symmetric matrices, the factorization

$$\mathbf{PAP}^T = \mathbf{LDL}^T, \qquad (11)$$

is more appropriate. For a stable decomposition in the indefinite case, the matrix **D** is block diagonal with blocks of order 1 or 2, and **L** is unit lower triangular.

Phases in Solution

In both the case of sparse and dense matrices, the factorization (6) is more expensive than the forward elimination and backsubstitution phases, (7) and (8) respectively. This is less significant if several sets of equations with the same matrix but differing right-hand sides need to be solved. The higher cost of the factorization can then be amortized over the cost of the multiple solutions. In the sparse case, a further distinction is important. Often much of the work concerning handling sparse data structures and choosing pivots can be performed once only for a particular matrix structure and the subsequent factorization of matrices with the same structure can be performed much more efficiently using information from this first factorization. In some cases, the differences in execution time can be quite dramatic as we illustrate in Table 3 where there is an order of magnitude difference in time for the three phases. The code `MA48` is a general sparse unsymmetric solver from the Harwell Subroutine Library and GRE 1107 is a test matrix from the Harwell-Boeing Sparse Matrix Collection. The ability to refactorize efficiently subsequent matrices is not present in all software packages but is very important particularly when solving nonlinear systems when the Jacobian will retain the same structure although the numerical values change.

[‡]An arrowhead matrix, **A**, has nonzero entries only in positions $a_{ii}, a_{i,n},$ and $a_{n,i}$, $i = 1, ..., n$

Table 3. Execution times (in seconds) for code MA48 for matrix GRE 1107 on a single processor of a CRAY Y-MP

First factorization	.66
Subsequent factorizations	.075
Back and forward substitution	.0068

Sparsity Preservation and Numerical Stability

In the sparse case, it is crucial that the permutation matrices of (6) are chosen to preserve sparsity in the factors as well as to maintain stability and many algorithms have been developed to achieve this. For general unsymmetric matrices the most popular method is called "Markowitz with *threshold pivoting*". Threshold pivoting ensures that pivots are within a certain threshold of the largest entry in the row or column. The threshold is often an input parameter and a typical value for it is 0.1. We control sparsity by choosing the pivot to be an entry satisfying the threshold criterion that has the fewest product of number of other entries in its row and column. The suggestion of using this sparsity control is due to Markowitz (1957). In the symmetric case, the Markowitz analogue is *minimum degree* where one chooses as pivot a diagonal entry with the least number of entries in its row.

Arithmetic Complexity

Although the LU factorization has a similar $\mathcal{O}(n^3)$ complexity[§] to matrix inversion for dense systems (and storage requirements of $\mathcal{O}(n^2)$), the complexity of the factorization process and storage requirements for sparse matrices depend on the structure and can be significantly less. For example the LU factorization of a tridiagonal matrix can be done in $\mathcal{O}(n)$ operations whereas the calculation of the inverse is at best $\mathcal{O}(n^2)$. Furthermore, the storage for the factors of a tridiagonal matrix are the same as the original matrix ($3n - 2$ reals) but the inverse of a tridiagonal matrix is dense. In fact, if we consider structure only and do not allow numerical cancellation, the inverse of an irreducible sparse matrix[¶] is always dense (Duff *et al* 1988). An archetypal example, is a five-diagonal matrix[‖] as obtained, for example, from the finite-difference discretization of a two-dimensional Laplacian. If the discretization has k grid points in each direction (so that the order of the matrix is k^2), the LU factorization would require $\mathcal{O}(k^6)$ if considered as a dense system but $\mathcal{O}(k^4)$ if considered banded (and the storage requirement reduced from k^4 to k^3). Although this figure is often quoted when comparing the complexity of LU factorization with other methods on such matrices, by using a nested dissection ordering algorithm, the work and storage can be reduced to $\mathcal{O}(k^3)$ and $\mathcal{O}(k^2 \log k)$ respectively. The bad news is that one can prove that, for a wide range of matrices including the five-diagonal one, this is asymptotically the best that can be done for any direct method based on LU factorization.

[§] See, however, the comments on Strassen's algorithm which follow.
[¶] A matrix is irreducible if it is not possible to reorder the matrix rows and columns so the reordered form has a nontrivial block triangular form.
[‖] By five-diagonal matrix, we mean a matrix that has nonzeros only in five diagonals.

Indirect Addressing and the Use of the BLAS

For arbitrarily structured sparse matrices, complicated data structures are needed for efficient execution (for example, Duff et al (1986)). Although this is hidden from the user of the sparse code, it can significantly affect the efficiency of the computation. Even for computers with hardware indirect addressing, access to data of the form A(IND(I)), I = 1, K carries a heavy penalty in terms of additional memory traffic and non-localization of data. When this is added to the fact that most loops are of the order of number of nonzero entries in a row rather than the order of the system, general sparse codes can perform particularly badly on some high performance computers relative to their dense counterpart. Much recent research on sparse direct techniques has been to develop algorithms and codes that use the same kernels as dense codes at the innermost loops.

The BLAS, or Basic Linear Algebra Subprograms, are well established standard operations on dense matrices and vectors, and include computations such as scalar products, solution of triangular systems, and multiplication of two matrices (Lawson et al 1979, Dongarra et al 1988 and 1990). The important thing is that the interface to each subprogram has been standardized and most vendors have developed optimized code for these kernels. For example, the matrix-matrix multiplication routine (_GEMM) performs at close to peak performance on many computers, even those with quite sophisticated architectures involving vector processing, memory hierarchy, caches etc. Much of the recent work in dense linear algebra has centred round the use of these kernels. However, in spite of some early work by Duff (1981) and others, it is only quite recently that it has become appreciated that these dense linear algebra kernels can be used effectively within direct methods for sparse matrices.

Table 4. Performance of _GEMM kernel in Mflop/s on a range of machines (single processor performance)

Machine	Peak	_GEMM
Meiko CS2-HA	100	88
IBM SP-2	266	213
Intel PARAGON	75	68
DEC Turbo Laser	600	450
CRAY 2	459	449
CRAY YMP	333	313

We show, in Table 4, the performance of the Level 3 BLAS kernel _GEMM on a range of computers with various floating-point chips and memory organizations. In many cases, this kernel attains about 90% or more of the peak performance of the chip and in every case more than 75% of peak is achieved. This remarkable performance is obtained by the fact that a (dense) matrix-matrix multiply performs $2n^3$ arithmetic operations but only requires $3n^2$ data references. This enables data that is brought into the memory hierarchy (say onto an on-chip cache) to be reused, thus amortizing the cost of transferring it from main memory or even further afield. Since the memory level closest to the floating-point unit can usually supply data at the same speed as the unit computes, we can then get close to the speed of the floating-point unit.

As a footnote to the complexity issue, we should record the fact that methods have been developed for multiplying dense matrices in $\mathcal{O}(n^\alpha)$ operations, where $\alpha < 3$, based on Strassen's method (Strassen 1969). _GEMM has been implemented using this algorithm (for example, Douglas et al (1992)) and so the use of this kernel in dense Gaussian elimination can reduce the complexity accordingly. Note that the added complexity of Strassen's algorithm and the need to pay more care to stability (Highman 1990) means that this is not the panacea to the "n^3" problem". Also the lowest value of α that has been so far obtained is 2.376 so that dense matrix computations still quickly become infeasible when the matrix order becomes very large.

The trick is now to develop sparse matrix techniques that can take advantage of these fast dense matrix kernels. Of course, it is possible just to solve the sparse system using a code for dense systems, and some people have advocated this approach arguing that the greater "peak" speed and memory of modern computers makes this feasible. I must stress that the complexity discussions we had earlier makes this really non-viable for all but the smallest sparse matrices. We illustrate the wide difference in execution times for sparse and dense codes on sparse matrices by the results in Table 5. Although there are now much faster machines than a CRAY Y-MP, the matrices used in this table are quite small by current standards.

Table 5. Comparison between MA48 (a sparse code) and LAPACK (SGESV) (a dense code) on a range of matrices from the Harwell-Boeing Sparse Matrix Collection. Times are for factorization and solution (in seconds on one processor of a CRAY Y-MP)

Matrix	Order	Entries	MA48	SGESV
FS 680 3	680	2646	0.06	0.96
PORES 2	1224	9613	0.54	4.54
BCSSTK27	1224	56126	2.07	4.55
NNC1374	1374	8606	0.70	6.19
WEST2021	2021	7353	0.21	18.88
ORSREG 1	2205	14133	2.65	24.25
ORANI678	2529	90158	1.17	36.37

Frontal and Multifrontal Methods

A more viable approach is to order the sparse matrix so that its nonzero entries are clustered near the diagonal (called bandwidth minimization) and then regard the matrix as banded, treating zeros within the band as nonzero. However, this is normally too wasteful as even the high computational rate of the Level 3 BLAS does not compensate for the extra work on the zero entries. A variable band format is used to extend the range of applicability of this technique. A related, but more flexible scheme, is the frontal method which owes its origin to computations using finite elements.

Here we assume that **A** is of the form

$$\mathbf{A} = \sum_{l=1}^{m} \mathbf{A}^{[l]}$$

where each element matrix $\mathbf{A}^{[l]}$ has nonzeros in only a few rows and columns and is normally held as a small dense matrix representing contributions to **A** from element l. If a_{ij} and $a_{ij}^{[l]}$ denote the (i,j)th entry of **A** and $\mathbf{A}^{[l]}$, respectively, the basic assembly operation when forming **A** is of the form

$$a_{ij} \Leftarrow a_{ij} + a_{ij}^{[l]}. \tag{12}$$

It is evident that the basic operation in Gaussian elimination

$$a_{ij} \Leftarrow a_{ij} - a_{ip}[a_{pp}]^{-1}a_{pj} \tag{13}$$

may be performed as soon as all the terms in the triple product (13) are *fully summed* (that is, are involved in no more sums of the form (12)). The assembly and Gaussian elimination processes can therefore be interleaved and the matrix **A** is never assembled explicitly. Variables that are internal to a single element can be immediately eliminated (called *static condensation*) and this can be extended to a submatrix from a set of elements, that is a sum of several element matrices. In this scheme, all intermediate working can be performed within a dense matrix, termed the *frontal matrix*, whose rows and columns correspond to variables that have not yet been eliminated but occur in at least one of the elements that have been assembled.

We can partition the frontal matrix, **F**, as:

$$\mathbf{F} = \begin{bmatrix} \mathbf{F}_{11} & \mathbf{F}_{12} \\ \mathbf{F}_{21} & \mathbf{F}_{22} \end{bmatrix} \tag{14}$$

where the fully summed variables correspond to the rows and columns of the block $\mathbf{F}_{11}$, from where the pivots can be chosen. The kernel computation in a frontal scheme is then of the form

$$\mathbf{F}_{22} \leftarrow \mathbf{F}_{22} - \mathbf{F}_{21}\mathbf{F}_{11}^{-1}\mathbf{F}_{12} \tag{15}$$

and can be performed using Level 3 BLAS. (We note that this expression is notational and the inverse of $\mathbf{F}_{11}$ is not explicitly calculated.)

The frontal method can be easily extended to non-element problems since any set of rows of a sparse matrix can be held in a rectangular array whose number of columns is equal to the number of columns with nonzero entries in the selected rows. A variable is regarded as fully summed whenever the equation in which it last appears is assembled. These frontal matrices can often be quite sparse but are suitable for computations involving Level 3 dense BLAS. A full discussion of the equation input can be found in Duff (1984).

If the frontal scheme is combined with an ordering to preserve sparsity and reduce the number of floating-point operations and if a new frontal matrix can be formed independently of already existing frontal matrices, we can develop a scheme that combines the benefits of using Level 3 BLAS with the gains from using sparsity orderings. This is developed in *multifrontal schemes* where the computation can be viewed as a tree, whose nodes represent computations of the form (15) and whose edges represent the transfer of data from the child to the parent (data of the form of the $\mathbf{F}_{22}$ matrices generated by (15)). Another approach that combines sparse ordering schemes with higher level dense BLAS is the supernodal approach (for example, Demmel *et al* (1995)).

Parallelization of Direct Methods

In the late 80's and early 90's, it was almost impossible to obtain research funding in linear algebra unless parallelism was mentioned in the abstract if not the title. The topic was also embraced by computer scientists who could theorize on what the complexity of elimination techniques might be on a range of hypothetical computers and exotic parallel metacomputers with $\mathcal{O}(n^p)$ processors. I daresay some of this work was useful as other than an intellectual exercise, but the last few years have seen a more mature study of realistic parallel algorithms that can be implemented on actual available computers and can (and are sometimes) even used by applications people or industry.

Although algorithms and software for the parallel solution of dense systems of equations have been developed (for example, the ScaLAPACK package of Blackford *et al* (1997), the irregularity of sparse matrix structures makes it much more difficult to efficiently parallelize methods for sparse equations. The PARASOL Project is an ambitious attempt in this direction.

PARASOL** is an ESPRIT IV Long Term Research Project (No 20160) for "An Integrated Environment for Parallel Sparse Matrix Solvers". The main goal of this Project, which started on January 1 1996, is to build and test a portable library for solving large sparse systems of equations on distributed memory systems. There are twelve partners in five countries, five of whom are code developers, five end users, and two software houses. The software is written in Fortran 90 and uses MPI for message passing. The solvers being developed in this consortium are: two domain decomposition codes by Bergen and ONERA, a multigrid code by GMD, and a parallel multifrontal method (called MUMPS) by CERFACS and RAL. The final library will be in the public domain.

It is common, when examining implementations on parallel computers, to stress the speedup, or how many times faster the application runs on multiple processors than a single processor, although it is recognized that a more important measure is the execution time relative to the fastest method on a uniprocessor. However, an often more important reason for parallel computation is the the benefit of having access to more memory. This is particularly true on machines which have memory entirely local to each processor and which use message passing to share data between processors. In this case, the more processors; the more memory. A side effect of this is that it may only be possible to run a large problem on several processors so a comparison with a uniprocessor code is inappropriate. Furthermore, the memory system may be inefficient when stressed (for example, because of memory paging) and so the speedup may be superlinear. We illustrate this amusing effect by some runs of a parallel multifrontal code from the PARASOL Project on a PARASOL test problem on an IBM SP2 and an SGI Origin 2000 in Table 6. The speedups on the Origin reflect the true parallelism of the software, whereas those on the SP2 include a memory effect.

ITERATIVE METHODS

In contrast to direct methods, iterative methods do not normally modify the matrix and do not form any representation of the inverse. Furthermore, most iterative methods do not require the matrix to be represented explicitly, sometimes it is sufficient to be able to form the product of the matrix with a vector, although this may restrict the preconditioning available (see later subsection).

Iteration techniques such as successive approximation have been around since the first days of scientific computing and the early iterative techniques for solving sparse linear equations were based on such approaches (Gauss-Seidel, SOR etc). While these

---

**For more information on the PARASOL project, see the web site at http://www.genias.de/parasol.

Table 6. Results for the symmetric version of the MUMPS code on CRANKSEG2.

Machine	Working processors	Time for factorization
SP2	16	1045.3
	24	457.3
	32	139.7
Origin	1	635.4
	2	411.0
	3	275.7
	4	220.4
	5	175.1
	6	158.3
	7	142.9
	8	135.7

methods are still used (for example, as smoothers in multigrid techniques), it is true to say that most modern methods for the iterative solution of sparse equations are based on Krylov sequences of the form

$$sp\{\mathbf{v}, \mathbf{A}\mathbf{v}, \mathbf{A}^2\mathbf{v}, ...\}, \tag{16}$$

with the approximate solution at each step the "best" solution that lies in the subspace of increasing dimension. What constitutes "best" determines which of the many methods is defined.

The residual at the i-th iteration of an iterative method can be expressed as

$$\mathbf{r}^{(i)} = P_i(\mathbf{A})\mathbf{r}^{(0)}, \tag{17}$$

where P_i is a polynomial such that $P_i(0) = 1$. If we expand $\mathbf{r}^{(0)}$ in terms of the eigenvectors of $\mathbf{A}$ we see that we want P_i to be small on the eigenvalues of $\mathbf{A}$ so that the spectrum of $\mathbf{A}$ is crucial in determining how quickly our method converges. For example, if there are many eigenvalues close to zero or if the spectrum is widely distributed, the degree of polynomial will have to be high in order to be small on all eigenvalues and so the number of iterations required will be large.

A major feature of most Krylov based methods for symmetric systems are that they can be implemented using short term recurrences which means that only a few vectors of length n need be kept and the amount of computation at each stage of the iterative process is modest. However, Faber and Manteuffel (1984) have shown that, for general matrices, one must either lose the cheap recurrences or lose the minimization property. Thus for the solution of general unsymmetric systems, the balance between these, added to the range of quantities that can be minimized and the differing norms that can be used, has led to a veritable alphabet soup of methods (Saad 1996), for example GMRES(k), CGS, Bi-CGSTAB(ℓ), TFQMR, FGMRES, GMRESR, ...

Preconditioning

The key to developing iterative methods for the solution of realistic problems lies in preconditioning, where by this we mean finding a matrix $\mathbf{K}$ such that

1. **K** is an approximation to **A**.

2. **K** is cheap to construct and store.

3. **Kx** = **b** is much easier to solve than the system (1).

We then solve the preconditioned system

$$\mathbf{K}^{-1}\mathbf{A}\mathbf{x} = \mathbf{K}^{-1}\mathbf{b}, \tag{18}$$

where we have chosen **K** so that our iterative method converges more quickly when solving equation (18) than equation (1). Lest this seem too much of a black art (which to some extent it is), if **K** were chosen as the product of the factors **LU** from an *LU* factorization (admittedly violating point 2 above), then the preconditioned matrix **B** = **K**$^{-1}$**A** would be the identity matrix and any sane iterative method would converge in a single iteration. ¿From our earlier discussion, we would like the preconditioned matrix **B** to have a better distribution of eigenvalues than **A**.

The preconditioning can also be applied as a right-preconditioning **AK**$^{-1}$ or as a two-sided preconditioning $\mathbf{K}_1^{-1}\mathbf{A}\mathbf{K}_2^{-1}$, when the matrix **K** can be expressed as a product $\mathbf{K}_1\mathbf{K}_2$. Common preconditioners include using the diagonal of **A** or a partial or incomplete factorization of **A**. A recent simple discussion of the merits of different forms of preconditioners and their implementation can be found in the report Duff and van der Vorst (1998) that is a preprint of a chapter in a forthcoming book Dongarra *et al* (1998).

A very interesting aspect of this is that convergence can occur in very few iterations if the eigenvalues are well clustered. This is, of course, true whether the matrix is dense or sparse. In the dense case, a direct method will require $\mathcal{O}(n^3)$ work whereas a single matrix vector multiplication only $\mathcal{O}(n^2)$. Thus, if our iterative method converges in only a few iterations, it may be a very attractive method for solving the dense system. An example where dense systems with well clustered eigenvalues are found is given by Rahola (1996). However, as in the case of sparse systems, preconditioning is normally needed to obtain a good eigenvalue distribution. This is less attractive in the dense case because another n^2 multiplications are required and, if it is required to solve for the preconditioning matrix, our subproblem is as hard as the original problem. It is sometimes possible, however, to develop sparse preconditioning matrices for the dense problem (for example, Alléon *et al* (1997)) so that the cost of this preconditioning is small compared to the multiplication by the original matrix. In some cases, the structure of the problem can be used to reduce the matrix-vector multiplication itself, for example by using multipole methods (Rahola 1998).

Parallelization of Iterative Methods

In contrast to direct methods, each step of an iterative method is relatively easy to parallelize since the only numerical computations involved are:

- **Av**

- $\mathbf{v}^T\mathbf{w}$

- $\mathbf{v} - \alpha\mathbf{w}$

and, for the preconditioning:

- $\mathbf{K}^{-1}\mathbf{v}$

The first operations can be performed with high efficiency on parallel computers, for example Erhel (1990), although scalar products require synchronization and much communication. The achilles heel for performance is, however, usually in the efficient implementation of the preconditioning, on which much research is still being done (see for example, Dongarra *et al* (1998)). We should add that it is vital that the preconditioner is effective in reducing the number of iterations because, if the convergence is so slow as to be meaningless, no amount of parallelism can make it viable.

DIRECT OR ITERATIVE OR ?

I am often asked whether it is better to use an iterative or a direct method to solve a set of linear equations. The answer is often quite simple, and not just because of the predilection of my research interest. One should use a direct method! This is even more true if you have several right-hand sides to solve with the same coefficient matrix. However, since your computer is unlikely to be blessed with an infinite amount of memory (unless you are a theoretician from the 80's), the range of problems for which such a technique is applicable is limited, significantly so if your underlying problem is three-dimensional. In such cases, one has to resort to an iterative method. However, since for all but the simplest cases, your chosen iterative method (even with a standard preconditioner) is unlikely to converge ... what do you do?

The answer is to use a technique that combines elements of both direct and iterative methods. Sophisticated preconditioners come into this category. The most obvious method to combine these approaches is the block Jacobi method where the matrix is treated as a block matrix, the solution of the subproblems corresponding to the diagonal blocks are performed using a direct method, and the system is solved using a block Jacobi algorithm. Clearly this inherits much of the parallelism of the point case but should have better convergence properties. Indeed a variant corresponding to using a block Jacobi scheme on the normal equations, termed the block Cimmino algorithm, has proven quite effective (for example, Arioli *et al* (1995)). As in the point case, we can sacrifice parallelism a little and gain faster convergence through the use of a block Gauss-Seidel method. The counterpart to block Cimmino is then block Kacmarz (for example, Bramley and Sameh (1992)).

In the framework of the solution of partial differential equations, a more general technique for combining direct and iterative methods is to use domain decomposition (Chan and Mathew (1994), Smith *et al* (1996)), where the problem is divided into separate domains (either overlapping or non-overlapping) and the local problems can be solved using direct methods, which of course is trivial to do in parallel since these subproblems are decoupled. In the case of non-overlapping domains, usually the main issue is the solution of the problem for the interface variables. It is normal to use an iterative method for this but the main problem is then to define a preconditioner, particularly if the matrix corresponding to the interface problem is not computed explicitly.

One way in which direct methods can be used as preconditioners for iterative methods are to perform only a partial factorization, as is the case for ILU(k) preconditioners, where a limited amount of fill-in is allowed to the matrix factors so that an incomplete factorization is performed. Another approach is to perform a "full" factorization but one of a reduced or simpler problem to the original, for example for a simpler differential equation than the original.

Multigrid methods have become very popular in recent years largely because of their provably optimal performance on simple elliptic problems. In these techniques, a sequence of grids is used. The solution is required on the finest grid. A few passes of a simple iterative method are performed and the residual is projected onto the next coarser grid. A correction is obtained by projecting back the solution of the residual equations on this coarser grid to the finer grid. Since the solution of the residual equations can be performed using an even coarser grid, multiple grids can be used. The essence of the method is that the simple iterative method (or smoother) tends to efficiently remove error components that vary as quickly as the grid size and the use of coarser grids enables the smoother components of the error to be reduced. While the individual sweeps of the smoothers can be parallelized, it is harder to parallelize across the grids but there has been some recent work on this, for example by the PARASOL Project mentioned earlier.

MATRIX DIAGONALIZATION

What physicists call matrix diagonalization, numerical analysts call the eigenproblem. In the case of small matrices, there are many techniques for both symmetric and unsymmetric matrices that involve computing transformations of the matrix to diagonal form usually in a two-step process, the first step transforming the matrix to tridiagonal or Hessenberg form, respectively. Although the use of high Level BLAS in these computations improves their efficiency, the transformations involved destroy sparsity in a way that is not normally as recoverable as in the case of LU factorization. Additionally, in the large sparse case, normally only a few eigenvalues and eigenvectors are required. Thus for large sparse matrices, we use other methods for obtaining the eigenvalues.

In fact the basis of eigensolution techniques for large sparse matrices is the same as we just discussed for the iterative solution of sparse equations, namely the Krylov sequence (16). Clearly, this is a generalization of the classical power method for computing eigenvalues and eigenvectors, but differs significantly because the previous powers are taken into account so that the size of the subspace increases as iterations are performed. Additionally, in the Krylov based methods, we are free to choose appropriate bases for the subspaces.

In our brief discussion that follows, we will be concerned with the standard eigenproblem (3), although we note that normally the generalized problem (2) is first converted to a standard one, for example on a matrix of the form $(\mathbf{A} - \sigma\mathbf{B})^{-1}\mathbf{B}$ or $\mathbf{L}^{-1}\mathbf{A}\mathbf{L}^{-T}$, when $\mathbf{B} = \mathbf{L}\mathbf{L}^T$ is symmetric positive definite.

Whereas for linear systems, we seek solutions of the form

$$\mathbf{Q}_k \mathbf{y}_k,$$

where $\mathbf{Q}_k = [\mathbf{q}_1 \mathbf{q}_2 .... \mathbf{q}_k]$ is the basis of the Krylov subspace of dimension k, for the eigenproblem, we solve a reduced eigenproblem for the projected subspace

$$\mathbf{Q}_k^* \mathbf{A} \mathbf{Q}_k. \qquad (19)$$

For the eigenproblem, we can again use preconditioning techniques to accelerate convergence. In this case, the subspaces are generated for a simple function of $\mathbf{A}$ so that the eigenvalues required are better represented in subspaces of the form (19) of low dimension.

Since the kernels of Krylov subspace methods for finding eigenvalues are exactly those listed for the iterative solution of linear equations at the end of the section on

Iterative Methods, eigenvalue calculations are also relatively easy to parallelize with possible bottlenecks in preconditioning.

A recently investigated phenomenon is that of non-normality (Chaitin-Chatelin and Frayssé 1996). Formally a matrix is normal if

$$\mathbf{A}^*\mathbf{A} = \mathbf{A}\mathbf{A}^*$$

and so all symmetric (or Hermitian) matrices are normal. The most important feature of normal matrices are that there exists a unitary matrix $\mathbf{Q}$ such that $\mathbf{Q}^*\mathbf{A}\mathbf{Q}$ is diagonal. This in turn means that the eigenproblem is well-conditioned in the sense that a small perturbation to the matrix produces a small perturbation of the eigenvalues. However, for non-normal but diagonalizable matrices the diagonalization $\mathbf{X}^{-1}\mathbf{A}\mathbf{X}$, implies nothing about the condition number, $\kappa(\mathbf{X})$, of $\mathbf{X}$, so that, by the Bauer-Fike theorem, if

$$\mathbf{X}^{-1}\mathbf{A}\mathbf{X} = \mathbf{D} = \text{diag}\{\lambda_1, ..., \lambda_n\}$$

and μ is an eigenvalue of the perturbed matrix $\mathbf{A} + \mathbf{E}$, then

$$\min_\lambda |\lambda - \mu| \leq \kappa(\mathbf{X})||\mathbf{E}||.$$

A similar result holds for non-diagonalizable matrices.

This means that even a matrix very close to $\mathbf{A}$ might have very different eigenvalues. This phenomenon can be viewed in a diagram containing contours of the resolvent

$$||(\mathbf{A} - \lambda\mathbf{I})^{-1}||$$

and pseudo-eigenvalues are defined by regions in the plane where the resolvent is large. There is now some debate about whether physical processes are governed by pseudo-eigenvalues rather than eigenvalues but certainly one should be very wary of making a qualitative judgement of a process or its stability on the strength of a single or few eigenvalues (Trefethen *et al* 1993). One should note that a complex symmetric matrix is not Hermitian and can be non-normal.

SOURCES OF SOFTWARE

The origins of much specialist work in numerical software stem from the requirement of computational physicists and chemists, and libraries and collections of software were developed to avoid duplication of effort and provide a sound base of portable codes. I would not advise the applications scientist or engineer to try to reverse this historical trend by coding his or her own matrix algorithm, even using a numerical recipes crib sheet.

There are many sources of software for sparse matrices, from the do-it-yourself approach of the Templates book for iterative methods (Barrett *et al* 1993) (a similar one for eigenproblems is forthcoming), to supported proprietary codes as are present in the Harwell Subroutine Library, http://www.dci.rl.ac.uk/Activity/HSL (HSL 1996). Some codes are available through netlib, http://www.netlib.org, and others from the Web pages of the researchers developing the code. The problem with the latter source is that, in addition to a lack of quality control, the researcher in question will often have no compunction against editing the code almost as you are downloading it.

We discuss sources of sparse matrix software in Duff (1997) and a recent report by Eijkhout (1998) discusses, in some detail, software for iterative methods for solving linear systems. The reports by Lehoucq and Scott (1996) discuss sparse eigenvalue software.

BRIEF SUMMARY

We have presented a few obvious, and hopefully some less obvious, comments on the solution of large sparse systems and large eigenproblems. We have emphasized the similarity between iterative solution techniques and matrix diagonalization procedures, and indicated briefly how direct and iterative methods can be combined to solve really large problems. We see considerable promise in both frontal and multifrontal methods on machines with memory hierarchies or vector processors and reasonable possibilities for exploitation of parallelism by multifrontal methods. A principal factor in attaining high performance is the use of dense matrix computational kernels, which have proved extremely effective in the dense case.

Finally, we have tried to keep the narrative flowing in this presentation by avoiding an excessive number of references. For such information, we recommend, for linear systems, the recent review by Duff (1997), where 215 references are listed. Saad (1992) has written a book on large-scale eigenproblems with an emphasis on computational aspects and more recent references on iterative solution of equations and eigenproblems can be found in Golub and van der Vorst (1996) and van der Vorst and Golub (1996) respectively.

ACKNOWLEDGEMENTS

I would like to thank my colleagues, Valérie Fraysse and Luc Giraud from CERFACS, and Nick Gould, John Reid, and Jennifer Scott from the Rutherford Appleton Laboratory for their comments on a draft of this paper and to Walter Temmerman of the Daresbury Laboratory for checking its accessibility to my main target audience.

REFERENCES

Alléon, G., Benzi, M. and Giraud, L. (1997), 'Sparse approximate inverse preconditioning for dense linear systems arising in computational electromagnetics', *Numerical Algorithms* **16**(1), 1–15.

Anderson, E., Bai, Z., Bischof, C., Demmel, J., Dongarra, J., DuCroz, J., Greenbaum, A., Hammarling, S., McKenney, A., Ostrouchov, S. and Sorensen, D. (1995), *LAPACK Users' Guide, second edition*, SIAM Press.

Arioli, M., Duff, I. S., Ruiz, D. and Sadkane, M. (1995), 'Block Lanczos techniques for accelerating the Block Cimmino method', *SIAM J. Scientific Computing* **16**(6), 1478–1511.

Barrett, R., Berry, M., Chan, T., Demmel, J., Donato, J., Dongarra, J., Eijkhout, V., Pozo, R., Romine, C. and van der Vorst, H., eds (1993), *Templates for the Solution of Linear Systems: Building Blocks for Iterative Methods*, SIAM, Philadelphia.

Blackford, L. S., Choi, J., Cleary, A., D'Azevedo, E., Demmel, J., Dhillon, I., Dongarra, J., Hammarling, S., Henry, G., Petitet, A., Stanley, K., Walker, D. and Whaley, R. C. (1997), *ScaLAPACK Users' Guide*, SIAM Press.

Bramley, R. and Sameh, A. (1992), 'Row projection methods for large nonsymmetric linear systems', *SIAM J. Scientific and Statistical Computing* **13**, 168–193.

Chaitin-Chatelin, F. and Fraysse, V. (1996), *Lectures on Finite Precision Computations*, SIAM Press, Philadelphia.

Chan, T. F. and Mathew, T. P. (1994), *Domain Decomposition Algorithms*, Vol. 3 of *Acta Numerica*, Cambridge University Press, Cambridge, pp. 61–143.

Demmel, J. W., Eisenstat, S. C., Gilbert, J. R., Li, X. S. and Liu, J. W. H. (1995), A supernodal approach to sparse partial pivoting, Technical Report UCB//CSD-95-883, Computer Science Division, U. C. Berkeley, Berkeley, California.

Dongarra, J. J., Du Croz, J., Duff, I. S. and Hammarling, S. (1990), 'A set of Level 3 Basic Linear Algebra Subprograms.', *ACM Trans. Math. Softw.* **16**, 1–17.

Dongarra, J. J., Du Croz, J. J., Hammarling, S. and Hanson, R. J. (1988), 'An extented set of Fortran Basic Linear Algebra Subprograms', *ACM Trans. Math. Softw.* **14**, 1–17.

Dongarra, J. J., Duff, I. S., Sorensen, D. C. and van der Vorst, H. A. (1998), *Numerical Linear Algebra for High-Performance Computers*, SIAM Press, Philadelphia.

Douglas, C. C., Heroux, M., Slishman, G. and Smith, R. M. (1992), GEMMW: A portable level 3 BLAS Winograd variant of Strassen's matrix-matrix multiply algorithm, Technical Report RC 18026 (79130), IBM T. J. Watson Research Centre, P. O. Box 218, Yorktown Heights, NY 10598.

Duff, I. S. (1981), The design and use of a frontal scheme for solving sparse unsymmetric equations, *in* J. P. Hennart, ed., 'Numerical Analysis, Proceedings of 3rd IIMAS Workshop. Lecture Notes in Mathematics 909', Springer Verlag, Berlin, pp. 240–247.

Duff, I. S. (1984), 'Design features of a frontal code for solving sparse unsymmetric linear systems out-of-core', *SIAM J. Scientific and Statistical Computing* **5**, 270–280.

Duff, I. S. (1997), Sparse numerical linear algebra: Direct methods and preconditioning, *in* I. S. Duff and G. A. Watson, eds, 'The State of the Art in Numerical Analysis', Oxford University Press, Oxford, pp. 27–62.

Duff, I. S. and van der Vorst, H. A. (1998), Preconditioning and parallel preconditioning, Technical Report RAL-TR-1998-052, Rutherford Appleton Laboratory.

Duff, I. S., Erisman, A. M. and Reid, J. K. (1986), *Direct Methods for Sparse Matrices*, Oxford University Press, Oxford, England.

Duff, I. S., Erisman, A. M., Gear, C. W. and Reid, J. K. (1988), 'Sparsity structure and Gaussian elimination', *SIGNUM Newsletter* **23**(2), 2–8.

Duff, I. S., Grimes, R. G. and Lewis, J. G. (1989), 'Sparse matrix test problems', *ACM Trans. Math. Softw.* **15**(1), 1–14.

Duff, I. S., Grimes, R. G. and Lewis, J. G. (1997), The Rutherford-Boeing Sparse Matrix Collection, Technical Report RAL-TR-97-031, Rutherford Appleton Laboratory. Also Technical Report ISSTECH-97-017 from Boeing Information & Support Services and Report TR/PA/97/36 from CERFACS, Toulouse.

Eijkhout, V. (1998), 'Overview of iterative linear system solver packages', *NHSE Review*. http://www.nhse.org/NHSEreview/98-1.html.

Erhel, J. (1990), 'Sparse matrix multiplication on vector computers', *Int J. High Speed Computing* **2**, 101–116.

Erisman, A. M. and Tinney, W. F. (1975), 'On computing certain elements of the inverse of a sparse matrix', *Communications of the Association for Computing Machinery* **18**, 177–179.

Golub, G. H. and van der Vorst, H. A. (1996), Closer to the solution: Iterative linear solvers, *in* I. S. Duff and G. A. Watson, eds, 'The State of the Art in Numerical Analysis', Oxford University Press, Oxford, pp. 63–92.

Higham, N. J. (1990), 'Exploiting fast matrix multiplication within the Level 3 BLAS', *ACM Trans. Math. Softw.* **16**, 352–368.

Higham, N. J. (1996), *Accuracy and Stability of Numerical Algorithms*, SIAM Press, Philadelphia.

HSL (1996), *Harwell Subroutine Library. A Catalogue of Subroutines (Release 12)*, AEA Technology, Harwell Laboratory, Oxfordshire, England. For information concerning HSL contact: Dr Scott Roberts, AEA Technology, 477 Harwell, Didcot, Oxon OX11 0RA, England (tel: +44-1235-432682, fax: +44-1235-432023, email: Scott.Roberts@aeat.co.uk).

Lawson, C. L., Hanson, R. J., Kincaid, D. R. and Krogh, F. T. (1979), 'Basic linear algebra subprograms for Fortran usage', *ACM Trans. Math. Softw.* **5**, 308–323.

Lehoucq, R. B. and Scott, J. A. (1996*a*), An evaluation of Arnoldi based software for sparse nonsymmetric eigenproblems, Technical Report RAL-TR-96-023, Rutherford Appleton Laboratory.

Lehoucq, R. B. and Scott, J. A. (1996*b*), An evaluation of subspace iteration software for sparse nonsymmetric eigenproblems, Technical Report RAL-TR-96-022, Rutherford Appleton Laboratory.

Markowitz, H. M. (1957), 'The elimination form of the inverse and its application to linear programming', *Management Science* **3**, 255–269.

Rahola, J. (1996), Efficient solution of dense systems of linear equations in electromagnetic scattering calculations, PhD Thesis, CSC Research Reports R06/96, Center for Scientific Computing, Department of Engineering Physics and Mathematics, Helsinki University of Technology.

Rahola, J. (1998), Experiments on iterative methods and the fast multipole method in electromagnetic scattering calculations, Technical Report TR/PA/98/49, CERFACS, Toulouse, France.

Saad, Y. (1992), *Numerical Methods for Large Eigenvalue Problems*, Manchester University Press, Manchester, UK.
Saad, Y. (1996), *Iterative methods for sparse linear systems*, PWS Publishing, New York, NY.
Smith, B., Björstad, P. and Gropp, W. (1996), *Domain Decomposition, Parallel Multilevel Methods for Elliptic Partial Differential Equations*, 1st edn, Cambridge University Press, New York.
Strassen, V. (1969), 'Gaussian elimination is not optimal', *Numerische Mathematik* **13**, 354–356.
Trefethen, L. N., Trefethen, A. E., Reddy, S. C. and Driscoll, T. A. (1993), 'Hydrodynamics stability without eigenvalues', *Science* **261**, 578–584.
van der Vorst, H. A. and Golub, G. H. (1996), 150 years old and still alive: Eigenproblems, *in* I. S. Duff and G. A. Watson, eds, 'The State of the Art in Numerical Analysis', Oxford University Press, Oxford, pp. 93–119.
Wilkinson, J. H. (1961), 'Error analysis of direct methods of matrix inversion', *J. ACM* **8**, 281–330.

ELASTIC ELECTRON COLLISION WITH CHIRAL AND ORIENTED MOLECULES

K.Blum, M.Musigmann, D.Thompson [+]

Institut für Theoretische Physik I
Universität Münster
Wilhelm Klemmstr.9
D-48149 Münster

[+] Department of Applied Mathematics
and Theoretical Physics
Queen's University
Belfast BT7 1NN Northern Ireland

INTRODUCTION

The interest in electron collisions from chiral molecules has been greatly stimulated by Farago[1,2]. In a series of papers he discussed experimental possibilities, and described some of the new "chiral effects" to be expected, if a beam of electrons passes through a chiral medium:

i.) the differential cross section for scattering of longitudinally polarised electrons depends on the handedness of electrons and molecules (that is, the spin-asymmetry should be different from zero),

ii.) a beam of unpolarised electrons becomes longitudinally polarised.

It has also been pointed out by Hegstrom[3] that chiral effects are to be expected in dissociation and rearrangement collisions. Furthermore, as is well known from optical studies[4], chiral phenomena can be produced in reactions with non-chiral but oriented molecules, if a "screw-sense" is defined by the geometry of the experiment[5].

A number of theoretical and numerical studies has been performed along these lines. On the experimental side, elastic collisions between electrons and randomly oriented chiral molecules have been performed by Kessler and his group[6,7,8].

They confirmed the existence of chiral effects, although the order of magnitude of 10^{-4} is very small.

On the theoretical side, a detailed analysis has been performed, classifying the expected chiral effects according to their space-time symmetries[9], and a series of detailed numerical calculations has been performed. The theoretical and experimental results have been reviewed recently[10].

A new development started with detailed numerical calculations with oriented molecules[9,11,12]. It has been found that the values of chiral observables can be increased by 2 - 3 order of magnitude for particular orientations.These results show that the "stereodynamic" of the collision plays an important role.

In the present paper we will review recent theoretical and numerical results. In section 2 some basic definitions will be given, and chiral observables will be introduced. In section 3 we will present data for chiral observables for HBr, H_2S_2, and $CHFClBr$ obtained by Smith et al.[12] within the "Continuum Multiple Scattering Method". In section 4 we will consider the "sterodynamics" of the collision in more detail, concentrating on diatomic molecules. After a brief discussion of how to produce and describe anisotropic molecular ensembles in the gasphase, general formulas will be given for the dependence of chiral observables on the orientation and alignment of the initial molecular sample. Numerical results will be presented which might be useful as a guide for future experiments.

SOME GENERAL FORMULAS AND DEFINITIONS

Let us consider elastic collisions from <u>fully oriented</u> (chiral or non-chiral) molecules with <u>axes fixed in space</u>. By this we mean that all molecules in the given target ensemble are pointing in one and the same direction with respect to a given "external" coordinate system. As "external" system we will choose the <u>collision system</u> where the incoming electron beam direction is taken as z-axis, and the scattering plane is choosen as xz-plane. The molecule-fixed rectangular system will be denoted by $\vec{e_i} = \vec{e_1}, \vec{e_2}, \vec{e_3}$. Throughout this paper we will discuss closed-shell molecules.

Let us consider a transition $\vec{k_0}\, m_0\, \vec{e_i} \to \vec{k_1}\, m_1\, \vec{e_i}$ where $\vec{k_0}$ ($\vec{k_1}$) denotes the wave vector of initial (scattered) electrons, and m_0 (m_1) the z-component of initial (final) electrons. The molecular axes $\vec{e_i}$ are assumed to be fixed during the collision. The indicated transition will be described in terms of scattering amplitudes $f(\vec{k_1}m_1, \vec{k_0}m_0, \vec{e_i}) \equiv f(m_1 m_0)$, normalised in such a way that its absolute square gives the corresponding differential cross section.

$$| f(m_1 m_0) |^2 = \sigma(m_1 m_0) \qquad (1)$$

An alternative set of amplitudes, denoted by g_0, g_1, g_2, g_3, have been introduced by Johnston et al.[9]. This set has definite transformation properties under space inversion and time reversal, and this has many advantages in general considerations. In particular, the important concept of "time-odd" ("false") chirality can be discussed in detail. This concept has first been introduced by Barron[4] for optical processes, and further developed and applied by Johnston et al.[9]. A detailed discussion of the importance of this concept can be found in the review by Blum and Thompson[10].

The amplitudes $f(m_1 m_0)$ (or the set g_i) can then be used to calculate observables of experimental interest. Of particular importance are those observables which vanish for non-chiral systems, and which are only non-zero for chiral systems (that is, if the target molecules are chiral, or if chirality is defined by the geometry of the experiment.) We give two examples:

i.) production of longitudinal polarisation $P'_\parallel$ for the scattered electrons for unpolarised initial electrons, that is, the component of the final spin polarisation vector with respect to $\vec{k}_1$. (Alternatively, P'_z and P'_x could be considered, that is, the components of the final spin polarisation vector with respect to the z- and x-axis respectively). All "in-plane" components $P'_\parallel, P'_z$, and P'_x vanish for non-chiral systems because of reflection invariance of the interaction with respect to the scattering plane.

ii.) the spin-asymmetry A, defined by the expression

$$A = \frac{\sigma(+) - \sigma(-)}{\sigma(+) + \sigma(-)} \quad (2)$$

where $\sigma(+)$ $(\sigma(-))$ denotes the differential cross section for completely longitudinally polarised initial electrons with $m_0 = +\frac{1}{2}$ $(-\frac{1}{2})$ with respect to z. For non-chiral systems, $\sigma(+)$ would be equal to $\sigma(-)$.

NUMERICAL PROCEDURE AND RESULTS FOR FULLY ORIENTED AND RANDOMLY ORIENTED MOLECULES

General scattering theories and computer programes for collisions with chiral molecules have been developed by D.Thompson and collaborators. Here we will briefly sketch the most recent form of the theory. The emphasis is here on collisions with polyatomic molecules containing more than one heavy atom. For details we refer to Smith et al.[12].

The computational model is an extension of the Continuum Multiple Scattering Method in which the molecular region is divided into "atomic" spherical regions centred on each nucleus. The interaction potential in each region is the sum of a spherically-symmetric spin-independent term and a spin-orbit term. The interaction potential between the "atomic" spheres inside the molecule is constant. Solutions are obtained in each region, and between the regions, enforcing continuity of the solutions across region boundaries leads to a K-matrix and a scattering amplitude, from which the chiral observables are obtained. Electron exchange and distortion effects have been taken into account. The method has been tested for HCl against a fuller treatment in which exchange and spin-orbit interactions are included exactly. In the following we will give some results for chiral and non-chiral molecules.

Results For HBr

At first we will present results for fully-oriented HBr-molecules (the experimentally more interesting case of partially-oriented diatomics will be discussed in section 4.) Here, chiral effects (non-vanishing results for $P'_\parallel$ and A) can only be expected if a screw-sense is defined by the geometry of the experiment. In other words, it must be possible to construct a pseudoscalar out of the given experimental parameters. Such a pseudoscalar is for example $[\vec{k}_1 \times \vec{k}_0] \cdot \vec{n}$ where the cross denotes the vector product. This pseudoscalar would be zero if $\vec{n}$ lies in the scattering plane. For homopolar molecules, no screw sense could be defined for $\vec{n}$ perpendicular to the scattering plane (see section 4). The electron spin "sees" this screw sense, and can react to it, and chiral effects can be produced.

Fig. 1 presents results of numerical calculations for HBr for the in-plane polarisation P'_z and the asymmetry A. The molecular axis is perpendicular to the scattering plane. The results show a considerable variation with scattering energy and angle.

Results For H_2S_2

H_2S_2 is a chiral molecule: One $S-H$-bond is rotated about $96,6°$ out of the plane of the two S-atoms and the other H-atom. Fig.2 shows numerical results for the longitudinal polarisation $P'_\parallel$ and asymmetry A at an orientation of $\alpha = \beta = 90°$ with respect to the collision system.(that is, the $S-S$-bond molecular z-axis is perpendicular to the scattering plane. The C_2-symmetry axis is choosen as x-axis of the molecular system).
Results for randomly oriented molecules are shown in fig. 3. As expected the magnitude is much less than for the oriented case, being of order 10^{-4}.

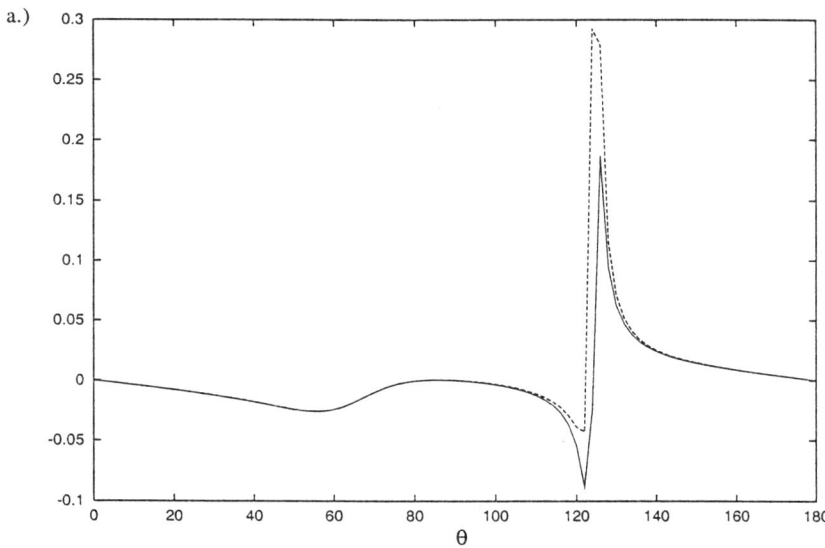
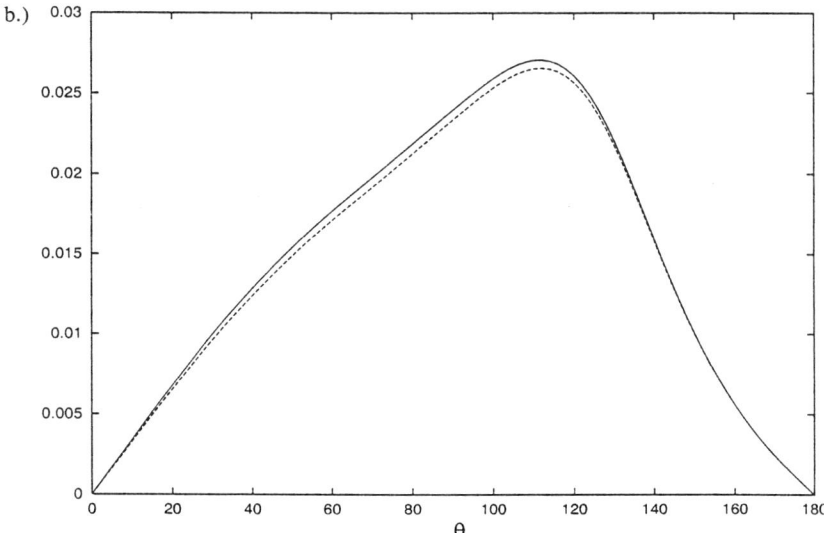

Figure 1. In-plane-polarisation P'_z (———) and asymmetry (– –) against scattering angle θ for HBr. Orientation : $\alpha = \beta = 90°$.
a.) 10eV, b.) 5eV.

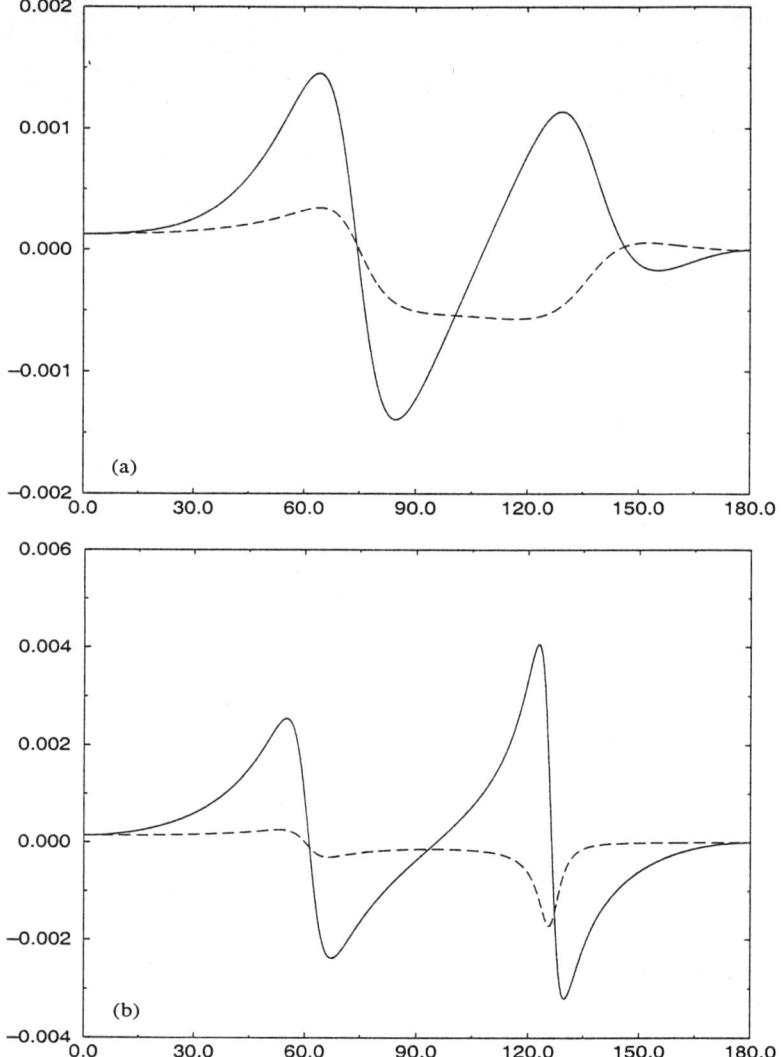

Figure 2. longitudinal polarisation (——) and asymmetry (– –) against scattering angle θ for H_2S_2. Orientation : $\alpha = \beta = 90°, \gamma = 0°$ a.) 10eV, b.) 5eV.

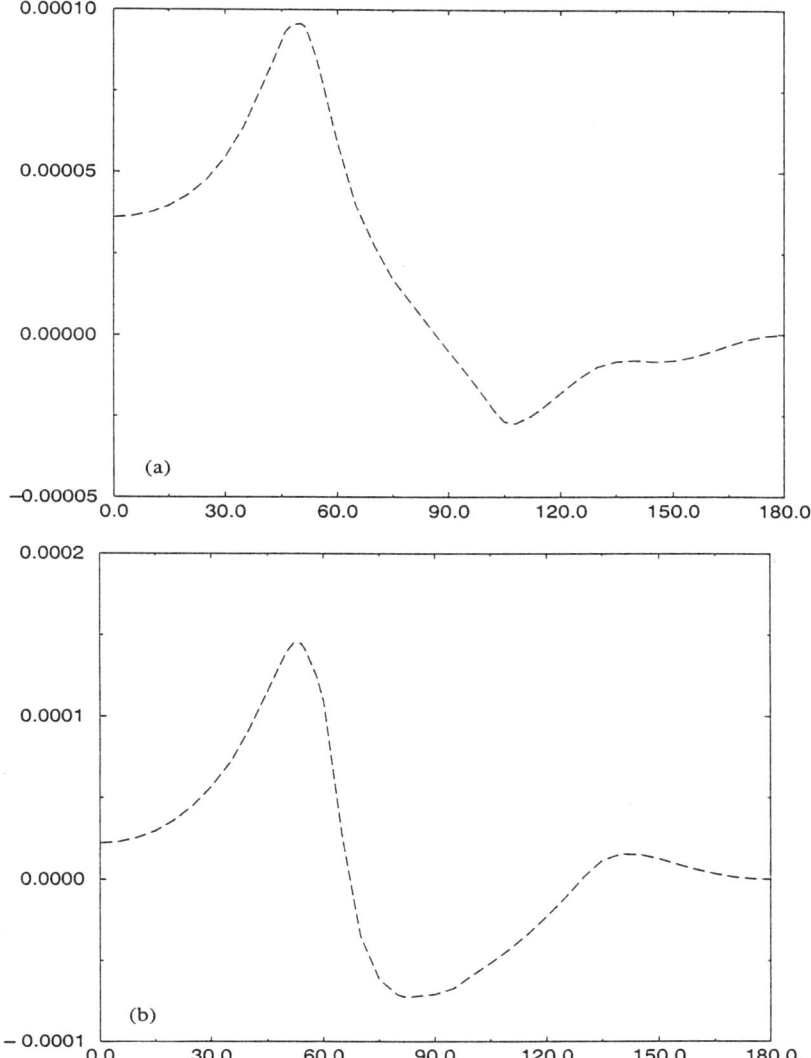

Figure 3. longitudinal polarisation for randomly oriented H_2S_2. a.) 10eV, b.) 5eV.

Results For $CHBrClF$

Results for the oriented molecule are shown in fig. 4. The effects, which in some cases reach magnitudes greater than 10%, are much larger than for H_2S_2, possibly due to the greater spin-orbit coupling. The sharp features at 60° in fig. 4b correspond to very small values of the cross section. However, there is a lot of cancellation in the average over all molecular orientations (fig. 5). Despite the presence of large Z nuclei the results have the same order of magnitude as for H_2S_2.

The dependence of $P'_{\|}$ and A on Z is therefore not straightforward but is a complicated one and depends on the molecular orientation. In order to unravel these relations a more systematic study of the stereodynamics of the collision is required. First attempts will be discussed in the following section.

COLLISIONS WITH PARTIALLY ORIENTED MOLECULAR ENSEMBLES

Description of Anisotropic Ensembles

The numerical results, presented in the preceding subsection, show that chiral effects are considerably larger for fully oriented molecules than for randomly oriented ones. The problem is with the experimental verification. Scattering from fully oriented molecules in the gasphase can not be directly observed. Experimentally feasible are collisions with molecular ensembles which have been partially oriented by optical pumping with linearly polarised laser light, by external static fields, or some other method which has been developed (see e.g. Scoles[13] for a review). Here we deal with molecular ensembles with an anisotropic axis distribution, but different molecules can point in different direction.

The first step is therefore to orient or align the molecules, the second step is to scatter projectiles from the prepared ensembles, using various experimental geometries. This method has been widely applied to study chemical reactions (see for example the recent review by Loesch[14]). Böwering and his group[15] have studied elastic electron collisions from anisotropic molecular ensembles at high energies and measured differential cross sections for unpolarised electrons.

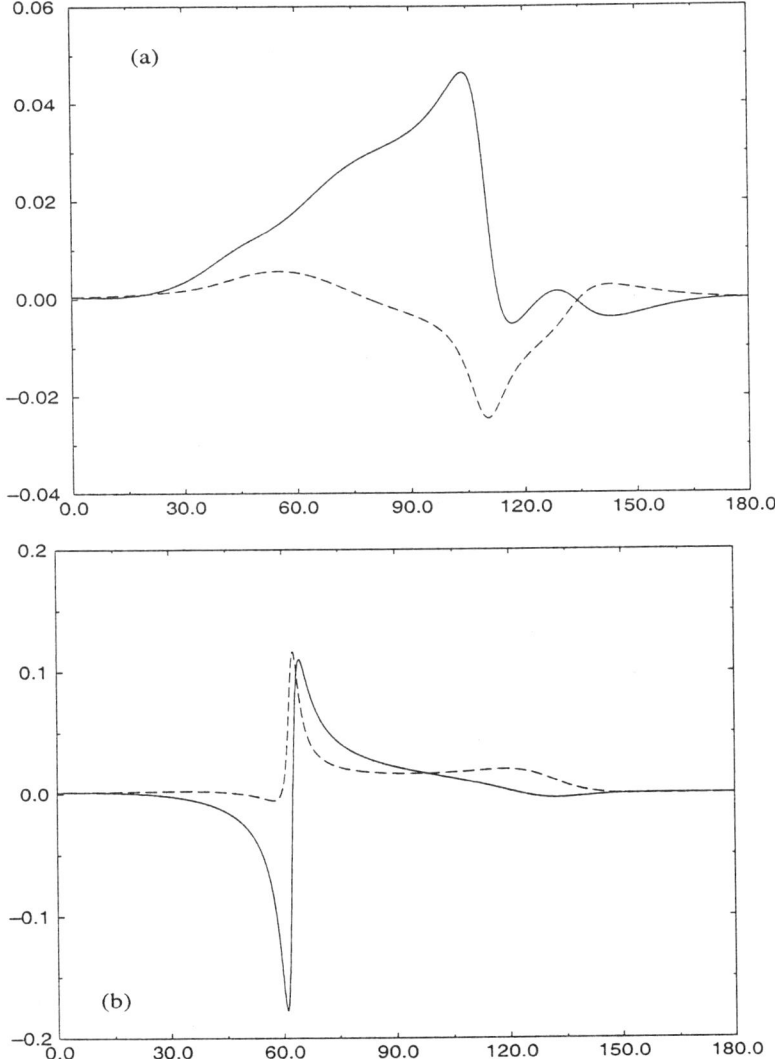

Figure 4. longitudinal polarisation (——) and asymmetry (– –) against scattering angle θ for $CHBrClF$. Orientation : $\alpha = \beta = 90°, \gamma = 0°$ a.) 10eV, b.) 5eV.

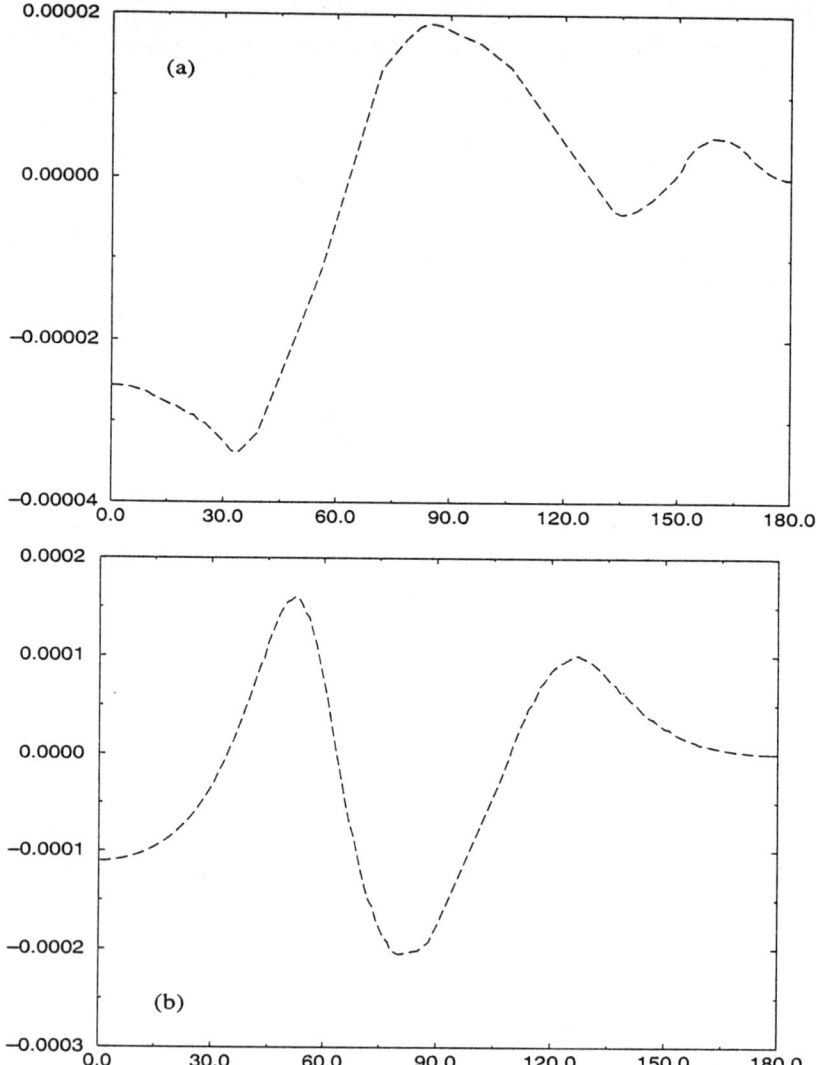

Figure 5. longitudinal polarisation for randomly oriented $CHBrClF$. a.) 10eV, b.) 5eV.

The general theory for elastic electron scattering from partially oriented molecules has been developed by Ostrawsky et al.[16] for diatomic molecules, and generalised to polyatomic molecules by Busalla and Blum[17,18]. Numerical results have been presented for N_2, CO and polyatomic molecules (in particular for cyclopropane C_3H_6). These results have been used to study the spatial aspects of electron-molecule collisions.

Recently this theory has been generalised to spin-dependent elastic collisions with polarised electrons in order to study in particular chiral effects. Here we will discuss the theory for <u>diatomic</u> molecules only. Collisions with partially oriented chiral molecules will be presented in a forthcoming publication[19].

Consider a molecular ensemble prepared with an anisotropic axis distribution by one of the mentioned methods. This preparation process will be described in a coordinate system called "director system". For example, for preparation by laser excitation or external fields, the direction of the relevant electric field vector will be taken as z'-axis of the director system. x' and y' can be choosen arbitrarily perpendicular to z, and perpendicular to each other. The axis distribution is than axially symmetric with respect to the "director" z'.

In general, the axis distribution of diatomic molecules can be characterised in terms of a distribution function $W(\alpha\beta)$ defined in such a way that $W(\alpha\beta)\sin\beta d\beta d\alpha$ is the probability of finding a molecule with axis $\vec{n}$ at a specific orientation fixed by polar angle β and azimuth angle α.

Let us concentrate on axially symmetric ensembles with z as symmetry axis. The distribution function is then independent of α and can be expanded in terms of Legendre polynomials $P_K(\cos\beta)$:

$$W(\beta) = \frac{1}{4\pi}\sum_K (2K+1) <P_K(\cos\beta)> P_K(\cos\beta). \tag{3}$$

Note that no integration over α has been performed in eq. (3). The coefficients $<P_K>$ are defined by the expression

$$<P_K(\cos\beta)> = \int_0^\pi d\beta \sin\beta W(\beta) \tag{4}$$

and are often called "order parameters". The more general case of polyatomic molecules will not be discussed here (see ref.[18] for details).

The upper limit of K in eq. (3) depends on the orientation process. In case of optical pumping by linearly polarised light it has been shown that only terms with $K = 0$ and $K = 2$ contribute because of dipole selection rules. If molecules in the gasphase are selected in states with sharp angular momentum J then $K_{max} = 2J$. Of experimental interest is the case of state-selected molecules with $J = 1$ and z-component $M = 0$ (see for example Loesch[14]). Here, the axis distribution is given by the absolute square of the corresponding spherical harmonic:

$$W(\beta) = \frac{1}{4\pi} \mid Y_{10}(\beta) \mid^2 = \frac{3}{4\pi} \cos^2 \beta$$

which can be written in a form similar to eq. (3):

$$W(\beta) = \frac{1}{4\pi}(<P_0> + 5 <P_2> P_2(\cos \beta)) \qquad (5)$$

with

$$<P_0> = 1, \quad <P_2> = \frac{2}{5}. \qquad (6)$$

We will return to this case below.

Collision With Anisotropic Molecular Ensembles. Steric Factors

Let us now consider elastic electron scattering from an anisotropic molecular ensemble. It will be assumed that the molecules are spinless and diatomic and in their electronic ground state. The general theory has been discussed in detail in ref.[20] (chapter 7) and ref.[17]. Here we will concentrate on spin-dependent collisions and will only give the essential results.

The differential cross section for a transition $\vec{k}_0\, m_0 \to \vec{k}_1\, m_1$ for molecules with axes pointing in a fixed direction β', α' will be denoted by $\sigma(\theta m_1 m_0 \alpha' \beta')$ where β' is the polar angle and α' the azimuth of the molecular axis in the collision frame. θ is the scattering angle.

Let δ be the polar angle of the "director" z' in the collision system and ε its azimuth angle. For example, if the anisotropic molecular sample has been prepared by optical pumping then z' would be the direction of the electric field of the laser light. The differential cross section for unpolarised initial electrons and for given δ and ε is given by the expression[16,20]

$$\sigma(\theta \varepsilon \delta) = \sum_{Kq} <P_K(\cos \beta)> Y^*_{Kq}(\delta \varepsilon) I_{Kq}(\theta) \qquad (7)$$

$$= \sum_{Kq} \sqrt{\frac{2K+1}{4\pi}} <P_K(\cos \beta)> d^{(K)}_{q0}(\delta) e^{i\varepsilon q} I_{Kq}(\theta) \qquad (8)$$

where in eq. (8) the spherical harmonics have been expressed in terms of the more convenient "small" Wigner functions $d^{(K)}_{q0}(\delta)$ (using the conventions of Zare[21]). The so-called <u>steric factors</u> I_{Kq} are defined by the expression

$$I_{Kq}(\theta m_1 m_0) = \int_0^{2\pi} d\alpha' \int_0^{\pi} d\beta' \sin \beta' W(\beta') \sigma(\theta \alpha' \beta', m_1 m_0) \qquad (9)$$

for spin-dependent cases. For unpolarised particles we define

$$I_{Kq}(\theta) = \frac{1}{2} \sum_{m_1 m_0} I_{Kq}(\theta m_1 m_0) \tag{10}$$

and these factors occur in eq. (8).

Eqs. (7) and (8) give a complete factorisation of the cross section. The order parameters $<P_K(\cos\beta)>$ describe orientation and alignment of the initial molecules in the director system x', y', z'. This is most convenient since all symmetry properties of the preparation process can easily be taken into account. The steric factors are defined in the collision system which is most convenient for numerical calculations. The steric factors contain all information on the dynamics of the collision. Finally, the spherical harmonics describe the geometry of the experiment.

For the in-plane polarisation P'_z for initially unpolarised electrons one obtains[19]

$$\sigma(\theta\delta\varepsilon)P'_z = 2 \sum_{K,q>0} \sqrt{\frac{2K+1}{4\pi}} <P_K(\cos\beta)> d^{(K)}_{q0}(\delta) \sin\varepsilon q' \tag{11}$$

$$\times \left[\Im I_{Kq}(\theta + \frac{1}{2} + \frac{1}{2}) - \Im I_{Kq}(\theta - \frac{1}{2} + \frac{1}{2}) \right]$$

where $\Im I_{Kq}(\theta m_1 m_0)$ denotes the imaginary part of the steric factors. We list some symmetry properties:

$$\begin{aligned}
I_{Kq}(\theta m_1 m_0) &= (-1)^q I^*_{K-q}(\theta m_1 m_0), & (12a) \\
I_{Kq}(\theta = 0, m_1 m_0) &= I_{Kq}(\theta = \pi, m_1 m_0) = 0 \quad \text{for} \quad q \neq 0, & (12b) \\
I_{Kq}(\theta m_1 m_0) &= I^*_{Kq}(\theta - m_1 - m_0). & (12c)
\end{aligned}$$

It follows that the parameters (eq. (10)) are real, and that all parameters with $q = 0$ are real. It should be noted that only terms with $q > 0$ contribute to eq. (11).

The expression for the anisotropy A (eq. (2)) is obtained by reversing the sign of the last term in eq. (11). We have $P'_z = A$ if the spin-flip terms $m_1 \neq m_0$ would vanish, and $P'_z = -A$ if the non-flip terms would be zero.

Results For Fully Oriented Molecules

It is important to note that the definition (9) of the steric factors can be inverted by multiplying both sides with a suitably choosen spherical harmonic, summing over K and q, and applying the orthogonality relation of the spherical factors. For initially unpolarised particles one obtains for the cross section:

$$\sigma(\theta \alpha' \beta') = \sum_K \sqrt{\frac{2K+1}{4\pi}} P_K(\cos\beta') I_{K0}(\theta) \tag{13}$$

$$+ 2 \sum_{K,q>0} \sqrt{\frac{2K+1}{4\pi}} d^K_{q0}(\beta') \cos\alpha' q \, I_{Kq}(\theta)$$

and for P'_z:

$$\sigma(\theta\alpha'\beta')P'_z = 2\sum_{K,q>0}\sqrt{\frac{2K+1}{4\pi}}d^K_{q0}(\beta')\sin\alpha'q \qquad (14)$$
$$\times\left[\Im I_{Kq}(\theta+\frac{1}{2}+\frac{1}{2}) - \Im I_{Kq}(\theta-\frac{1}{2}+\frac{1}{2})\right].$$

Eqs. (13) and (14) describe all spatial aspects of the collision process. For example, we can recover the general results given in section 2. It follows from eq. (14) that P'_z vanishes if the molecular axis lies in the scattering plane ($\alpha' = 0$) since all sinus functions are zero. For $\theta = 0$ and $\theta = \pi$ the steric factors, occurring in eq. (14), vanish because of condition (12b). Hence, the pseudoscalar $[\vec{k}_1 \times \vec{k}_0]\cdot\vec{n}$ must be different from zero.

For homonuclear molecules it can be shown that all steric factors with K odd vanish. If the molecular axis is then perpendicular to the scattering plane ($\alpha' = \beta' = \frac{\pi}{2}$) then P'_z vanishes since $\sin q\frac{\pi}{2} = 0$ for q even, and $d^{(K)}_{q0}(\frac{\pi}{2}) = 0$ for $K + q$ odd.

It should be noted that results for the Longitudinal Polarisation $P'_\parallel$ can be obtained by substituting $I_{Kq}(\theta\lambda_1 m_0)$ for $I_{Kq}(\theta m_1 m_0)$ in eqs. (11) and (14) where λ_1 is the helicity of the scattered electrons.

It follows from eqs. (13) and (14) that the steric factors contain all information on the stereodynamics of the collision. The experimental or numerical determination of the full set of all relevant steric factors should therefore be the main goal. In principle, the steric factors can be obtained by using eqs. (9) or (11), and by measuring $\sigma(\theta\delta\varepsilon)$ for a sufficient number of angles δ and ε. So far, experiments have been performed for simple chemical reactions with molecules prepared in states with $J = 1$, $M = 0$ (eqs. (5),(6)), and the steric factors $I_{2q}(\theta)$ have been determined (14). In this case the order parameters are known (eq.(6)), but only steric factors with $K = 2$ can be obtained experimentally, and this gives only limited information on the stereodynamics.

In order to guide experimentalists the full set of steric factors should be calculated numerically for cases of interest.

A first systematic study has been performed recently for elastic collisions with CO and C_3H_6 (Busalla et al.[18]). The main result is that terms up to $K \approx 4$ contain the dominant contributions, the steric factors decrease with increasing K and increasing q. The results have then been used to unravel the stereodynamics of the collision.

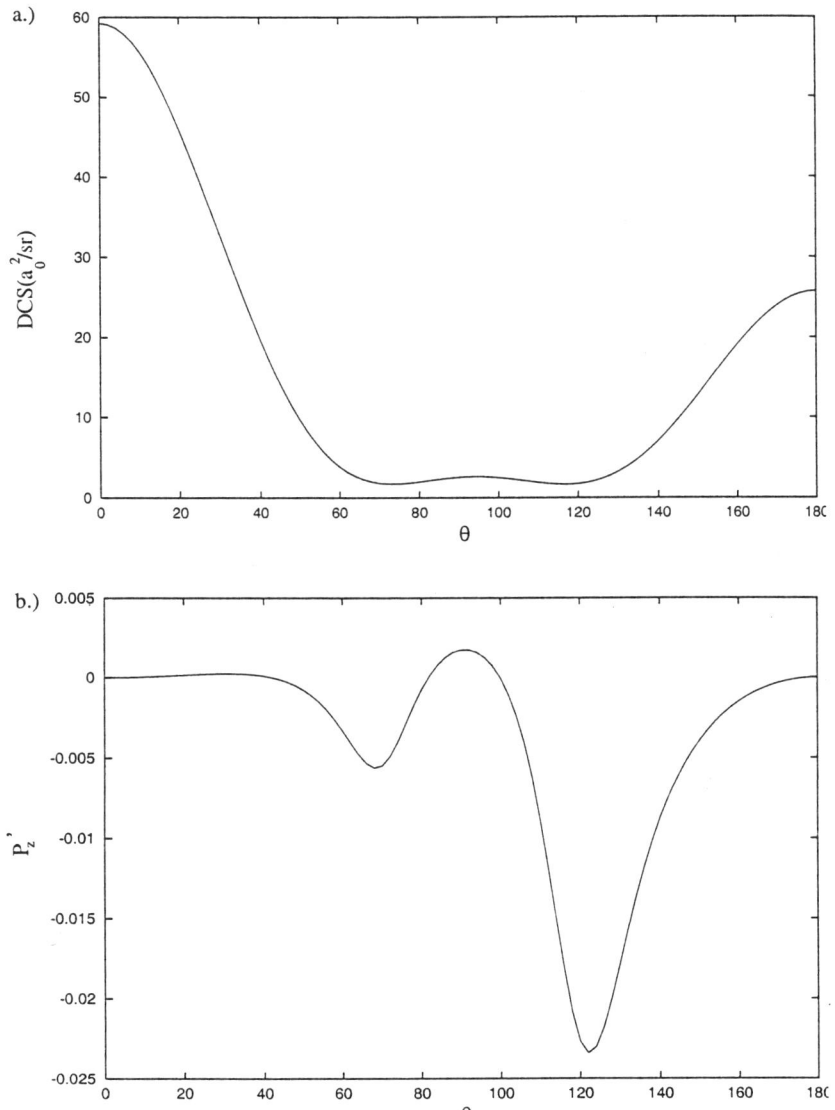

Figure 6. Differential cross section (fig.a) and in-plane polarisation P'_z (fig.b) against scattering angle θ for $E = 10eV, \delta = 90°, \varepsilon = 45°$

Special Cases and Numerical Results

Applying the theoretical model and computer code, described in section 3, we have calculated the spin-dependent steric factors for collisions with HBr for $5eV$ and $10eV$ up to $K = 10$. Again we have found that these parameters decrease with increasing K and q. In addition, the steric factors for spin-flip are 2-3 orders of the magnitude smaller than the non-flip factors. As a consequence, there is only a small difference between P'_z and asymmetry A.

From an experimental point of view the steric factors with $K = 2$ are of special interest. In order to get some idea about the magnitude of the effects we have calculated cross section and spin polarisation for elastic collisions with HBr, assuming that the molecules have been prepared in states with $J = 1$, $M = 0$ by optical pumping with linearly polarised laser light. The relevant order parameters are given by eq. (6), and eqs. (13) and (14) reduce to the expressions:

$$\sigma(\theta\varepsilon\delta) = \frac{1}{\sqrt{4\pi}} I_{00} + \sqrt{5} <P_2> \left[I_{20}(\theta)\frac{1}{2}(3\cos^2\delta - 1) \right. \quad (15)$$

$$\left. - I_{21}(\theta)\sqrt{6}\sin\delta\cos\delta\cos\varepsilon + I_{22}(\theta)\sqrt{\frac{3}{2}}\sin^2\delta\cos 2\varepsilon \right]$$

and

$$\sigma(\theta\varepsilon\delta)P'_z = 2\sqrt{\frac{5}{4\pi}} <P_2> \quad (16)$$

$$\times \left[-\sqrt{6}\sin\delta\cos\delta\sin\varepsilon \left(\Im I_{21}(\theta + \frac{1}{2} + \frac{1}{2}) - \Im I_{21}(\theta - \frac{1}{2} + \frac{1}{2}) \right) \right.$$

$$\left. + \sqrt{\frac{3}{2}}\sin^2\delta\sin 2\varepsilon \left(\Im I_{22}(\theta + \frac{1}{2} + \frac{1}{2}) - \Im I_{22}(\theta - \frac{1}{2} + \frac{1}{2}) \right) \right]$$

where explicit expressions for the d-functions have been inserted.

Let us assume that the director axis (eg. direction of the electric field of the laser) is perpendicular to the scattering plane ($\delta = \frac{\pi}{2}$). In this case the second term in eq. (15), and the first term in eq. (16) vanish, and only the imaginary parts of the steric factors with $K = 2$, $q = 2$ contribute to P'_z, which becomes maximal for $\varepsilon = \frac{\pi}{4}$. The director axis lies then in the xy-plane of the collision system with azimuth angle $\frac{\pi}{4}$ to the x-axis.

We have calculated P'_z for $\delta = \frac{\pi}{2}$, $\varepsilon = \frac{\pi}{4}$ for $E = 10eV$ as a function of the scattering angle θ. The results are shown in fig 6. The results for P'_z are considerable smaller than in fig. 1b due to the averaging over the axis distribution.

REFERENCES

1. P.S. Farago: J.Phys. B13, L567 (1980)
2. P.S. Farago: J.Phys. B14, L743 (1981)
3. R.A. Hegstrom: Nature 297, 643 (1982)

4. L.D. Barron: "Molecular Light Scattering", Cambridge University Press, Cambridge (1982)
5. K. Blum, D. Thompson: J.Phys. B22, 1823 (1989)
6. S. Mayer, J. Kessler: Phys. Rev. Lett. 74, 4803 (1995)
7. S. Mayer, C. Nolting, J. Kessler: J.Phys. B29, 349 (1996)
8. C. Nolting, S. Mayer, J. Kessler: J.Phys. B30, 5491 (1997)
9. C. Johnston, K. Blum, D. Thompson: J.Phys. B26, 965 (1993)
10. K. Blum, D. Thompson: "Advances in At. Mol. Opt. Physics" (eds.:B. Bederson, H. Walther), Academic Press, New York (1997)
11. D. Thompson, M. Kinnin: J.Phys. B28, 247 (1995)
12. I. Smith, D. Thompson, K. Blum: J.Phys. B, in press
13. G. Scoles: "Atomic and Molecular Beam Methods", Oxford Univ. Press, Oxford (1988)
14. H.J. Loesch: Ann. Rev. Phys. Chem. 46, 255 (1995)
15. M. Volkmer, Ch. Meier, M. Fink, N. Böwering: Phys. Rev. A56, 1690 (1997)
16. C. Ostrawsky, K. Blum, C. Gillan: J.Phys. B28, 2269 (1995)
17. A. Busalla, K. Blum: "Novel Aspects of Electron-Molecule Scattering" (ed.:K. Bedler), World Scientific Publ. (1997)
18. A. Busalla, K. Blum, T. Beyer, B. Nestmann: J. Phys. B, to be published
19. M. Musigmann, K. Blum, D. Thompson: to be published
20. K. Blum: "Density Matrix Theory and Applications", 2.ed. Plenum Press, New York (1996)
21. R.N. Zare: "Angular Momentum", A Wiley-Interscience Publication, (1988)

POSITRON SCATTERING BY ATOMS

B.H. Bransden

Department of Physics
University of Durham
South Road
Durham, DH1 3LE
UK

1. INTRODUCTION

It is a great pleasure to contribute to this meeting in honour of Phil Burke. We first met some forty-four years ago when we used to attend Professor Sir Harrie Massey's research group meetings on the theory of few nucleon collisions. At the time Phil was a research student at U.C.L and I was working in the Physics Department at Belfast and we saw each other about once a term. In those forty-four years Phil has made enormous contributions to collision theory, and particularly to the theory of atomic and nuclear collisions, but in paying tribute to his work I should like to acknowledge the key role Phil has played in the development in the UK of computational science through his persuasive and skillful advocacy in the Research Councils which I often had occasion to see first hand.

Today I shall talk about some aspects of positron collision with atoms mainly at energies above the positronium (Ps) formation threshold, not attempting anything like a comprehensive review but picking out some particular topics of interest.

Taking positron scattering by atomic hydrogen as the prototype of a positron collision, the following processes are possible:

$e^+ + H(1s) \rightarrow e^+ + H(1s)$ elastic scattering

$\rightarrow Ps(nl) + H^+$ positronium formation
  lowest threshold at 6.8 eV
$\rightarrow e^+ + H(nl)$ excitation
  lowest threshold at 10.2 eV
$\rightarrow e^+ + e^- + H^+$ ionization
  threshold at 13.65 eV
$\rightarrow H^+ + \gamma$ radiation

The coupling to the annihilation channels is weak so that these can be ignored in calculating collision cross sections. However the annihilation rate is proportional to the probability of finding the positron and electron at the same position and its measurement provides a stern test for any theory.

The first theoretical investigations of the elastic scattering of positrons by atoms were started by Ore[1] in 1949 and continued by Massey and Moussa[2] in 1958, and at about the time Phil Burke was working on n-d scattering, the first calculations of positronium formation in e^+ - H and e^+ - He collisions were carried out by Massey and Mohr[3] using the first Born approximation. For atomic hydrogen the formation cross section was found to rise steeply from threshold to a maximum of about $4\pi a_o^2$ at 14 eV, thereafter decreasing steadily to about πa_o^2 at 32 eV. At that time no cross section measurements existed, but the development of monoenergetic beams of positrons [4, 5, 6, 7] in the 1970's has been a key stimulus in the development of the theory of the various positron collision process. In addition to the intrinsic interest of positrons which were the first anti-particles discovered, the theory has received additional stimulus from applications in astrophysics,[7] and also in surface and condensed matter physics[8] where the positron can be used as an important probe.

A schematic diagram showing the behaviour of cross sections for the e^+ - H(1s) system is shown in Fig 1.

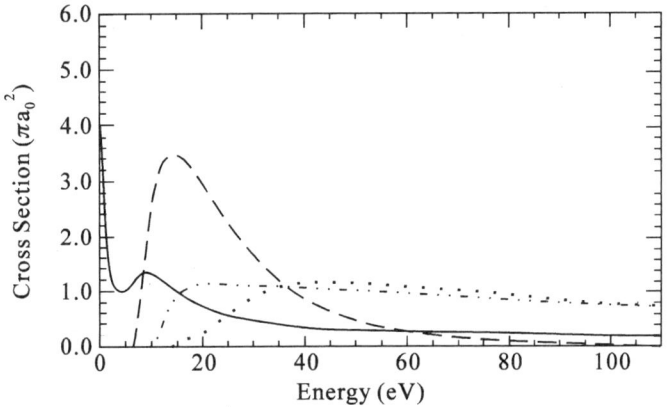

Figure 1. Schematic diagram of cross sections for e^+-H(1s) scattering; elastic scattering solid line, Ps formation dashed curve, excitation dot-dashed curve, ionization dotted curve.

In accounting for the characteristics of the cross sections shown in Fig. 1, several qualitative points must be kept in mind:

(1) There is no exchange interaction between the positron and the target electrons.
(2) The static interaction is repulsive, in contrast to the case of electrons. The dipole polarization potential remains attractive and this results in a kind of Ramsauer minimum in the elastic scattering at low energies.
(3) The odd and even Born amplitudes are of opposite sign with the result that in the total cross section for e^+ scattering is less than that for e^- scattering, up to energies at which the

first Born approximation is accurate, when the e$^+$ and e$^-$ total cross sections coalesce, which is shown in Fig 2 for e$^{\pm}$ - He scattering.

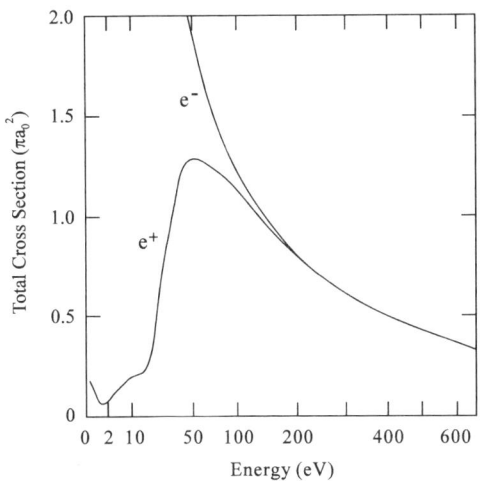

Figure 2. The total cross sections for e$^+$-He and e$^-$-He scattering.

2. LOW ENERGY SCATTERING AND THE VARIATIONAL METHOD

The first accurate calculations for positron scattering were made for elastic scattering by H(1s) below the Ps formation threshold using the Kohn and inverse Kohn variational methods with a Hylleraas trial fuctions,[9, 10, 11, 12] later extended to the calculation of the K-matrix in the region between the Ps formation and excitation thresholds by Humberston[13] and Brown and Humberston.[14] Similar detailed and accurate calculations have been carried out for He,[15, 16] Li[17] and H$_2$[18] targets.

3. TWO-CENTRE COUPLED CHANNEL MODELS

As the energy is raised above the excitation threshold the number of open channels increases so that very detailed variational calculations become impracticable and the most extensive calculations have been based on two centre expansions of the wave function in terms of target and positronium eigenstates and pseudostates. For scattering by atomic hydrogen the boundary conditions for excitation can be expressed in terms of the e^+-p coordinate x and the e^--p coordinate r, and for the rearranged Ps-p system the e^+-e^- coordinate R and s the coordinate joining the centre of mass of the positronium to the proton (Fig 3).

The total wave function can then be expanded approximately as:

$$\Psi = \sum_{i=1}^{I} \phi_i(r) F_i(x) + \sum_{n=1}^{N} \psi_n(R) G_n(s) \tag{1}$$

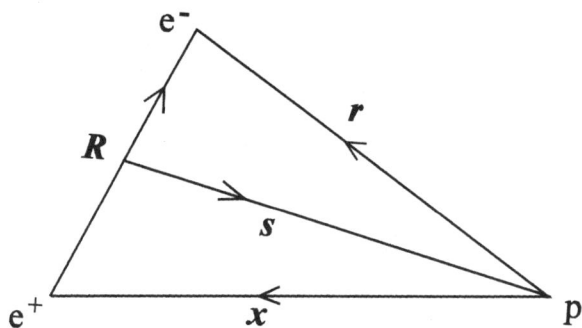

Figure 3.   Jacobi coordinates (x,r) or (s,R) for e^+-H scattering.

In the first group of terms the basis functions $\phi_i(r)$ are hydrogenic bound state functions together with pseudostates which represent the omitted bound states and the continuum. Similarly the basis functions $\psi_i(R)$ represent positronium states and pseudostates. Although pseudostates can be found in different ways, for example by the use of eigendifferentials, in the present context the method introduced by Phil Burke and his co-workers[19,20] has been widely employed in which the functions of ϕ_i and ψ_n diagonalize the target and positronium internal Hamiltonians, giving (in atomic units)

$$\langle \phi_i | -1/2\, \nabla_r^2 + V_{e^-p} | \phi_j \rangle = \delta_{ij} \varepsilon_i$$

$$\langle \psi_n | -\nabla_R^2 + V_{e^+e^-} | \psi_m \rangle = \delta_{nm} \eta_n \tag{2}$$

To carry out the diagonalizations the functions ϕ_i and ψ_n are expressed as combinations of some suitable set of L^2 functions. A popular choice has been Slater functions but these have the limitation that linear dependence problems make it difficult to use many more than 10-15 functions for each angular momentum. This problem can be overcome by choosing the L^2 functions to be members of an orthogonal set. Suitable choices are Sturmian functions[21] which are orthogonal with respect to $1/r$ or the orthonormal set introduced by Stelbovics.[22]

$$f_i(r) = [\lambda i!]^{\frac{1}{2}} [(2\ell + 2 + i)!]^1 (\lambda r)^{\ell+1} \exp(-\lambda r / 2) L_i^{2\ell+2}(\lambda r) \tag{3}$$

In this case the diagonalizations can be carried out using up to 100 or more functions. This set has an additional advantage. Stelbovics[22] following earlier work by Yamani and Reinhardt[23] showed that the hydrogenic continuum functions have an exact expansion in terms of this set and that if the expansion is truncated the resulting pseudostates generate an integration over the hydrogenic continuum of Gaussian type. Another possibility is to employ sets of Gaussian functions,[24] in which case the evaluation of matrix elements of potentials between basis functions on different centres can be carried out analytically. This may be very useful when more complicated reactions are studied, for example the interaction of positronium or positronium hydride with atoms.

By inserting the expansion (1) into the Schrödinger equation coupled integro-differential equations are obtained for the channel functions $F_i(x)$ and $G_j(s)$ which can be put in the form

$$(\tfrac{1}{2}\nabla_x^2 + \tfrac{1}{2}k_i^2)F_i(x) = \sum_{j=1}^{I}\langle\phi_i|V_{e^+p}(x) + V_{e^+e^-}(R)|\phi_j\rangle F_j(x)$$

$$+ \sum_{n=1}^{N}\langle\phi_i|H - E|\psi_n G_n(s)\rangle$$

$$i = 1,2,...I$$

$$(\tfrac{1}{4}\nabla_s^2 + \tfrac{1}{4}\bar{k}_n^2)G_n(s) = \sum_{m=1}^{N}\langle\psi_n|V_{e^+p}(x) + V_{e^+p}(r)|\psi_m\rangle G_m(s)$$

$$+ \sum_{i=1}^{I}\langle\psi_n|H - E|\phi_i F_i(x)\rangle$$

$$n = 1,2,..N \qquad (4)$$

with

$$E = \tfrac{1}{2}k_i^2 + \varepsilon_i = \tfrac{1}{4}\bar{k}_n^2 + \eta_n \qquad (5)$$

In the second set of equations representing the Ps-H$^+$ channels the direct local interactions vanish unless the states n and m are of different parity, and the static interaction between a charge and a positronium atom always vanishes. The couplings between the e$^+$-H and the Ps-H$^+$ channels are non-local and non-separable. The non-local kernels are of importance for suprisingly large vaues of x or s, extending up to 30-50 au. The equations are to be solved with the usual scattering boundary conditions.

After a partial wave decomposition sets of radial equations are found for each J, which can be written together as

$$\left(\frac{d^2}{dy^2} + \frac{\ell_p(\ell_p+1)}{y^2} + K_p^2\right)f_p(y)$$

$$= \sum_{p'=1}^{N+I}\left[V_{pp'}(y)f_{p'}(y) + \int_0^\infty dy' \, K_{pp'}(y,y')f_{p'}(y')\right] \qquad (6)$$

It is of interest to consider some of the numerical methods which have been employed to solve these equations.

(a) In their paper[25] on n-d scattering Phil Burke and Harry Robertson were faced with a single channel equation of this kind. They represented the intergral over the kernel by an integration rule with up to thirty equally spaced points and expressed the second derivation by a central difference formula using the same pivitol points

$$f''(y_n) \approx \tfrac{1}{h^2}[f(y_{n-1}) - 2f(y_n) + f(y_{n+1})] \qquad (7)$$

thus obtaining a set of simultaneous linear equations. The equations were solved using a programme written by Harry Robertson[26] for the pilot ACE machine developed at the National Physical Laboratory. This must be the first example in the UK that a collision problem was solved by an electronic machine and must be the origin of Phil's life-long quest for increased computing capacity!

(b) Somewhat later in 1967 Jundi and I[27] were studying the two state system

$$e^+ + H(1s) \rightarrow Ps(1s) + H^+$$

and to avoid the inconvenience of using pivotal points at equal intervals, we converted the equations to integral equations using the free particle Green's function:-

$$f_p(y) = s_{\ell_p}(k_p y)\delta_{pi} + \sum_{p'} \int_0^\infty dy g_{\ell_p}(y,y')\left[V_{pp'}(y')f_{p'}(y') + \int_0^\infty dy'' K_{pp'}(y',y'')f_{p'}(y'')\right] \qquad (8)$$

On introducing numerical integration rules a system of linear equations is obtained which can be solved by standard methods.

(c) Another possible technique is to solve a sequence of coupled differential equations by iterating the non-local kernel terms. In 1971, Dirks and Hahn[28] showed that a straight forward iteration diverged, but they devised a scheme to converge the iteration sequence which enabled them to obtain accurate results for the two-state problem, however the method is certainly less robust than those based on integral equations.

(d) Rather than solving the integral equations in configuration space a different approach, introduced in this context by Basu et al[29] in 1976, and which has proved to be very efficient, is to construct the corresponding coupled equations in momentum space for the half-off shell K or T matrices. These have the general form

$$T_{pp'}(k,k_o) = V_{pp'}(k,k_o) - \pi^{-1}\sum_{p'}\int_0^\infty k'dk' \frac{V_{pp''}(k,k')T_{p''p'}(k',k_o)}{E + i\varepsilon - \tfrac{1}{2}k'^2 - \varepsilon_p} \qquad (9)$$

where $V_{pp'}(k,k')$ is an off-shell Born approximation matrix element. Once again the equations can be solved using standard linear algebraic methods after allowing for the singularity at $E = \tfrac{1}{2}k'^2 + \varepsilon_p$. This method appears to be very robust and has been widely exploited.

(e) Coming to the present decade the R-matrix method has been developed for e^+ scattering by Phil Burke and his collaborators[30,31], and this method has also been exploited with great success by James Walters and coworkers.

(f) Finally I should like to mention that Cliff Noble and I have been investigating the use of a least squares approach[32], originally suggested by Merts and Collins[33] for coupled differential equations. It is based on representing the radial functions $f_i(y)$ over a region 0 < y < y_{max} by an expansion in basis functions, such as shifted Legendre functions. The coefficients of the expansion satisfy linear equations found by requiring the radial functions to satisfy the coupled equations in a least squares sense.

Application of the Coupled Equation Method

The e^+ - H System. The very first coupled channel calculations for the e^+-H system were made by Burke, Schey and Smith[34] in a 1s-2s-2p close coupling approximation in order to calculate the elastic scattering cross section and the n = 2 excitation cross sections in the energy range up to 54 eV. The positronium formation channel was omitted on the grounds that at low energies < 6.8 eV Cody and Smith had shown that the inclusion of the 2p state providing 65% of the dipole polarizability was more important than virtual positronium formation. In fact Bray and Stelbovics[35], and Higgins et al [30], have shown that below the positronium formation threshold a single centre expansion with a large basis can reproduce the results of the accurate variational calculations. However the convergence of a two centre expansion is much more rapid[36]. Above threshold the Ps formation cross section becomes very large and the Ps channels must be represented in the wave function. When the Ps formation cross section becomes small at high energies a single centre expansion can be employed again. The early two-state approximations H(1s), Ps(1s), were extended in energy range to 8-200 eV by Basu and coworkers[29] in 1976, and later in 1990 Mukherjee et al[37] employed enhanced basis sets H(1s,2s,2p), Ps(1s) and $H(1s,2s,2\bar{p})$, Ps(1s) which suggested that indeed the elastic and excitation cross sections were influenced by the positronium channel. This was confirmed by Ghosh and Darewych[38] employing an optical potential method. These calculations and others using a six state basis[39] H(1s,2s,2p), Ps(1s,2s,2p) confirm that capture is mainly to the ground state of positronium and that the contribution of the $\ell = 0$ partial wave is very small, the major contributions coming from the partial waves with $\ell = 1$ to $\ell = 4$.

The major advance in the last few years has been the ability to carry out large basis set calculations either by the T-matrix[40] or R-matrix methods[41,42]. As an example in Figures 4 and 5, the Ps formation and ionization cross sections of Kernoghan et al[42] are shown using a (30,3) basis. The rotation (a,b) refers to the number of states and pseudostates on the hydrogen and the Ps centres respectively. Similar calculations by Mitroy[40] with a (28,3) basis are in good agreement with those shown. The calculated cross sections are fair agreement with the experimental measurements. Calculations with basis sets of different size show that if too few pseudostates are employed pseudoresonances appear below the thresholds of the pseudostate channels, for example in the (9,9) basis calculations of Kernoghan et al[41], however when a suitable average cross section is calculated agreement with the larger basis calculations is good. An important result is that below threshold calculations[40] with large two-centre basis sets agree well with the accurate variational phase shifts. This agreement is maintained above threshold up the n = 2 excitation threshold of the target where the cross sections can be compared with the variational results of Humberston[13] and Brown and Humberston[14].

Other targets

Coupled channel calculations allowing for Ps formation have been carried out using the R-matrix method for the alkali metals using a one-electron approximation. This work has been reviewed recently by Walters et al[46]. For alkali targets positronium formation is exothermic and the cross section is large up to energies of the order ~ 30eV after which the positronium channel has little influence on the cross sections for elastic scattering, excitation and ionization. In the case of Li excellent agreement can be obtained with the accurate variational calculations of Watts and Humberston[17] at low energies, provided pseudostates are included in the expansion to represent the continuum.

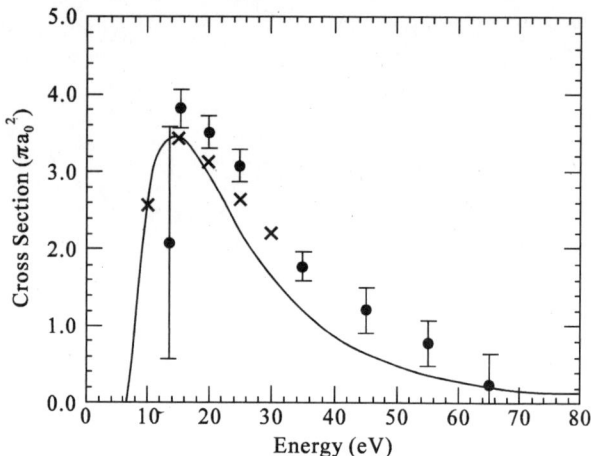

Figure 4. The total Ps formation cross section for e^+-H(1s) collisions. Calculated cross sections: —— two centre coupled channel results with (30,3) basis,[42] xx calculation with hyperspherical basis.[43] Experimental cross section ⊕ from Weber et al.[44]

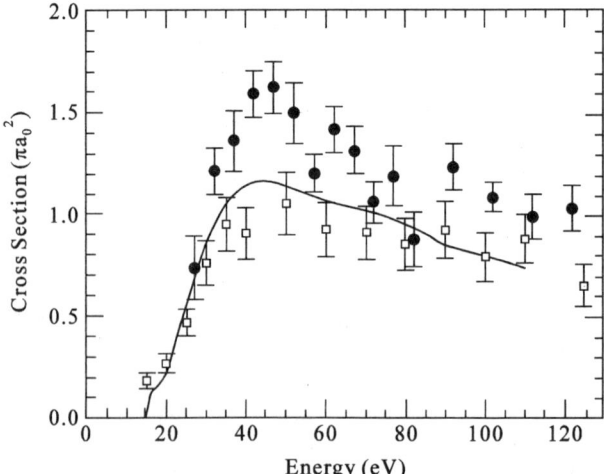

Figure 5. The cross section for ionization of H(1s) by e^+ impact. Calculated cross sections: —— two centre coupled channel results with (30,3) basis.[42] Experimental cross sections ⊕ from Jones et al[45], ⊕ from Weber et al.[44]

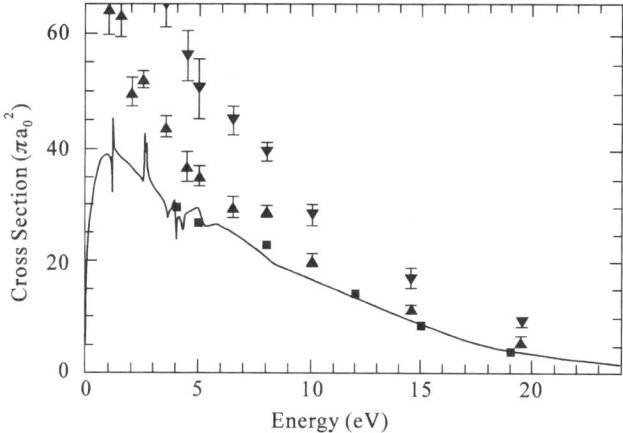

Figure 6. The Ps formation cross section for e^+-Na collisions. A (5,6) close coupling calculation[47] ———; upper and lower limits of measured cross section[50] ▼, ▲.

Using the T-matrix approach, extensive calculations of e^+-Ne scattering have been made by Ryzhikh and Mitroy[47] who employed Hartree-Fock frozen core wave functions for the target. A (5,6) close coupling basis was employed consisting of Na(3s,3p,4s,3d,4p) and Ps(1s,2s,2p,3s,3p,3d) states with no pseudostates. The calculated cross sections agree closely with those obtained with a similar model by McAlinden et al[48]. Calculations for lithium by McAlinden et al. seem to show that for the alkalis the continuum has a comparitively small effect. In Figs 6 and 7 the cross sections of Ryzhihh and Mitroy for Ps formation and the total cross section are compared with the experimental data of the Detroit group[50, 51, 52]. The agreement is satisfactory for the total e^+ scattering cross sections but less so for the Ps formation cross section. Confidence in the theory is enhanced by the corresponding results for K[53] for which agreement with experiment is obtained both for the formation and total cross sections.

Coupled channel calculations have also been carried out for e^+-He scattering, a more difficult system because of the exchange interaction between the electrons on different centres in the rearranged system Ps + He^+. Earlier work was based on a one-electron approximation[54] but this has been avoided in the recent calculation of Adhikari and Ghosh[55] which uses a basis of He(1sn$l$), Ps(1s), Ps(2s) Ps(2p) state with nl = 1s,2s,2p,3ps,3p. They obtained reasonable agreement with the Ps formation cross section measured by Overton et al[56], and very good agreement with the total cross section data[57, 58, 59] see Figs. 8 and 9.

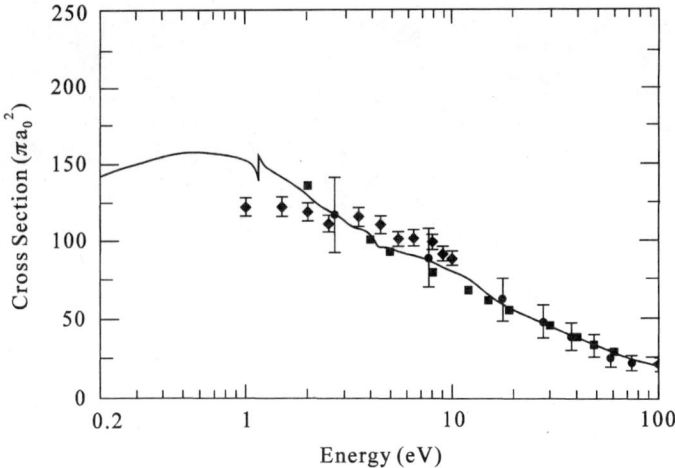

Figure 7. The total cross section for e^+-Na scattering. A (5,6) close coupling calculation[47] ——; experimental cross sections ● from Kwan et al[51], ◆ from Kauppila et al.[52]

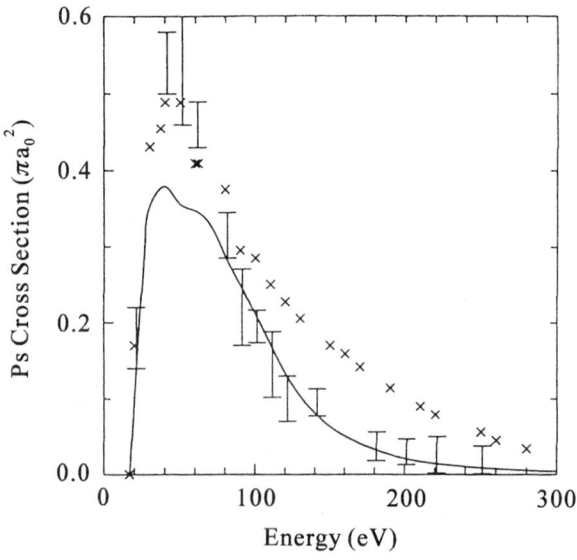

Figure 8. The cross section for Ps formation in e^+-He collisions. Theoretical cross section[55] ——; Experimental cross section xxx Fromme et al.[59], I Overton et al.[53] (Reproduced with permission of Elsevier Science from reference [55].)

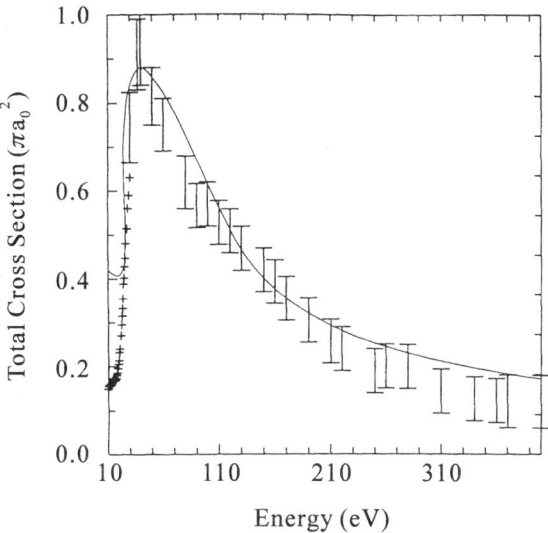

Figure 9. The total cross section for e$^+$-He scattering. Theoretical cross section[55] —— ; Experimental cross section +++ from Stein et al[57], I I from Griffith and Heyland[58] and from Fromme et al.[59] (Reproduced with permission of Elsevier Science from reference [55].)

Back in 1954 Massey and Mohr started the study of positronium interactions with atoms, pointing out that the main mechanism at low energies was electron exchange. Subsequently calculations were carried out in the static exchange approximation at low energies for scattering by hydrogen and helium and estimates of the influence of the van der Waals interaction were made[60,61,62]. It is now possible to carry out calculations with more realistic basis sets. Detailed coupled channel calculations for H-Ps have been published.[63]

4. COUPLED CHANNEL CALCULATIONS WITH A HYPERSPHERICAL BASIS

The expansion of the wave function for an e$^+$-atom system in terms of states or pseudostates on the atomic and Ps centres is not the only possibility. As in heavy particle scattering "molecular orbital" rather than atomic orbital basis sets can be envisaged. The analogous expansion to the heavy particle molecular orbital expansion can be formulated with the use of hyperspherical coordinates which were introduced by Archer et al[64] for e$^+$-H scattering and more recently been employed by Zhou and Lin[65] and in a slightly different form by Igarashi and Toshima[43].

In terms of the Jacobi coordinates x and r, or s and R (see Fig 3), a hyperradius can be defined as

$$y^2 = x^2 + r^2 = 2s^2 + R^2/2 \tag{10}$$

Corresponding hyperangles ϕ_H or ϕ_{Ps} are then defined by

$$\tan\phi_H = \frac{r}{x} \text{ and } \tan\phi_{Ps} = \frac{R}{2s} \tag{11}$$

for the arrangements $e^+ + H$ and $Ps + H^+$ respectively. The total wave function satisfies the equation

$$\left[-\frac{1}{2}(\frac{d^2}{dy^2} + \frac{5}{y}\frac{d}{dy}) + H_{ad} - E\right]\Psi = 0 \tag{12}$$

where

$$H_{ad} = \frac{\Lambda^2}{2y^2} + V_{e^+p} + V_{e^-p} + V_{e^+e^-} \tag{13}$$

and

$$\Lambda^2 = -\frac{1}{\sin^2\phi\,\cos^2\phi}(\frac{d}{d\phi}\sin^2\phi\cos^2\phi\frac{d}{d\phi}) + \frac{\mathbf{L}'^2}{\cos^2\phi} + \frac{\mathbf{L}^2}{\sin^2\phi} \tag{14}$$

In (14), for the arrangement $e^+ + H$, $\phi = \phi_H$ and $\mathbf{L}'$ and $\mathbf{L}$ are the angular momentum operators for r and x, while for the arrangement $Ps + H^+$, $\phi = \phi_{Ps}$ and $\mathbf{L}'$ and $\mathbf{L}$ are the angular momentum operators for R and s. Igarashi and Toshima expand Ψ in terms of the eigenfunctions $\psi_i(y,\Omega)$ of $H_{ad}(y)$, in which y is a fixed parameter. The basis functions are taken to be a mixture of hyperspherical harmonics and Slater orbitals.

Coupled equations are found using the expansion

$$\Psi = \sum_i y^{\frac{5}{2}} F_i(y)\psi_i(y,\Omega) \tag{15}$$

The equations obtained are coupled differential equations rather than integro-differential equations. In the asymptotic region the solution is projected on the representation in Jacobi coordinates to extract the K or T-matrices.

Between the Ps threshold and the n = 2 excitation threshold of atomic hydrogen (the Ore gap) the elastic scattering and Ps formation cross sections, in $e^+ + H(1s)$ collisions, computed by Zhou and Lin[65] agree almost exactly with the variationally determined values of Brown and Humberston[14]. The values determined by Igarashi and Toshima[43] are also close to the variational values. Between the excitation threshold and 30 eV, the Ps formation cross section has also been computed by Igarashi and Toshima. Their results which are included in Fig 4 agree well with the coupled channel calculations and with the experimental data.

5. POSITRONIUM FORMATION AT HIGH ENERGIES

At energies for which Ps formation is improbable it is reasonable to look for pertubation models. The first order Born approximation for the Ps formation amplitude in hydrogen is

$$T_{fi} = \langle\Phi_f|V_f|\Phi_i\rangle = \langle\Phi_f|V_i|\Phi_i\rangle \tag{16}$$

where

$$V_i = V_{e^+p}(x) + V_{e^+e^-}(R), \quad V_f = V_{e^+p}(x) + V_{e^-p}(v) \tag{17}$$

and Φ_f and Φ_i are the unperturbed wave functions in the initial and final states. As in heavy particle charge exchange, in the limit of high energies the non-relativistic scattering amplitude is not given by the first order terms (16) but by the second order terms. Of the four second order terms containing $V_{e^+p}G_0V_{e^+p}$, $V_{e^-p}G_0V_{e^+p}$, $V_{e^+p}G_0V_{e^+e^-}$ and $V_{e^-p}G_0V_{e^+e^-}$ the last two are the most important and correspond to a classical double scattering (Thomas scattering). Shakeshaft and Wadhera[66] showed that if the final Ps state is of even parity these two terms cancel at high energies and the cross section decreases like E^{-6} characteristic of the first order Born and distorted wave approximations. On the other hand if the final Ps state is odd the two terms are of the same sign and dominate the cross sections which behaves as $E^{-5.5}$. As a consequence capture into the 2p state is predicted to be the dominant process as $E \to \infty$. However this behaviour only sets in at very large energies and even up to 8 keV for which explicit (but approximate) second Born approximation calculations have been made by Basu and Ghosh[67] capture into the n=2 level is small compared with ground state capture. This result was confirmed by a complete calculation of the second Born approximation by Igarashi and Toshima.[68] Their results show that both the first and second Born approximations to the Ps formation cross section agree quite well with the measured cross section above 20 eV, see Fig. 10. This does not imply that the first Born approximation gives an accurate representation of the scattering

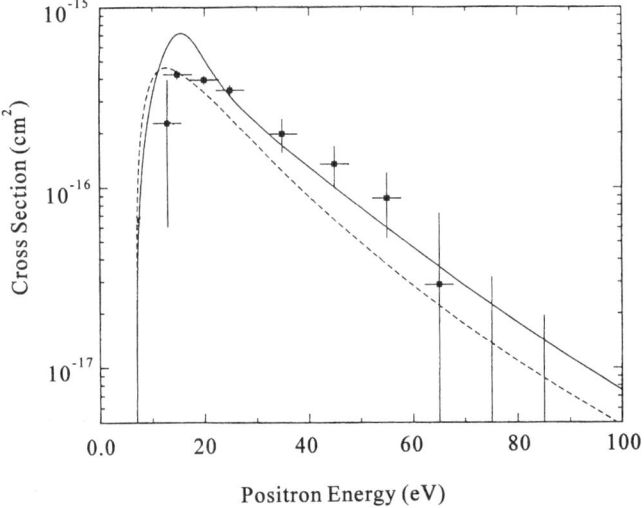

Figure 10. The positronium formation cross section in e⁺-H(1s) collisions in the first and second Born approximations; —— Second Born approximation[68], - - - First Born approximation. The theoretical cross sections are the sum of capture into the 1s,2s and 2p states. The experimental data is from Sperber et al.[69]

amplitude because the angular distributions away from the forward peak are very different. The situation in this respect is similar to that found for charge exchange in ion-atom scattering.

In heavy particle charge exchange the Continuum Distorted Wave (CDW) approximation introduced by Cheshire[70] and developed by Crothers[70] has been shown to be successful in providing a description of the reaction at high energies, suggesting that this kind of model might be equally successful when applied to the Ps formation reaction. At present calculations for H and He targets have been carried out[72] using unsymmetical forms of the CDW and eikonal approximations which exhibit a significant post-prior discrepancy up to energies of the order of 200 eV, see Figure 11, pointing the need in the future to develop symmetrical distorted wave models satisfying time reversal invariance.

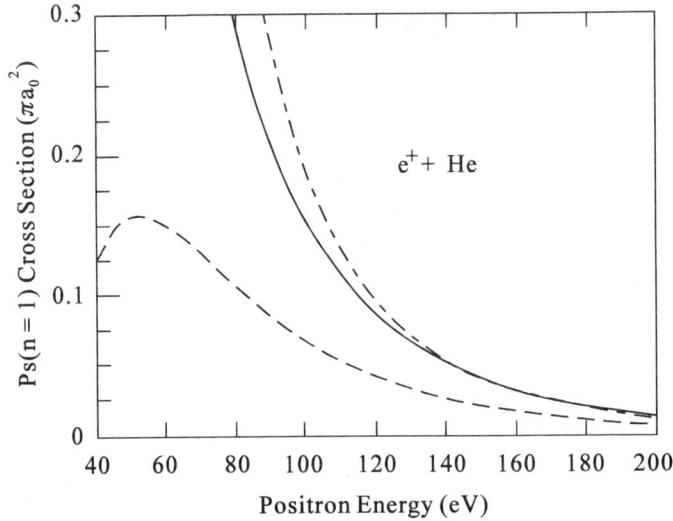

Figure 11. Total cross sections for $e^+ + He(1s^2) \rightarrow Ps(n=1) + He^+(1s)$, —— Born approximation; —·— CDW (Post); - - - CDW (Prior) from ref. [72].

6. CONCLUSIONS

Because of the technical difficulties in dealing with the rearrangement channel the topic of e^+-atom scattering developed rather slowly from its beginnings in the 1950s. However during the last few years coupled channel models have been shown to be successful in providing accurate cross sections for the interaction of positrons with one and two electron atoms and progress is being made on the very interesting topic of positronium scattering. In this short review it has not been possible to cover many important topics, in particular the application of optical potential methods[73,74] which may point the way to the study of more complicated systems.

REFERENCES

1. A. Ore, Univ. I Bergen Arbok Natur. Rekke 9 and 12 (1949).
2. H.S.W. Massey and A.H.A. Moussa, *Proc. Phys. Soc. (London).* 71:38 (1958).
3. H.S.W. Massey and C.B.O. Mohr, *Proc. Phys. Soc. (London).* 57: 695 (1954).
4. T.C. Griffith and G.R. Heyland, *Phys. Rep. C.* 39: 169 (1978).
5. M. Charlton, *Rep. Prog. Phys.* 48: 737 (1985).
6. M. Charlton and G. Laricchia, *J. Phys. B.* 23: 1045 (1990).
7. R.J. Drachman, in *Positron Scattering in Gases,* eds. J.W. Humberston and M.R.C. McDowell, p. 203, Plenum Press, New York (1984).
8. R.N. West, *Positron Studies of Condensed Matter.* Taylor and Francis, London (1974).
9. C. Schwartz, *Phys. Rev.* 124: 1468 (1961).
10. R.L. Armstead, *Phys. Rev.* 171: 91 (1968).
11. A.K. Bhatia, A. Temkin, R.J. Drachman and H. Eiserike, *Phys. Rev. A.* 3: 1328 (1971).
12. A.K. Bhatia, A. Temkin and H. Eiserike, *Phys. Rev. A.* 9: 219 (1974).
13. J.W. Humberston, *Cand. J. Phys.* 60: 591 (1982).
14. C.J. Brown and J.W. Humberston, *J. Phys. B.* 17: L423 (1984); 18: L401 (1985).
15. J.W. Humberston, *Adv. At. Mol. Phys.* 22: 101 (1979).
16. P. van Reeth and J.W. Humberston, *J. Phys. B.* 30: L95 (1997).
17. M.S.T. Watts and J.W. Humberston, *J. Phys. B.* 25: L491 (1992).
18. E.A.G. Armour, D.J. Baker and M. Plummer, *J. Phys. B.* 23: 3057 (1990).
19. P.G. Burke and T.C. Webb, *J. Phys. B.* 3: L131 (1970).
20. P.G. Burke and J.E.B. Mitchell, *J. Phys. B.* 6: 320 (1973).
21. M. Rotenburg, *Adv. Atomic Mol. Phys*. 6: 233 (1970).
22. A.T. Stelbovics, *J. Phys. B.* 22: L159 (1989).
23. H.A. Yamani and W.P. Reinhardt, *Phys. Rev. A.* 11: 11 (91975).
24. R.N. Hewitt, C.J. Noble and B.H. Bransden, *J. Phys. B.* 23: 4185 (1990).
25. P.G. Burke and H.H. Robertson, *Proc. Phys. Soc. A.* 70: 777 (1957).
26. H.H. Robertson, *Proc. Camb. Phil. Soc.* 52: 538 (1956).
27. B.H. Bransden and Z. Jundi, *Proc. Phys. Soc.* 92: 880 (1967).
28. J.F. Dirks and Y. Hahn, *Phys. Rev. A.* 3: 310 (1971).
29. M. Basu, G. Banerji and A.S. Ghosh, *Phys. Rev. A.* 13: 1381 (1976).
30. K. Higgins, P.G. Burke and H.R.J. Walters, *J. Phys. B.* 23: 1345 (1990).
31. K. Higgins and P.G. Burke, *J. Phys. B.* 26: 4269 (1993).
32. B.H. Bransden and C.J. Noble, *J. Phys. B* (submitted, 1998).
33. A.L. Merts and L.A. Collins, *J. Phys. B.* 18: L29 (1985).
34. P.G. Burke, H. Schey and K. Smith, *Phys. Rev.* 129: 1258 (1963).
35. I. Bray and A.T. Stelbovics, *Phys. Rev. A.* 48: 4787 (1993).
36. J. Mitroy and K. Ratnavilu, *J. Phys. B.* 28: 287 (1995).
37. M. Mukherjee, M. Basu and A.S. Ghosh, *J. Phys. B.* 23: 757 (1990).
38. A.S. Ghosh and J.W. Darewych, *J. Phys. B.* 24: L269.
39. J. Mitroy and A.T. Stelbovics, *J. Phys. B.* 27: 3257 (1994).
40. J. Mitroy, *J. Phys. B.* 29: L263 (1996).
41. A.A. Kernoghan, M.T. McAlinden and H.R.J. Walters, *J. Phys. B.* 28: 1079 (1995).
42. A.A. Kernoghan, D.J.R. Robinson, M.T. McAlinden and H.R.J. Walters, J. Phys. B 29: 2089 (1996).
43. A. Igarashi and N. Toshima, Phys. Rev. A. 50: 232 (1994).
44. M. Weber, A. Hofmann, A. Raith, W. Sperber, F. Jacobsen and K.G. Lynn. *Hyper. Ints*. 89: 221 (1994).
45. G.O. Jones, M. Charlton, J. Slevin, G. Larricchia, A. Kövér, M.R. Poulsen and S. Nic Chormaic, *J. Phys. B.* 26: L483 (1993).

46. H.R.J. Walters, A.A. Kernoghan, M.T. McAlinden and C.P. Campbell, in *Photon and Electron Collisions with Atoms and Molecules*, eds. P.G. Burke and C.J. Joachain, p. 313, Plenum Press, New York (1997).
47. G. Ryzhik and J. Mitroy, *J. Phys. B.* 30: 5545 (1997).
48. M.T. McAlinden, A.A. Kernoghan and H.R.J. Walters, *Hyper. Ints.* 89: 161 (1994).
49. M.T. McAlinden, A.A. Kernoghan and H.R.J. Walters, *J. Phys. B.* 30: 1543 (1997).
50. S. Zhou, S.P. Parikh, W.E. Kauppila, C.K. Kwan, D. Lin, A. Surdutovich and T.S. Stein, *Phys. Rev. Lett.* 73: 236 (1994).
51. C.K. Kwan, W.E. Kauppila, R.A. Lukaszaw, S.P. Parikh, T.S. Stein, Y.J. Wan and M.S. Dababneh, *Phys. Rev. A.* 44: 1620 (1991).
52. W.E. Kauppila, C.K. Kwan, T.S. Stein and S. Zhou, *J. Phys. B.* 27: L551 (1994).
53. M.T. McAlinden, A.A. Kernoghan and H.R.J. Walters, *J. Phys. B.* 29: 535 (1996).
54. R.N. Hewitt, C.J. Noble and B.H. Bransden, *J. Phys. B.* 25: 557 (1992).
55. S.K. Adhikari and A.S. Ghosh, *Chem. Phys. Lett.* 262: 460 (1996).
56. N. Overton, R.J. Mills and P.G. Coleman, *J. Phys. B.* 26: 3951 (1993).
57. T.S. Stein, W.E. Kauppila, V. Pol, J.H. Smart and G. Jenson, *Phys. Rev. A.* 17: 1600 (1978).
58. T.C. Griffith and G.R. Heyland, *Phys. Repts.* 39: 169 (1978).
59. D. Fromme, G. Kruse, W. Raith and G. Sinapius, *Phys. Rev. A.* 33, 727 (1986).
60. P.A. Fraser, *Proc. Phys. Soc.* 78: 333 (1961); 79: 721 (1962).
61. P.A. Fraser and M. Kraidy, *Proc. Phys. Soc.* 89: 533 (1966).
62. B.H. Bransden and M.I. Barker, *J. Phys. B.* 1: 1109 (1968).
63. C.P. Campbell, M.T. McAlinden, F.G.R.S. MacDonald and H.R.J. Walters, *Phys. Rev. Lett.* 80: 5097 (1998).
64. B.J. Archer, G.A. Parker and R.T. Pack, *Phys. Rev. A.* 41: 1303 (1990).
65. Y. Zhou and C.D. Lin, *J. Phys. B.* 27: 5065 (1994).
66. R. Shakeshaft and J.M. Wadhera, *Phys. Rev. A.* 22: 968 (1980).
67. M. Basu and A.S. Ghosh, *J. Phys. B.* 21: 3439 (1988).
68. A. Igarashi and N. Toshima, *Phys. Rev. A.* 47: 2386 (1993).
69. W. Sperber, D. Becker, K.G. Lynn, W. Raith, A. Schwab, G. Sinapius, G. Spicher and M. Weber, *Phys. Rev. Lett.* 68: 3690 (1992).
70. I.M. Cheshire, *Proc. Phys. Soc.* 84: 89 (1964).
71. D.S.F. Crothers, J. Phys. B. 15: 2061 (1982).
72. B.H. Bransden, C.J. Joachain and J.F. McCann, *J. Phys. B.* 25: 4965 (1992).
73. F.A. Gianturco, A. Jain and J.A. Rodriguez-Ruiz, *Phys. Rev. D.* 48: 4321 (1993).
74. I.E. McCarthy, K. Ratnavelu and Y. Zhou, *J. Phys. B.* 26: 2733 (1993).

FROM POSITRON TO POSITRONIUM SCATTERING

G Laricchia, AJ Garner and K Paludan

Department of Physics and Astronomy
University College London
London WC1E 6BT, UK

INTRODUCTION

Although still considered somewhat exotic particles, positrons (e^+) and positronium (Ps, the electron-positron bound state) are currently employed in the exploration of a range of fundamental phenomena in atomic[1] and condensed matter[2] physics, astrophysics[3] as well as in the diagnostics of the electronic and structural properties of industrially important materials[4] and of living biological systems[5]. The considerable growth of positron techniques and their applications has been enabled by symbiotic developments in the understanding of the basic processes involved in the interactions of positrons with matter and the ensuing improvements in the efficiency for their production and manipulation.

Early studies of positron interactions with matter were simply carried out by implanting β^+ particles from nuclear decay into some medium and measuring their annihilation characteristics (e.g. rates, energies of the annihilation quanta or the angles between them). Whilst a wealth of knowledge continues to be gleaned in this way (e.g. density fluctuations and cluster formation in fluids[6], vacancies and defects in solids[7]), the degree of experimental control and versatility has been considerably enhanced with the advent of quasi-monoenergetic positron beams[1,2,4].

Among the first experiments to be performed with slow positron beams were measurements of total cross-sections from easily accessible atomic and molecular systems[8,9]. As an example, the results for helium are shown in Fig. 1 where they are compared with corresponding electron (e^-) measurements[8]. Differences in the interactions of these two projectiles with the atom are manifested by the striking divergence between the cross-sections below ~200eV. The polarization and static interactions are both attractive and add up in the case of electrons, whilst they have opposite signs for positrons (the static interaction being the repulsive one) and tend to reduce the overall scattering probability. The exchange interaction is, of course, only active for e^-. The sharp increase in the positron cross-section near the threshold for Ps formation, indicated in the figure by the arrow, was an early suggestion of the importance of this process in positron-atom collisions[10]. At higher kinetic energies, as the influences of exchange (for electrons) and Ps formation (for positrons) diminish, the static

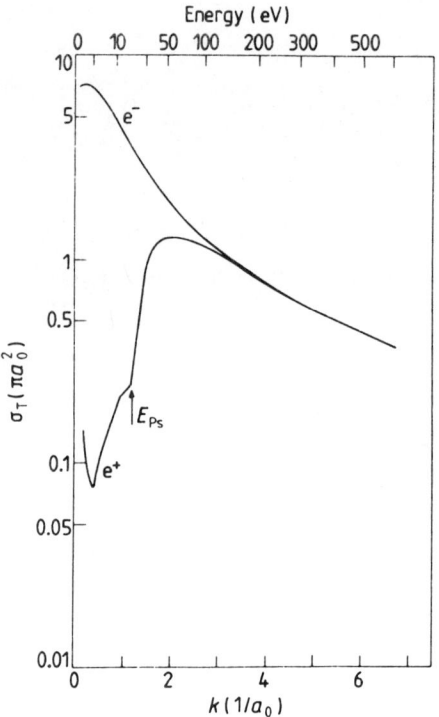

Figure 1. Total cross-section for $e^{\pm}$ - He scattering [8].

interaction becomes dominant and the cross-sections of the two projectiles ultimately merge. Qualitatively, the tendencies of the positron total cross-sections discussed above for helium are typical of many targets[8,9], including atomic hydrogen (where excellent agreement has recently been achieved between theory[11] and experiment[12]). A different behaviour is, however, exhibited by the cross-sections from some molecules[8], for example O_2, where strong coupling between channels might be responsible for the smooth variation of the total cross-section across the thresholds for Ps formation, excitation and ionization[13,14]. In the case of alkali atoms, Ps formation is an exoergic reaction and the cross-section for Ps(1s) formation diverges as the impact energy tends to zero[15]. As an illustration of this atomic family, calculations for rubidium[16] are shown in Fig. 2 where they are compared with experiment[17]. The peak in the total cross-section around 6eV (also present in potassium and cesium) has been traced to contributions from excited state Ps[18,19].

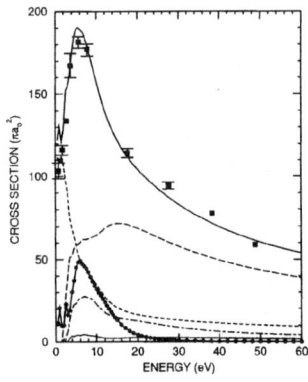

Figure 2. Total cross-section for e^+ - Rb scattering:- full curve, theory [16]; circles, experiment[17].

CURRENT STATUS

As indicated in Table 1, among the specific processes which may occur in positron collisions, those which have thus far received most attention from experimentalists are positronium formation and direct ionization. In the study of both processes, signal to background levels have often been improved by setting up coincidences between two (or more) of the final state particles (or their products). Recently, direct measurements[20,21,22] of Ps - atom (molecule) total cross-sections have also become possible through the realization of quasi-monoenergetic Ps beams[23]. A brief survey of results is given below. Further details may be found in other recent reviews[24,25].

Positronium Formation

Depending on the relative spin orientation of its constituents, positronium can be formed in a triplet (s=1) state (called ortho- or o-Ps) or singlet (s=0) state (called para- or p-Ps). The states are characterized by different annihilation properties, as summarized in Table 2. Positronium is structurally equivalent to hydrogen and, with half the reduced mass, its Bohr energy levels are approximately half those of hydrogen, so that its binding energy in a state of principal quantum number n is approximately $6.8eV/n^2$. The large magnetic moment of the positron (658 times that of the proton) and the presence of QED effects such as virtual annihilation render fine and hyperfine separations markedly different from the corresponding hydrogen values[26].

Table 1. Current status in e^+ - atom (molecule) collision experiments. Q denotes the cross-section for the particular process, $d\Omega$ the solid angle, E the e^+ energy, E_{Ps} and E_i the Ps formation and direct ionization thresholds respectively.

Process	Measured cross-sections
$e^+ + A \to$ all	Q_t for many targets, incl. H and the alkali
elastic scattering $e^+ + A \to e^+ + A$	some Q_{el} and $dQ_{el}/d\Omega$
target excitation $e^+ + A \to e^+ + A^*$	scarce
electron capture $e^+ + A \to Ps + A^+$	Q_{Ps} ($1 \le$ (E-E_{Ps}) $\le 10^2$ eV) incl.H and the alkali and some $dQ_{Ps}/d\Omega$
direct ionization $e^+ + A \to e^+ + ze^- + A^{z+}$	Q_i^{z+} ($1 \le$ (E-E_i) $\le 10^3$ eV) incl. H; some $dQ_i^+/d\Omega$, $d^2Q_i^+/d\Omega_1 dE_1$, $d^3Q_i^+/d\Omega_1 d\Omega_2 dE_1$
annihilation $e^+ + A \to 2\gamma + A^+$	over a very restricted energy range
formation of compounds $e^+ + A \to (eA)^+ + h\nu$	scarce
$e^+ + AB \to (PsA) + B^+ + h\nu$	scarce

Table 2. Summary of ground state Ps properties.

name	state	lifetime (s)	annihilation quanta	
o-Ps	$^3S_1$	142×10^{-9}	odd	(3)
p-Ps	$^1S_0$	125×10^{-12}	even	(2)

With a few exceptions discussed below, the formation of Ps in positron-atom/molecule collisions, represented by the reaction

$$e^+ + A \rightarrow Ps + A^+ \qquad (1)$$

has been measured integrated over all the quantum states of the Ps atom and the ion. Some of the experimental methods used (and inherent assumptions) are summarized in Table 3. Cross-sections are available for all the inert atoms, some molecules, atomic hydrogen and some alkali atoms[1]. Although, experimental determinations of Q_{Ps} from atomic hydrogen[12,31] are in good agreement with most recent theoretical determinations[11,34,35], for more complex targets the level of agreement among experiments (and between these and theory) is variable, despite the similarity of many of the methods. As indicated in Table 3, with the exception of the work of ref. 27 (which disagree strongly with all other measurements but is thought to have been affected by systematic errors), all experimental (and indeed all theoretical) determinations of Q_{Ps}, assume that annihilation is negligible in comparison to reaction (1). Indeed annihilation (and its coupling to scattering channels) has been neglected when calculating scattering parameters. Recently, however, the influence of virtual processes on the annihilation probability has been considered[36,37,38] and found to be significant especially in the vicinity of energy thresholds for inelastic processes. It might be worthwhile therefore to re-consider how well the scattering of positrons by atoms can really be described (or measured) without considering this absorbing ion-producing channel explicitly, especially in energy regions where many thresholds are presents.

As indicated in Table 3, in addition to the assumption of negligible annihilation, some works also rely on the complete detection of scattered positrons. Discrepancies of the type shown in Fig. 3, among various experimental determinations for helium[39] (at high energies) and for rubidium[17] (at low energies) have been interpreted in terms of inefficient collection of e^+ scattered to large angles. In the figure, the contrasting quantum state distribution predicted for the two targets can be seen. Positronium formation in the first excited state has been observed in binary collisions with neon, argon and molecular hydrogen through the detection of the 5.1eV Lyman-α photon from the (2P-1S) transition in delayed coincidence with an annihilation photon[40]. A maximum yield of ~6×10^{-2} Ps(n=2) per stopped e^+ was found at 16eV incident energy on molecular hydrogen. Contributions from excited state Ps have also been observed in an

Table 3. Experimental methods and assumptions used in the determination of Q_{Ps}.

detect	no annihilation	all e^+	reference
3 γ-rays (o-Ps)	no	no	27
no e^+ in final state (all Ps)	yes	yes	e.g. 28,29
A^+ and no e^+ (all Ps)	yes	yes	e.g. 30,31
A^+ below E_i (all Ps)	yes	no	e.g. 32,33
2 γ-rays (p-Ps and quenched o-Ps).	yes	yes	e.g. 12,29

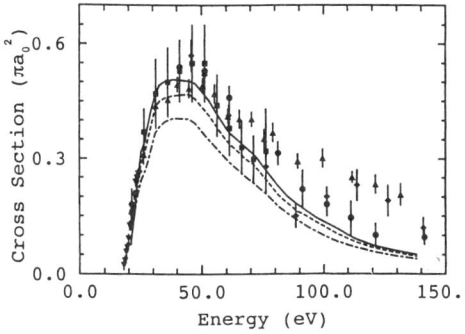
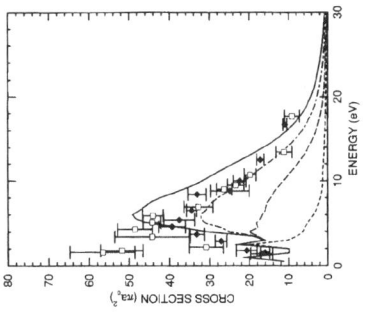

Figure 3. Comparisons of measurements of the Ps formation cross-sections for He (left) and Rb (right) with theory: for He (dash-dot n=1, full curve all n) [39], Rb (short dash n=1, full curve all n) [17].

angular differential study[41] but further studies, especially in view of the results shown in Fig. 3, would be highly desirable.

An enhancement of Ps formation simultaneous to ionic excitation has been observed in CO_2[42]. In this study, a 3.5eV photon was detected in delayed coincidence with an annihilation photon originating from Ps formation. Because of the close proximity between the C-state of CO_2 which lies at 10.56 eV above the ground state, and the threshold for Ps formation simultaneous to ionic excitation at 10.50eV, the enhanced Ps formation has been associated with a quasi-excitation of the molecule, followed by the capture of the excited electron by the near-stationary positron, with the remnant ion absorbing the difference in binding energies[43]. As well as producing electronically excited ions, Ps formation may leave the remnant ion in a vibrationally excited state[44]. This has been deduced through the observation of HeH^+ produced by positron bombardment of a gaseous mixture of helium and molecular hydrogen where it is thought that the proton is transferred from the vibrationally excited H_2^+.

The behaviour of $dQ_{Ps}/d\Omega$ has been studied primarily at small forward angles to assess, as discussed further below, the degree of collimation for positronium beam production[23]. Studies over a larger angular range are available for argon and krypton[45] where discrepancies exist with available theory[46]. The angular dependence of transfer ionization (e^+, Ps e^-) has also been investigated for the same gases[45] whilst a total suppression of this channel has been reported for helium and neon[47].

Direct ionization

During the last decade, experimental data for equi-velocity e^+, e^-, proton, p^+, and antiproton, p^-, impact ionization have been successfully used to disclose simple mechanisms involved in ionization processes[48,49]. The reaction represented by

$$e^+ + A \rightarrow e^+ + me^- + A^{m+} \qquad (2)$$

has been studied with m=1,2,3 by measuring energy loss spectra[50]; detecting an ion in coincidence with a positron[51,52,53,54,55] or detecting an ion in coincidence with an ejected electron[55,56]. Recently new experimental data for single and double ionization of the rare gases neon, argon, krypton, and xenon by p^- and e^+ impact have been obtained[57,58,59] and compared with existing data for e^- and p^+ impact[60]. Direct single ionization cross-sections for all four projectiles had previously been measured for atomic hydrogen, helium and molecular hydrogen[61]. From the investigation of targets with a low-atomic number (Z), a certain degree of understanding of single and multiple ionization had emerged. For high-Z atoms, the static interaction between the projectile

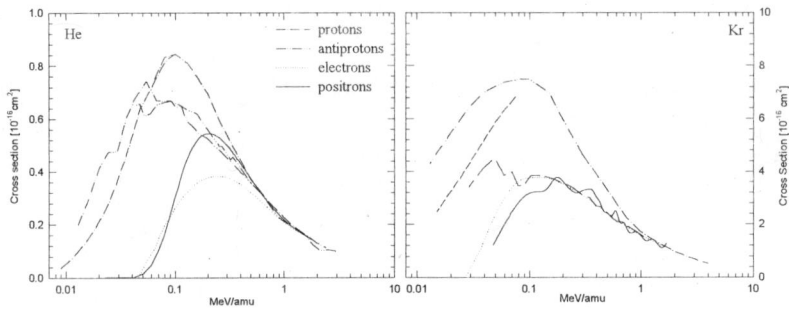

Figure 4. Direct single ionization cross-sections for ($e^\pm$, $p^\pm$) - He and Kr. The data shown are measurements from refs. 59, 60 and references therein.

and the undistorted target appears to play an important role in the case of the lighter projectiles. In Fig. 4 single ionization cross-sections for e^+, e^-, p^+, and p^- impact on helium and krypton are shown as examples of a low-Z and a high-Z target respectively. For krypton, the p^+ cross-section exceeds the p^- cross-section up to 70% whereas there is hardly any difference between the e^+ and e^- cross-sections above 200 keV/amu. This is in contrast to the low-Z targets where the difference at intermediate velocities between the e^+ and e^- cross-sections is very similar to that between the p^+ and p^- cross-sections. Both differences are ascribed to the polarization of the target electron cloud by the incoming projectiles which increases the cross-section of the positive projectiles over those of their negative antiparticles[53].

The different relationship between the cross-sections for krypton is interpreted as being due to the increased Coulomb interaction between target nucleus and projectile, the tendency of which is to attract negative projectiles towards the high-electron density of the target atom whereas the positive projectiles are repelled[60]. For high-Z targets, this interaction appears to be strong enough to deflect the light projectiles and, in the case of e^+, counteract the polarization effect at intermediate velocities. As the heavy projectiles are more difficult to deflect their trajectories remain unaffected in this velocity range and the difference between the e^+ and e^- cross-sections becomes significantly smaller than that between the p^+ and p^- cross-sections.

The Wannier law for the threshold behaviour of the direct ionization cross-section continues to attract much attention[62]. In the case of positrons, measurements[63] for helium and molecular hydrogen have found an exponent of around 2 in the range 1-3eV above threshold in contrast with the value of 2.651 predicted by theory[64]. However in a recent theoretical investigation[65], anharmonic terms in the three particle potential around the Wannier configuration have been revealed to be more important for positron than electron impact ionization. Although good agreement has been found up to 10eV above threshold between this "extended threshold" theory and experiment, the theoretical findings entail stricter demands on positron beam quality, than previously thought, for this type of studies.

For double ionization the most prominent effect of moving to high-Z targets is that inner-shell ionization followed by Auger-decay becomes increasingly important at high velocities[60,66]. For the low-Z targets the relationship between the double ionization cross-sections for e^+, e^-, p^+, and p^- impact at high velocities is understood in terms of interference between participating double ionization mechanisms[67, 68]. As just one mechanism, inner-shell ionization, becomes dominant, the effect of interference is reduced and the relationship between the four double ionization cross-sections is seen

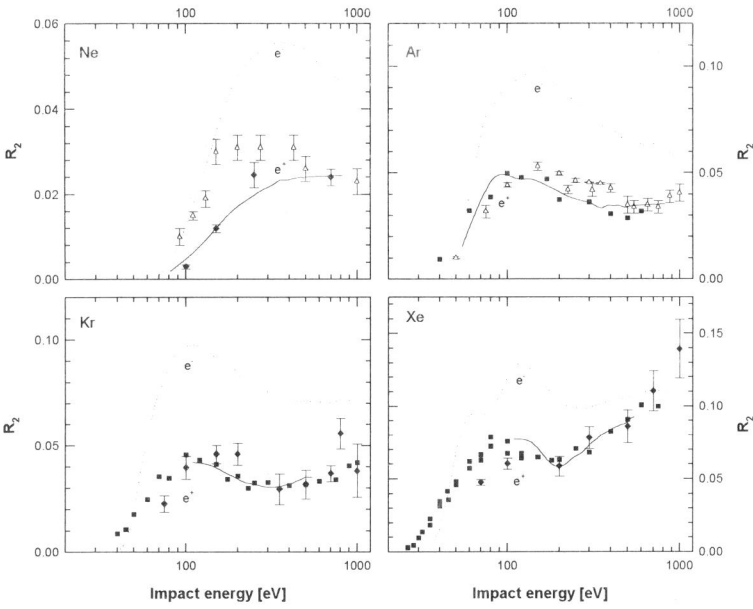

Figure 5. Ratios of double-to-single ionization cross-sections for $e^{\pm}$ from inert atoms. From ref. 57.

to change and become progressively similar to the relationship between underlying inner-shell cross-sections.

In the past, mass and charge effects have been highlighted by considering the ratio between double- and single-ionization cross-sections, R_2 [69]. Figure 5 shows R_2 values for electron and positron impact on neon, argon, krypton and xenon, as well as a prediction for e^+ impact, based on the equation $R_2^{dir}(e^+) = R_2(e^-)R_2^{dir}(p^+)/R_2(p^-)$ (where $R_2^{dir} = Q^{++}_{dir}/Q^+_{dir}$). This expression is found to be generally in good agreement for all targets. This implies that trajectory effects cancel out in the ratio R_2 and therefore are of similar importance in single and double ionization. It likewise seems that exchange effects are either not important or are of similar importance in single and double ionization by e^- impact over the energy region where good agreement is found between measured and predicted values of $R_2^{dir}(e^+)$.

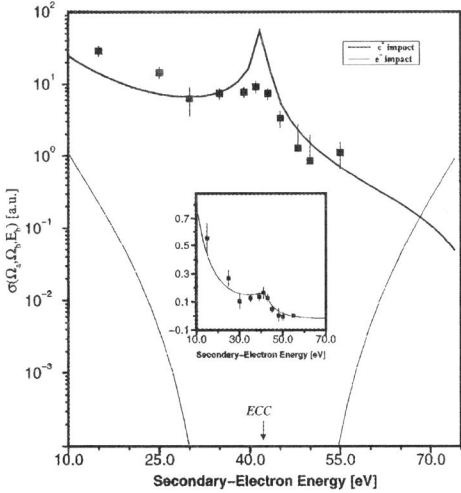

Figure 6. Triply differential cross-section for $e^{\pm}$(100eV) - H_2. The inset shows experimental results for positrons[78] compared with theory[79] convoluted with the experimental resolutions.

Progress has also been made in differential studies of positron impact ionization[70]. A primary motivation has been the investigation of positronium formation into a low-lying continuum state, corresponding to the process known in ion-atom collisions as Electron Capture to the Continuum (ECC). The phenomenon is manifested by a cusp-like peak in the electron energy spectrum at 0° at an energy corresponding to the same velocity as that of the scattered projectile. In the case of a light projectile such as the e^+, theoretical descriptions are complicated by the large deflections suffered by the projectile and discrepancies exist among their predictions[71,72]. The first doubly differential measurements [73,74] were performed on an argon target and the energy of both ejected electrons and scattered projectiles were investigated. No cusp was observed in agreement with classical trajectory Monte Carlo calculations[75] but, whilst the ejected electron spectra was well described by this theory, significant deviations were observed in the scattered projectile spectra[75]. These discrepancies remain unresolved[76]. The first triply differential cross-section with e^+ has recently been measured[77] for molecular hydrogen. These measurements have revealed the influence of positronium formation into a low lying continuum state. As shown in Fig. 6, the measurements are in good accord with a recent calculation in which an asymmetric capture probability to continuum states has been traced as originating from the deflection of the positron by the two-centre potential formed by the ion and ejected electron in the final state[78].

Ps beam and scattering

A beam of Ps atoms [39] may be produced through the charge-exchange reaction of Eq. 1. The angular and energy distributions of the Ps beam depend on the medium (gaseous or solid) from which the e^+ beam scatters, the geometry (transmission or reflection) and the characteristics of the e^+ beam itself (angular divergence, energy spread, etc.). It might be worth noting that at the typical velocities employed in atomic scattering experiments, the beam consists essentially of o-Ps because of its comparatively longer lifetime. The gas method is as efficient as that using solids and is the best characterized, producing a Ps beam with a relatively narrow kinetic energy spread which depends on the relative production in excited states or simultaneous to other inelastic processes.

A schematic diagram of the Ps beam operated at UCL[20-22] is given in Fig. 7. A monoenergetic positron beam is magnetically guided into the "Ps production" gas cell. An electrostatic element repels charged particles emerging from this cell whilst the exiting Ps is allowed to traverse a second "Ps scattering" cell. The Ps atoms can be detected at the end of the beamline by a channel-electron-multiplier array and/or γ-ray counters. Helium, argon and molecular hydrogen have been investigated with respect to their efficiency for collimated Ps production and, among them, molecular hydrogen has been found to be the best[21].

With this apparatus, the total cross-section of Ps scattering from argon[20], helium[21], molecular hydrogen[21] and molecular oxygen[22] have been measured. Figure 8 shows the results for helium and molecular hydrogen[21] together with corresponding calculations. In both cases the measurements display a broad peak at the lower energies before a gentle decrease at higher energies. In the case of helium, significant discrepancies are found between the experimental results and the calculations performed within a coupled pseudo-state formalism (with no exchange) for target elastic collisions and a First Born Approximation (FBA) for target inelastic processes[79]. Better agreement is found between experiment and the target-elastic total cross-section calculations[80] performed using a three-Ps state close coupling approximation for elastic scattering and n=2 excitation and a Born approximation with exchange for n≥3 excitations and ionization. Also shown is a static-exchange calculation[81] for the elastic scattering cross-section.

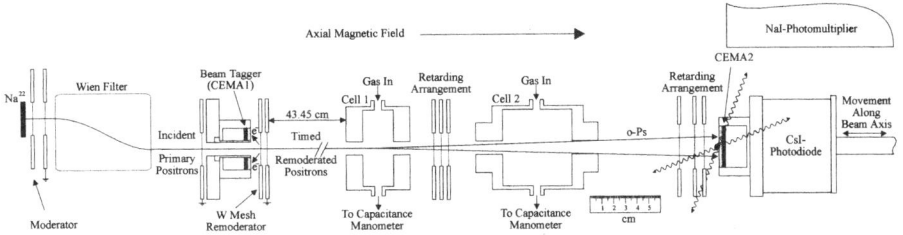

Figure 7. Schematic diagram of the positronium beam in operation at UCL[22].

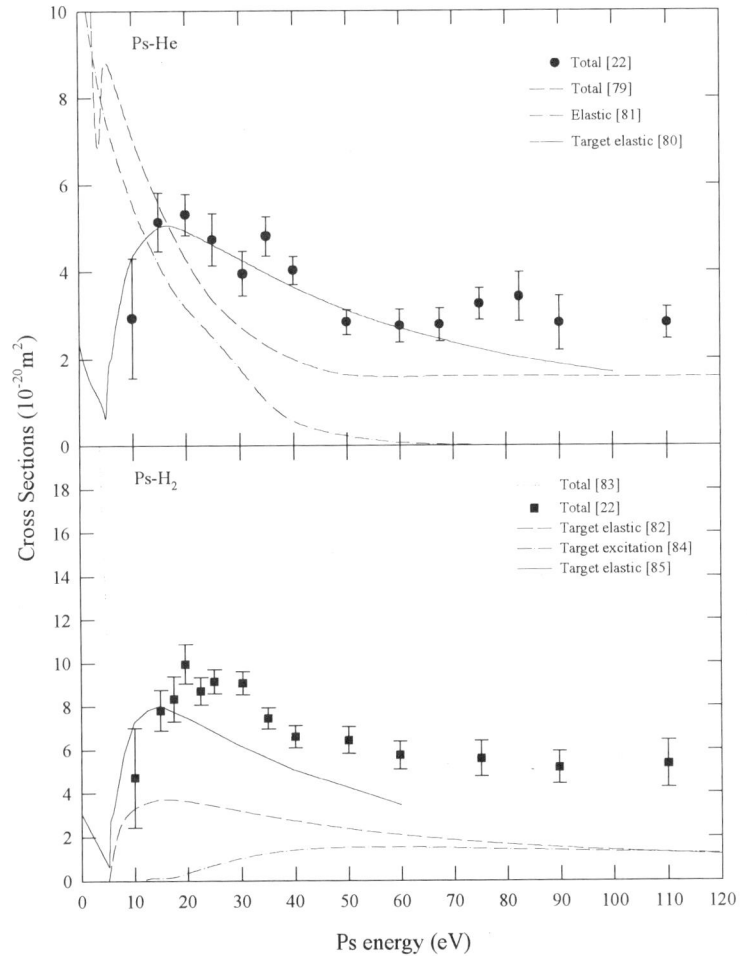

Figure 8. Cross-sections of positronium scattering from He and H_2.

In the case of molecular hydrogen, the experimental results are compared with those of a target-elastic FBA calculation[82] which neglects elastic scattering and even parity state transitions of the positronium-excitation channel. Also shown are the results for the sum of elastic scattering and vibrational excitation cross-sections calculated within the context of multichannel scattering theory[83]. The electronic excitation of H_2 from the ground state to B $^1\Sigma_u^+$ and b $^3\Sigma_u^+$ has been calculated within the framework of the FBA by considering discrete excitation of Ps up to n=6 and including break up[84]. Better agreement with experiment is found with the results of a three-Ps state coupled channel calculation[85] for target elastic total cross-section which includes the effects of exchange using a nonlocal model exchange potential.

CONCLUSIONS AND OUTLOOK

Collisions involving positron and positronium are contributing to the understanding of general atomic physics phenomena as well as illuminating specific processes and interactions such as annihilation and exchange. It is hoped that the considerable effort presently aimed towards the production of large ensembles of cold positrons[86] will result in beams of extremely well defined energy ($\leq kT$) which should enable the detailed study of near-threshold phenomena, resonances, annihilation and positronic compounds.

ACKNOWLEDGMENTS

We wish to thank the Engineering and Physical Sciences Research Council, The Royal Society and NATO for supporting positron research at UCL and express our gratitude and best wishes to Prof. Phil Burke, FRS whose contributions, and that of his group, have had a major impact for the advancement of the subject.

REFERENCES

[1] e.g. "Low energy positron and positronium physics" HH Andersen, EAG Armour, JW Humberston and G Laricchia, eds; Nucl Instr Meth B 143 (1998) and references therein
[2] e.g. PJ Schultz and KG Lynn Rev Mod Phys 60 (1988) 701
[3] e.g. RJ Drachman in ref. 1 p.1
[4] e.g. "Slow-positron beam techniques for solids and surfaces" WB Waeber, M Shi and AA Manuel, eds; Appl Surf Science 116 (1997)
[5] e.g. IEEE Trans Nucl Science 44 (1997); Adv Drug Delivery Rev 26 (1997)
[6] "Positron annihilation studies" SC Sharma ed; World Scientific, Singapore; (1987)
[7] "Positron annihilation" YC Jean, M Eldrup, DM Schrader, RN West, eds; Mat Sc Forum 255-257 (1997)
[8] WE Kauppila and TS Stein Adv At Mol Opt Phys 26 (1990) 1
[9] e.g. A Zecca, GP Karwasz and RS Brusa Nuovo Cimento 19 (1996) 1
[10] KF Canter, PG Coleman, TC Griffith and GR Heyland J Phys B 5 (1972) L167
[11] AA Kernoghan, DJR Robinson, MT McAlinden and HRJ Walters J Phys B 29 (1996) 2089
[12] S Zhou, H Li, WE Kauppila, CK Kwan and TS Stein Phys Rev A 55 (1997) 361
[13] G Laricchia, J Moxom and M Charlton Phys Rev Letts 70 (1993) 3229
[14] Y Katayama, O Sueoka and S Mori J Phys B 20 (1987) 1645
[15] e.g. MST Watts and JW Humberston J Phys B 25 (1992) L491
[16] AA Kernoghan, MT McAlinden and HRJ Walters J Phys B 29 (1996) 3971

[17] SP Parikh, WE Kauppila, CK Kwan, RA Lukaszew, D Prybyla, TS Stein and S Zhou Phys Rev A 47 (1993) 1535
[18] HRJ Walters, AA Kernoghan and MT McAlinden in "The Physics of Electronic and Atomic Collisions" ed LJ Dube, JBA Mitchell, JW McConkey and CE Brion (New York; AIP) (1995) 397
[19] RN Hewitt, CJ Noble and BH Bransden J Phys B 26 (1993) 3661
[20] N Zafar, G Laricchia, M Charlton and A Garner Phys Rev Lett 76 (1996) 1595
[21] AJ Garner, G Laricchia and A Özen J Phys B 29 (1996) 5961
[22] AJ Garner, A Özen and G Laricchia (1998) in ref. 1 p.155
[23] G Laricchia in "Positron Spectroscopy of Solids" Proceedings of the International School of Physics <<Enrico Fermi>> vol. 125 A Dupasquier and AP Mills, eds (IOS; Amsterdam) (1995) p401
[24] W Raith in "Photonic, electronic and atomic collisions" F Aumayr and H Winter, eds (World Scientific; Singapore) (1998) 341
[25] G Laricchia and M Charlton in "Positron beams and their applications" PG Coleman ed, (World Scientific, Singapore) (1998) in press
[26] S Berko and HN Pendleton Ann Rev Nucl Part Sci 30 (1980) 543
[27] M Charlton, G Clark, TC Griffith and GR Heyland J Phys B 16 (1983) L465
[28] LS Fornari, LM Diana and PG Coleman Phys Rev Lett 51 (1983) 2276
[29] TS Stein, J Jiang, WE Kauppila, CK Kwan, H Li, A Surdutovich and S Zhou Can J Phys 74 (1996) 313
[30] D Fromme, G Kruse, W Raith and G Sinapius Phys Rev Letts 57 (1986) 3031
[31] V Kara, K Paludan, J Moxom, P Ashley and G Laricchia in ref. 1 p. 94
[32] J Moxom, G Laricchia and M Charlton J Phys B 26 (1993) L367
[33] WE Meyerhof and G Laricchia J Phys B 30 (1997) 2221
[34] K Higgins and PG Burke J Phys B 24 (1991) L343
[35] J Mitroy J Phys B 29 (1996) L263
[36] G Laricchia and C Wilkin Phys Rev Letts 79 (1997) 2241; Nucl Instr Meth B 143 (1998) 135
[37] EP da Silva, JSE Germano and MAP Lima Phys Rev Letts 77(1998) 1028
[38] P Van Reeth and JW Humberston J Phys B 31 (1998) L231
[39] CP Campbell, MT McAlinden, AA Kernoghan and HRJ Walters Nucl Instr Meth B 143 (1998) 41
[40] G Laricchia, M Charlton, G Clark and TC Griffith Phys Lett A109 (1985) 97
[41] N Zafar, G Laricchia, M Charlton and TC Griffith J Phys B 24 (1991) 4461
[42] G Laricchia, M Charlton and TC Griffith J Phys B 21 (1988) L227
[43] G Laricchia and J Moxom Phys Lett A174 (1993) 255
[44] J Xu, LD Hulett, J Moxom, W Wu, S Datz and DM Schrader Phys Rev A 49 (1994) R4373
[45] T Falke, W Raith, M Weber and U Wesskamp J Phys B 28 (1995) L505; T Falke, T Brandt, O Kuhl, W Raith and MJ Weber Phys B 30 (1997) 3247
[46] MT McAlinden and HRJ Walters Hyperfine Interactions 84 (1994) 407
[47] H Bluhme, H Knudsen, JP Merrison and MR Poulsen Phys Rev Letts 81 (1998) 73
[48] H Knudsen and JF Reading Phys. Rep. 212 (1992) 107
[49] DR Schultz, RE Olson and C0 Reinhold J Phys B24 (1991) 521
[50] S Mori and O Sueoka J Phys B 27 (1994) 4349
[51] D Fromme, G Kruse, W Raith and G Sinapius Phys Rev Letts 57 (1986) 3031
[52] H Knudsen, L Brun-Nielsen, M Charlton and MR Poulsen J Phys B 23 (1990) 3955
[53] FM Jacobsen, NP Frandsen, H Knudsen, U Mikkelsen and DM Schrader J Phys B 28 (1995) 4691; FM Jacobsen, NP Frandsen, H Knudsen and U Mikkelsen J Phys B 28 (1995) 4675
[54] J Moxom, P Ashley and G Laricchia Can Jour Phys 74 (1996) 367

[55] Á Kövér, G Laricchia and M Charlton J Phys B 27 (1994) 2409
[56] J Moxom, G Laricchia, M Charlton, GO Jones and Á Kövér J Phys B 25 (1992) L613
[57] V Kara, K Paludan, J Moxom, P Ashley and G Laricchia J Phys B 30 (1997) 3933
[58] K Paludan, H Bluhme, H Knudsen, U Mikkelsen, SP Møller, E Uggerhøj and EJ Morenzoni Phys B 30 (1997) 3951
[59] K Paludan, G Laricchia, P Ashley, V Kara, J Moxom, H Bluhme, H Knudsen, U Mikkelsen, SP Møller, E Uggerhøj and EJ Morenzoni Phys B 30 (1997) L581
[60] RD DuBois, LH Toburen and ST Manson Phys. Rev. A 29 (1984) 70; RD DuBois and ST Manson Phys. Rev. A 35 (1987) 2007; ST Manson and RD DuBoisJ. Physique C 9 (1987) 263; E Krishnakumar and SK SrivastavaJ Phys. B 21 (1988) 1055; P Nagy, A Skutlanz and V Schmidt J Phys B 13 (1980) 1249; VV Afrosimov, Yu A Mamaev, MN Panov and NV FedorenkoSov. Phys.-Tech. Phys. 14 (1969) 109
[61] GO Jones, M Charlton, J Slevin, G Laricchia, Á Kövér, MR Poulsen and SN Chormaic J Phys B 26 (1993) L483; H Knudsen, U Mikkelsen, K Paludan, K Kirsebom, SP Møller, E Uggerhøj, J Slevin, M Charlton and E Morenzoni Phys Rev Lett 74 (1995) 4627; LH Andersen, P Hvelplund, H Knudsen, SP Møller, JOP Pedersen, S Tang-Petersen, E Uggerhøj, K Elsener and E Morenzoni Phys Rev A 41 (1990) 6536; P Hvelplund, H Knudsen, U Mikkelsen, E Morenzoni, SP Møller, T Worm and E Uggerhøj J Phys B 27 (1994) 925
[62] MY Kuchiev and VN Ostrovsky Phys Rev A 58 (1998) 321; PV Grujic Commnents At Mol Phys 33 (1997) 351 and references therein
[63] P Ashley, J Moxom and G Laricchia Phys Rev Lett 77 (1996) 1250
[64] H Klar J Phys B 14 (1981) 4165
[65] W Ihra, JH Macek, F Mota-Furtado and PF O'MahonyPhys Rev Letts (1997) 78 4027
[66] S Helms, U Brinkmann, J Deiwiks, R Hippler, H Schneider, D Segers and J Paridaens J Phys B 28 (1995) 1095
[67] JH McGuire Phys Rev Letts 49 (1982) 1153
[68] LH Andersen, P Hvelplund, H Knudsen, SP Møller, AH Sørensen, K Elsener, K-G Rensfelt and E Uggerhøj Phys Rev A 36 (1987) 3612
[69] M Charlton, L Andersen, L Brun-Nielsen, BI Deutch, P Hvelplund, FM Jacobsen, H Knudsen, G Laricchia, MR Poulsen and J Pedersen J Phys B 21 (1988) L545; M Charlton, L Brun-Nielsen, BI Deutch, P Hvelplund, FM Jacobsen, H Knudsen, G Laricchia and MR Poulsen J Phys B 22 (1989) 2779
[70] Á Kövér, G Laricchia and R Finch Nucl Instr Meth B 143 (1998) 100 and references therein
[71] A Bandypadhyay, K Roy, P Mandal and NC Sil J Phys B 27 (1994) 4337
[72] DR Schultz and CO Reinhold J Phys B 23 (1990) L9
[73] J Moxom, G Laricchia, M Charlton, GO Jones and Á Kövér J Phys B 25 (1992) L613
[74] Á Kövér, G Laricchia and M Charlton J Phys B 26 (1993) L575; J Phys B 27 (1994) 2409
[75] RA Sparrow and RE Olson J Phys B 27 (1994) 2647
[76] Á Kövér, RM Finch, M Charlton and G Laricchia J Phys B 30 (1997) L507
[77] Á Kövér and G Laricchia Phys Rev Lett 80 (1998) 5309
[78] J Berakdar Phys Rev Lett 81 (1998) 1393
[79] MT McAlinden, FRGS MacDonald and HRJ Walters Can Jour Phys 74 (1996) 434
[80] PK Biswas and SK Adhikari Phys Rev A (1998) in press
[81] NM Sarkar and AS Ghosh J Phys B 30 (1997) 4591
[82] PK Biswas and AS Ghosh Phys Letts A 223 (1996)173
[83] M Comi, GM Prosperi and A Zecca Nuovo Cimento 2 (1983) 1347
[84] PK Biswas and SK Adhikari J Phys B 31 (1998) L315
[85] PK Biswas and SK Adhikari J Phys B 31 (1998) L737
[86] e.g. CM Surko in "Photonic, electronic and atomic collisions" F Aumayr and H Winter, eds (World Scientific; Singapore) (1998) 341

EMBEDDING AND R-MATRIX METHODS AT SURFACES

J.E. Inglesfield
Department of Physics and Astronomy,
University of Wales Cardiff,
Cardiff, CF2 3YB, United Kingdom

INTRODUCTION

There are many examples of solving the Schrödinger equation in which space can be divided into two regions. In electron scattering from atoms and molecules, there is an inner region in which the electron interacts strongly with the correlated many-electron system of the atom or molecule, and an outer region where the electron propagates in free space in which the source and detector are situated. The R-matrix method[1] takes advantage of this division of space, surrounding the inner region by the R-matrix sphere within which a many-electron calculation can be performed. This gives the R-matrix on the spherical surface, from which the scattering properties of the electron in the outer region can be found. An analogous division of space can be useful when the one-electron Schrödinger equation is to be solved for a finite – and usually complicated region – joined onto an extended substrate, for example the surface of a solid. An embedding potential can be added on to the Hamiltonian for the surface region[2], so that the wavefunctions in this region are automatically matched on to the substrate wavefunctions: the Schrödinger equation is then solved explicitly only in the surface region. The embedding potential, which is defined over the boundary between the region of interest and the substrate, can be found from the substrate Green function. In this paper I shall describe the embedding method and its applications in surface physics, and some recent applications of the R-matrix method in electron scattering from solids.

I shall begin by deriving the embedding method as a variational principle for trial wavefunctions defined only over the region of interest. There are links between the embedding method and R-matrix theory, not just in the division of space into two regions but also in the mathematical structure. This will be discussed in the subsequent section.

Two surface applications of the embedding method will then be given. The first is the study of the Rydberg series of electron states bound by the image potential[3], where the embedding method enables both the effect of the surface and substrate, and the long-range image tail, to be treated very accurately. The second example is the study of Xe adsorbed on the surface of Ag, in particular the way that the rare gas Xe atoms behave like a thin layer of dielectric when a static uniform electric field is applied[4]. I shall then go on to present a quite different application, to the problem of electrons confined by some hard wall potential[5]. If the confinement geometry is complex, this boundary condition problem can be difficult to solve, but in the embedding method the hard wall potential is replaced by an embedding potential on the boundary and the problem can be solved using straightfoward basis set methods. This confinement problem is not of purely theoretical interest – for example, the cylindrical confinement problem is relevant to confinement in a quantum dot, and confinement along a pipe can be applied to nanotubules.

R-matrix methods have only recently been applied to condensed matter physics[6], to the theory of low energy electron energy-loss spectroscopy from NiO. In these experiments electrons with energies of typically tens of eV are scattered off the surface of transition metal oxides[7], and the dipole-forbidden intra-atomic d-d excitations on the Ni^{2+} ions can be measured. These are ideal for studying by the R-matrix technique, as they involve localised excitations, and there is a wealth of experimental data available. In the final section of the paper I shall describe the calculation of inelastic electron scattering from NiO. A single scattering approach, in which the Ni^{2+} ion is put into an effective crystal field, gives a reasonable description of the experimental data. The way in which this work can be extended will also be described.

THE EMBEDDING METHOD

In the embedding method[2], we consider the region of interest, region I, joined on to the substrate region II, and derive a variational principle for a trial function ϕ defined explicitly *only* in region I. The boundary condition that the wavefunction must be matched in amplitude and derivative on to the solution of the Schrödinger equation in region II is replaced by extra boundary terms in the Hamiltonian for region I involving the embedding potential.

To derive the embedding variational principle we must first notionally extend the trial function ϕ into region II. We do this with the exact solution ψ of the Schrödinger equation in region II, at some energy ϵ, which matches in *amplitude* on to ϕ over the boundary surface S separating I and II. In Hartree atomic units, with $e = \hbar = m = 1$, the expectation value of the Hamiltonian is then given by:

$$E = \frac{\int_I d\mathbf{r}\phi H\phi + \epsilon \int_{II} d\mathbf{r}\psi\psi + \frac{1}{2}\int_S d\mathbf{r}_S \phi \left(\frac{\partial \phi}{\partial n_S} - \frac{\partial \psi}{\partial n_S}\right)}{\int_I d\mathbf{r}\phi\phi + \int_{II} d\mathbf{r}\psi\psi}. \tag{1}$$

The first integral in the numerator is the expectation value of the Hamiltonian through region I, the region of interest; the second term is the contribution from region II. The final term in the numerator, an integral over the interface S, contains the difference in normal derivatives on either side of S (measured outwards from I) and comes from the kinetic energy operator in the Hamiltonian. Next we use the following results which can be derived from Green's theorem:

$$\frac{\partial \psi(\mathbf{r}_S)}{\partial n_S} = -2 \int_S d\mathbf{r}'_S G_0^{-1}(\mathbf{r}_S, \mathbf{r}'_S)\psi(\mathbf{r}'_S) \tag{2}$$

$$\int_{II} d\mathbf{r}\psi\psi = -\int_S d\mathbf{r}_S \int_S d\mathbf{r}'_S \psi(\mathbf{r}_S)\frac{\partial G_0^{-1}(\mathbf{r}_S, \mathbf{r}'_S)}{\partial \epsilon}\psi(\mathbf{r}'_S) \tag{3}$$

– here G_0^{-1} is the surface inverse of the Green function in II satisfying the zero normal derivative boundary condition. The value of ψ on the interface which appears in the right-hand integrals of (2) and (3) can be replaced by the value of ϕ, as ϕ and ψ match in amplitude, and then substituting into (1) gives us the embedding variational principle:

$$E = \frac{\int_I d\mathbf{r}\phi H\phi + \frac{1}{2}\int_S d\mathbf{r}_S \phi \frac{\partial \phi}{\partial n_S} + \int_S d\mathbf{r}_S \int_S d\mathbf{r}'_S \phi \left(G_0^{-1} - \epsilon \frac{\partial G_0^{-1}}{\partial \epsilon}\right)\phi}{\int_I d\mathbf{r}\phi\phi - \int_S d\mathbf{r}_S \int_S d\mathbf{r}'_S \phi \frac{\partial G_0^{-1}}{\partial \epsilon}\phi}. \quad (4)$$

This expression gives us E purely in terms of the trial function ϕ defined in region I and on S; it is a genuine variational principle, giving an upper limit to the energy, and by minimising E with respect to variations in ϕ and energy parameter ϵ, we obtain the solution of the Schrödinger equation in I which matches in amplitude and derivative on to the solution in II.

To obtain the energy eigenvalues and eigenstates of the system we substitute a basis set expansion of ϕ into (4):

$$\phi(\mathbf{r}) = \sum_i a_i \chi_i(\mathbf{r}). \quad (5)$$

E is stationary with respect to variations in the coefficients a_i when these satisfy the eigenvalue equation:

$$\sum_j H_{ij} a_j = E \sum_j O_{ij} a_j, \quad (6)$$

where the Hamiltonian matrix elements are given by:

$$H_{ij} = \int_I d\mathbf{r}\chi_i(\mathbf{r})\left[-\frac{1}{2}\nabla^2 + V(\mathbf{r})\right]\chi_j(\mathbf{r}) + \frac{1}{2}\int_S d\mathbf{r}_S \chi_i(\mathbf{r}_S)\frac{\partial \chi_j(\mathbf{r}_S)}{\partial n_S} + \int_S d\mathbf{r}_S \int_S d\mathbf{r}'_S \chi_i(\mathbf{r}_S)\left[G_0^{-1}(\mathbf{r}_S, \mathbf{r}'_S) + (E-\epsilon)\frac{\partial G_0^{-1}(\mathbf{r}_S, \mathbf{r}'_S)}{\partial \epsilon}\right]\chi_j(\mathbf{r}'_S), \quad (7)$$

and O_{ij} is the overlap matrix:

$$O_{ij} = \int_I d\mathbf{r}\chi_i(\mathbf{r})\chi_j(\mathbf{r}). \quad (8)$$

We should note that the basis functions are not generally orthogonal through region I – they should be chosen with sufficient flexibility to give the required values of $\phi(\mathbf{r}_S)$ and $\partial\phi(\mathbf{r}_S)/\partial n_S$ rather than satisfying a homogeneous boundary condition over S.

The structure of the embedding method is clear from (6) and (7). The matrix elements of the original Hamiltonian $(-\frac{1}{2}\nabla^2 + V)$ are evaluated through region I; the second term on the right-hand side of (7) involving the normal derivative ensures hermiticity. The final double integral in H_{ij} is the matrix element of the embedding potential G_0^{-1} evaluated at energy ϵ plus the energy derivative term to give G_0^{-1} at energy E, to first order in $(E-\epsilon)$. What started off life in (4) as the normalization of the trial function in region II gives the energy correction to G_0^{-1}. Of course to proceed accurately, an initial value of ϵ is chosen, and the resulting energy eigenvalue E used to evaluate G_0^{-1} in the next iteration, the process being repeated until self-consistency is achieved with $E = \epsilon$. The relationship between the normalization and the energy derivative of the effective potential (3) also occurs in pseudopotential theory, and in many-body Green function theory, where the quasiparticle normalization involves the energy derivative of the self-energy.

At energies in the substrate continuum, the embedding potential is complex, and it broadens the discrete states of a finite region I into a continuous spectrum. In the continuum it is generally convenient to calculate the Green function $G(\mathbf{r}, \mathbf{r}'; E)$ in region I, and this can also be expanded in terms of the basis functions:

$$G(\mathbf{r}, \mathbf{r}'; E) = \sum_{ij} G_{ij}(E)\chi_i(\mathbf{r})\chi_j(\mathbf{r}'). \tag{9}$$

The Green function matrix is given in the usual way by:

$$\sum_k (H_{ik} - EO_{ik})G_{kj} = \delta_{ij}, \tag{10}$$

but as the embedding potential is evaluated at energy E – the energy at which we require the Green function – the energy derivative term in (7) disappears. It is easy to show from the residue of the poles of this Green function that the correct normalization of the states is automatically preserved.

The embedding potential G_0^{-1} can be found in a number of ways. The easiest way to find it in surface calculations is to divide the substrate into layers, and then use the layer-KKR method to find the reflection coefficients of the substrate[8]. The embedding potential replacing the substrate is related directly to the reflection matrix. Unfortunately the layer-KKR method is restricted to a muffin-tin form of substrate potential, in which an electron feels a spherically symmetric potential inside "muffin tins" centred on each atom and a flat interstitial potential in between the atoms. The zero-derivative Green function for region II with the full potential can in fact be found very accurately, by using (10) with the embedding potential in the Hamiltonian (7) set to zero. The way that this is done in practice is first to calculate G_0 for one layer of the substrate. This is then built up layer by layer into G_0 for the full semi-infinite substrate using techniques borrowed from multiple scattering theory[9]. Other ways of finding G_0 are more appropriate in other geometries. For example to obtain the embedding potential for embedding an impurity into a solid, the Green function for the infinite solid can first be found from a band-structure calculation, and then the Green function with zero derivative on the impurity boundary calculated from this using the matching Green function method[10].

EMBEDDING AND R-MATRIX THEORY

There is considerable overlap between R-matrix theory and the embedding method – both are based on Green functions, after all. Region II in embedding corresponds to the inner region in R-matrix theory; these are the regions which in some sense are eliminated and replaced by the embedding potential and the R-matrix respectively. Although we have referred to region II as the substrate, and in the examples we give in the following sections this region is extended, the embedding method can just as well be used to eliminate a finite region II. To provide an obvious analogue with R-matrix theory, the core of an atom can be replaced by an embedding potential in a band-structure calculation. In embedding, region I is the region we concentrate on, where we need to know the wavefunctions, charge density and density of states; correspondingly in R-matrix theory the outer region is where the scattering experiments take place.

The R-matrix describes processes in which the target atom or molecule, of which the N-electron wavefunctions are completely contained within the R-matrix sphere, scatters an electron from the external region either elastically or inelastically[1]: in inelastic scattering the target state changes. It is convenient to define channel functions $\bar{\Phi}_i$, which consist of an N-electron target wavefunction coupled to the angular momentum and spin of the scattering electron[11]. The full $(N+1)$-electron wavefunction of the system, Ψ, can then be projected on to this channel function, integrating over all the electron and spin coordinates except the radial coordinate r of the scattering electron, to give a one-electron function F_i:

$$F_i(r) = \langle \bar{\Phi}_i | \Psi \rangle'. \tag{11}$$

The prime on the bracket indicates the restricted integration; the usual definition of the one-electron function has a factor r on the right-hand side, but we omit this to make the analogy with the embedding method more obvious. The function $F_i(r)$ is the wavefunction of the scattering electron in the i'th channel, when the target is in a particular state coupled to particular values of angular momentum and spin of this electron. The R-matrix relates the amplitudes and derivatives of these scattering wavefunctions on the R-matrix sphere, at radius a:

$$F_i(a) = \sum_j R_{ij}(E) \left. \frac{dF_j}{dr} \right|_{r=a}. \tag{12}$$

This is just the inverse of (2), apart from the extra channel indices in this case, establishing the link between R-matrix theory and embedding.

By comparison between (12) and (2) we see that the R-matrix is essentially the zero-derivative Green function G_0 at the sphere boundary. A complication compared with embedding is that the system inside the R-matrix sphere is a many-electron system, and first the $(N+1)$-electron Green function is found, satisfying:

$$(H_{N+1} - E)G_0(\mathbf{r}_1, \mathbf{r}_2, \ldots, \mathbf{r}_{N+1}; \mathbf{r}'_1, \mathbf{r}'_2, \ldots, \mathbf{r}'_{N+1}) =$$
$$\delta(\mathbf{r}_1 - \mathbf{r}'_1)\delta(\mathbf{r}_2 - \mathbf{r}'_2) \ldots \delta(\mathbf{r}_{N+1} - \mathbf{r}'_{N+1}), \tag{13}$$

with the zero-derivative boundary condition applying to each coordinate. Projecting this on to channel functions to the left and to the right as in (11) gives us a one-electron Green function:

$$\mathcal{G}_{ij}(r, r') = \langle \bar{\Phi}_i | G_0 | \bar{\Phi}_j \rangle', \tag{14}$$

and from Green's theorem:

$$F_i(a) = \frac{a^2}{2} \sum_j \mathcal{G}_{ij}(a, a) \left. \frac{dF_j}{dr} \right|_{r=a}. \tag{15}$$

This equation relates the Green function (14) to the R-matrix defined by (12). It is equivalent to (2), the difference in sign on the right-hand side coming from the trivial difference in normal derivative – into region II in embedding and outward from the R-matrix sphere here. The factor of a^2 in (15) comes from the integration over the surface of the R-matrix sphere.

We have discussed in the last section how the Green function satisfying the zero-derivative boundary condition is found in embedding calculations. In the R-matrix method a spectral representation is generally used for the Green function – this makes it very easy to work at different energies. The states satisfying the zero-derivative boundary condition are the eigenstates of the Hamiltonian $(-\frac{1}{2}\nabla^2 + V)$ plus the normal derivative term (the first two terms in (7)); the normal derivative term in (7) is the Bloch operator of R-matrix theory[12]. These eigenstates are usually expanded in a basis set of functions with the zero-derivative boundary condition so that the normal derivative term disappears. With a finite basis set of say $\mathcal{N}$ functions, the first $\mathcal{N}$ poles of the R-matrix are given accurately, but the effect of omitted poles can be important for many applications and the Buttle correction[13], derived from the Green function for a model problem, is used to correct for these. If a basis set satisfying *arbitrary* boundary conditions at $r = a$ is used, the lowest $\mathcal{M}$ poles (say) of the R-matrix are well reproduced and the remaining $\mathcal{N} - \mathcal{M}$ poles are spread out to simulate the infinity of other poles in the exact R-matrix[14]. The Buttle correction is then no longer necessary. This is the same as the choice of basis functions in the embedding method.

EMBEDDING CALCULATIONS OF SURFACE ELECTRONIC STRUCTURE

The guiding principle in electronic structure calculations of solids is the use of the three-dimensional periodicity of perfect crystals – Bloch's theorem tells us that the electron wavefunctions can then be labelled by the Bloch wavevector **k** which gives the change in phase in going from one unit cell to the next. Unfortunately the surface destroys the symmetry in the perpendicular direction and only two-dimensional periodicity parallel to the surface remains, with a surface wavevector **K**. This is usually circumvented by considering an array of slabs, each typically 5-9 atomic layers thick and separated by a thick layer of vacuum, thus restoring the periodicity in the third dimension[15]. This procedure is useful, because many properties such as charge density, work function and surface energy are quite local, and are not much affected by the second surface of the slab as long as this is at least the screening length away. However the spectrum of states is affected by the slab geometry – in the slab the states at fixed surface wavevector are discrete, whereas the states at the surface of a semi-infinite solid can be discrete (the localised surface states) or continuous (bulk states hitting the surface). This is important for interpreting electron spectroscopies like photoemission.

The embedding method allows us to calculate the electronic structure of the surface of a true semi-infinite solid[8]. Region *I* consists of the top layer or two of atoms, plus the vacuum region, and region *II* consists of the substrate, the perfect bulk crystal. As usual in condensed matter physics, the calculations are carried out in the framework of density functional theory[16] – the electrons satisfy a one-electron Schrödinger equation containing the Hartree potential and a local exchange-correlation potential, in addition to the potential due to the atomic nuclei. The Hartree and exchange-correlation potentials depend self-consistently on the electron density, and as the density in region *II* is essentially unchanged from the perfect bulk crystal, self-consistency is found within region *I*. As discussed previously, the embedding potential to replace region *II* is either found using the layer-KKR method, or using embedding ideas to build up the zero-derivative bulk Green function G_0 layer by layer. Within region *I* any convenient basis set can be used, and in our work we use linearised augmented plane waves (LAPWs).

A calculation where embedding really comes into its own is the study of image states[3], electronic states located mostly outside the solid in which the electron is bound to the surface by its image potential:

$$V(z) \sim -\frac{1}{4|z - z_0|}. \qquad (16)$$

Here $|z - z_0|$ is the distance of the electron from the image plane; the factor 4 in the denominator comes from the electron-image distance (one factor of 2), and the fact that the image potential is induced by the electron itself (another factor of 2). If the energy of the state lies in a bulk band-gap at a particular value of the surface wavevector, the electron cannot penetrate the crystal, and it cannot leak into the vacuum because its energy lies below the vacuum zero – hence we have an image potential surface state.

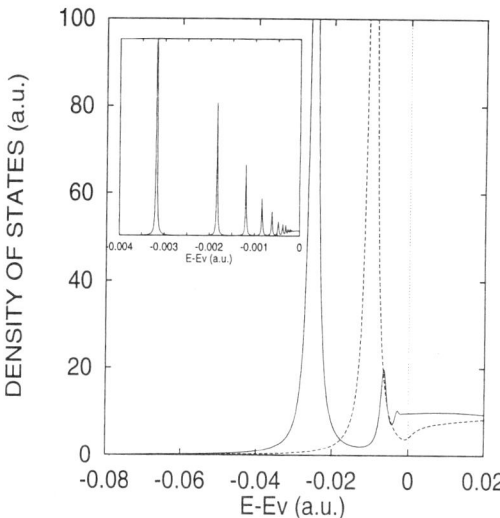

Figure 1: Density of states at **K** = 0 in the near-surface region, at the Ni(001) surface. Energy is measured from the vacuum zero, with an imaginary part of 0.001 a.u. The solid line shows the full calculation in which the electrons feel the long-range image potential, and the dashed line shows results with the local density approximation potential.

These states form a Rydberg series, and as their energies lie above the Fermi energy they are explored by techniques such as inverse photoemission[17]. Their main interest lies in the fact that they probe the effective potential near the surface. To describe image states using embedding, region I consists of the near-surface region just outside the surface, bounded on the one side by the surface atomic layer, and on the other side by the vacuum in which the electron accurately feels the asymptotic image potential (16). These are replaced by embedding potentials on the sides of the near-surface region. In the near-surface region the potential felt by the electron is taken to be an interpolation between the self-consistent density functional potential (which in the local density approximation does not contain the image potential), and (16).

Results for image states on Ni(001) are shown in figure 1, which gives the density of states in the near-surface region at **K** = 0. A small imaginary part (0.001 a.u.) has been added to the energy, to broaden the states – in reality there are Auger broadening mechanisms which are not included in the calculation. Two states are clearly resolved in this figure, but a remarkable feature is the continuous – almost constant – local density of states across the vacuum threshold. This comes from the broadening of the discrete image states so that they merge continuously with the states above threshold. By working at a much smaller imaginary part (inset of figure 1) many more Rydberg states can be resolved, but the threshold itself is invisible. The physical reason for this is that with the long-range image potential, an electron never reaches the true vacuum.

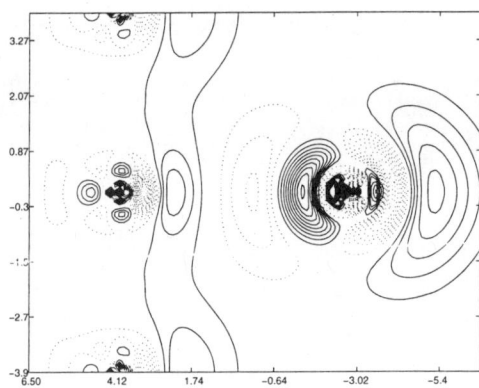

Figure 2: Contour plot of the screening charge on Ag(001) with a Xe overlayer, induced by a field of 0.01 a.u. The Ag is to the left of the figure (at $z = 3.9$ a.u.) and the Xe to the right (at $z = -2.75$ a.u.). Solid lines show where electronic charge is lost, and dashed lines where electrons are gained.

We now turn to a calculation of the screening of a static external electric field at a metal surface[4]. Embedding is very suitable for this study, because the bulk continuum is included, and this acts as a reservoir of electrons to supply the screening charge. The top layer or two of atoms, plus a restricted region of vacuum, constitute region I; it is usually unnecessary to extend region I deeper into the bulk, because of the near-perfect screening in metals of the perturbation due to the surface and the external field. Figure 2 gives the screening charge at the Ag(001) surface with an overlayer of Xe in an external field $\mathcal{E}$ of 0.01 a.u. – the figure shows the charge density difference with and without the field. The screening is perfect to within a few percent, by which we mean that the total screening charge from the calculation is equal to the classical value of $\mathcal{E}/4\pi$ per unit area. The screening charge itself is mostly on the very top of the top layer of Ag, and one of the surprises of this type of calculation is just how effective metallic screening really is. The Xe atoms behave like a classical dielectric, and are polarised. The centre of gravity of the screening charge averaged over the surface, in the limit of small field, corresponds to the classical image plane z_0 (16). This calculation gives an outward shift in z_0 on adsorption of the Xe overlayer of 4.31 a.u., in remarkable agreement with the value of 4.06 a.u. obtained from classical dielectric theory.

EMBEDDING METHOD FOR QUANTUM CONFINEMENT

The embedding method can be used to find the eigenstates of quantum systems confined by an effectively infinite potential barrier[5]. There have been several papers on this type of problem in recent years[18], for example finding the solutions to the Schrödinger equation for an atom confined inside cylindrical or spherical cavities. Advances in nanostructure fabrication techniques mean that this is not of purely theoretical interest

Table 1: Ground state energy in a.u. of H atom displaced 0.5 a.u. off centre in a spherical cavity of radius 3 a.u. The same basis set is used in both the embedding method and Brownstein's method[18], with $M = N$. V is taken to be 1.8×10^9 a.u.

Method	$N = 2$	$N = 4$	$N = 6$
Embedding	−0.31730	−0.41323	−0.41389
Brownstein	−0.44906	−0.41013	−0.41389

– for example the case of spherical confinement is relevant to an impurity in a quantum dot. To apply embedding to this type of problem, the confined region becomes region I, and the region outside where the potential V is very large is region II. The advantage of embedding once again is that any convenient basis can be used to solve the Schrödinger equation in the confinement region, the embedding potential on its boundary forcing the wavefunction amplitude to go (almost) to zero as required by confinement.

The large potential in region II leads to a simplification of the embedding formalism, and the embedding potential becomes local and energy-independent:

$$G_0^{-1} \approx \sqrt{\frac{V}{2}} \delta(\mathbf{r}_S - \mathbf{r}'_S). \tag{17}$$

The energy-independence means that the variational principle (4) simplifies, becoming:

$$E = \frac{\int_I d\mathbf{r}\, \phi H \phi + \int_S d\mathbf{r}_S \left\{ \frac{1}{2} \phi \frac{\partial \phi}{\partial n_S} + \sqrt{\frac{V}{2}} \phi^2 \right\}}{\int_I d\mathbf{r}\, \phi \phi}. \tag{18}$$

The boundary condition which this imposes on ϕ is that:

$$\frac{\partial \phi}{\partial n_S} = -\sqrt{2V}\phi \tag{19}$$

over S, which for large V and well-behaved functions means:

$$\phi(\mathbf{r}_S) \approx 0, \tag{20}$$

as we require. In practice very large values of V can be used, so the error in the eigenvalue due to leakage into the substrate can be made as small as we like.

As an example of this embedding approach we consider a H atom placed 0.5 a.u. off centre in a spherical cavity of radius 3 a.u., a problem previously treated using other approaches, in particular by Brownstein[18] who developed a new stationary principle for the zero-amplitude boundary condition. Following Brownstein we use the basis functions:

$$u_{\alpha,\beta}(r,\theta) = e^{-r} r^\alpha \cos^\beta(\theta), \tag{21}$$

where r and θ are the radial and polar coordinates relative to the atom at the origin, and $\alpha = 0, 1, \ldots (N-1)$, $\beta = 0, 1, \ldots (M-1)$. Our results are shown in table 1, and we see that the embedding method converges to the same ground state energy as Brownstein's method, with satisfactorily few basis functions. The confining potential used is $V = 1.8 \times 10^9$ a.u., so that there is negligible error due to leakage.

An advantage of embedding shared with Brownstein's method is that there is no need to construct basis functions which implicitly satisfy the boundary conditions – this can be difficult or impossible for complicated geometries. Instead, the boundary condition is imposed as a variational constraint. An advantage over Brownstein is that embedding gives a true minimum principle, as we can see from the table. On the other hand, Brownstein's stationary principle only involves volume integrals, whereas (18) requires a sometimes tedious surface integral over S. It is perhaps surprising that the embedding method converges so well with the enormous potentials used in region II. The reason is that the wavefunctions are only constrained to be near-zero on the boundary S, and there are no constraints at all on the wavefunction outside region I – in fact outside region I the ϕ which minimizes (18) is meaningless.

Recently we have been applying this method to treat the conductance properties of electron waveguides, in particular the way that electrons travel round corners[19].

R-MATRIX STUDY OF ELECTRON ENERGY-LOSS SPECTROSCOPY FROM NiO

The R-matrix method can be thought of as many-body embedding, with an electron in the outer region able to excite the many-electron system contained within the R-matrix sphere. We have recently applied the method to study the energy losses of low energy electrons scattered off NiO, in particular the losses due to d-d excitations on the Ni^{2+} ions[6,20]. The requirement that the excitations should be localised, preferably on a single ion to make the calculation tractable at this stage, means that we cannot yet study the electron excitations across the energy gap between the valence and conduction bands. Nevertheless, the d-d excitations are of particular interest because this type of excitation in the cuprate superconductors may play a role in the superconductivity[21]. The experiments, in which electrons with an energy of typically 20 - 100 eV are scattered off the surface in the same geometry as in low-energy electron diffraction, show a wealth of angle-, spin-, polarization- and energy-dependence.

To apply R-matrix theory to NiO, we put the Ni^{2+} ion on which the inelastic scattering occurs into a crystal field. This is similar in spirit to previous d^8 crystal field calculations of the Ni^{2+} target by itself[22]. The crystal field potential due to the octahedral coordination has the form:

$$V_c(r,\theta,\phi) = \beta r^4 \left[Y_{40} + \left(\frac{5}{14}\right)^{1/2} (Y_{44} + Y_{4-4}) \right] + V_M. \tag{22}$$

This is added on to a single-configuration Hartree-Fock calculation for Ni^{2+}, in which the Coulomb interaction is scaled by 0.7 to represent the effects of hybridization with the ligand orbitals. The parameter β is fitted graphically so that the N-electron target calculation reproduces the d-d excitation energies measured in electron energy loss as closely as possible (table 2), and V_M is a constant energy shift due to the Madelung potential, fitted to Hartree-Fock band-structure calculations. The values of β and V_M are 0.0418 a.u. and 0.75 a.u. respectively.

The R-matrix (12) is calculated with channel functions which are symmetry adapted to the octahedral environment, $\bar{\Phi}^{\Gamma}_{i(p_1)p_2 hl}$. Here the superscript Γ denotes the quantum numbers of the compound state of the target + scattering electron: P, the cubic representation; M_P, the state in the representation; S, the spin; M_S, the component of spin; and π, the parity. The subscript i refers to the target state with symmetry p_1; p_2 is the symmetry of the scattering electron with angular momentum l, and h indicates that different p_2's can be built up from an l. Next the T-matrix is found from the R-matrix, and from this the scattering from the ion can be immediately calculated. In inelastic scattering from NiO, the inelastic scattering event on a particular Ni^{2+} ion is sandwiched between multiple elastic scattering off all the other ions. We have not yet included these processes in the calculation, and we consider only the single inelastic

Table 2: Ground state and excited states of 3d-3d excitations in NiO. The excitation energies are from the EELS experiments[7] and are compared with our crystal field results. Energies are in eV.

Symmetry	EELS	Theory
$^3A_{2g}$	0.00	0.00
$^3T_{2g}$	1.10	1.05
^1E_g	1.60	1.70
$^3T_{1g}$	1.70	1.75
$^1T_{2g}$	2.75	2.70
$^1A_{1g}$	2.81	2.80
$^3T_{1g}$	3.00	3.13
$^1T_{1g}$	3.55	3.28
^1E_g	–	4.06
$^1T_{2g}$	–	4.12
$^1A_{1g}$	–	7.04

scattering. In the cubic representation, the total cross section averaged over incident angles and electron and target spins, for scattering from target state i to i', is then given by:

$$\bar{\sigma}(i' \leftarrow i) = \frac{4\pi^3}{p_i^2} \sum_{PS} \sum_{(p_2 hl; p_2' h'l')} \frac{N_P(2S+1)}{2N_i(2S_i+1)} |\langle \bar{\Phi}^\Gamma_{i'(p_1')p_2'h'l'} |t| \bar{\Phi}^\Gamma_{i(p_1)p_2 hl} \rangle|^2. \quad (23)$$

Here t represents the single-centre T-matrix, p_i is the momentum of the incident electron, and the summations are over the symmetry and spin of the compound system, and over the quantum numbers of the scattering electron before and after scattering. N_i and N_P are the degeneracies of the target state and the compound state in their representations p_1 and P, and $(2S_i + 1)$ and $(2S + 1)$ are the corresponding spin degeneracies. Similar expressions can be found for the differential cross sections, but they have a more complicated structure.

Results for differential cross sections as a function of loss energy are shown in figure 3, for two incident electron energies (20 and 50 eV) and two scattering geometries. The cross sections are averaged over the spin of the scattering electron and also the target spin – the multi-domain structure of NiO means that the Ni^{2+} spins are effectively randomly orientated. To compare with the experimental results of Gorschlüter and Merz[7], the discrete losses of our theory are broadened by a combination of Lorentzian and Gaussian broadening. From the comparison in figure 3 it is clear that the relative intensities are in fair agreement with the experimental findings. The major difference is that our calculations do not include the excitations across the band gap, responsible for the rise in the experimental loss spectra above about 3.5 eV, nor do they include the surface peak observed experimentally at 0.6 eV. However, we are able to identify contributions from different excitations to overlapping peaks in the loss spectrum – thus the peak at 1.7 eV is due to ^1E_g and $^3T_{1g}$ losses, and the peak at 3.2 eV due to $^3T_{1g}$ and $^1T_{1g}$ losses. As the scattering geometry is changed from specular to off-specular, our atomic model reproduces correctly the observed variations in peak intensities: the structure at 1.7 eV is raised due to an increase in both the ^1E_g and $^3T_{1g}$ intensity, and at 2.8 eV a strong shoulder appears, mainly due to $^1A_{1g}$.

These comparisons show that a description of the scattering process in terms of one single inelastic scattering at the Ni^{2+} ion seems to be adequate for explaining

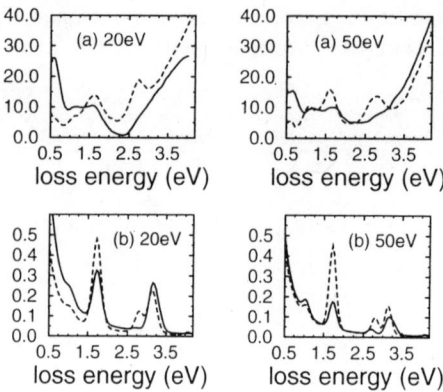

Figure 3: Spin-averaged differential cross sections of NiO(001) for two incident electron energies (20, 50 eV) and two scattering geometries. The experimental spectra are shown in (a), from Gorschlüter and Merz[7] (arbitrary units are used for the y-axis); the calculated spectra are in (b). The solid curves correspond to specular scattering at $\theta_i = 45°$, and the dashed curves to off-specular with $\theta_i = 45°, \theta_f = 10°$.

the spectral shape. The role of multiple scattering should not be neglected, however, especially for the case of strong forward inelastic scattering in near-specular scattering geometry, and for explaining the variation in the surface loss peak with scattering geometry.

The formalism to include multiple scattering has fortunately been developed already in the context of inelastic electron scattering from surface vibrations[23]. What is crucial is to establish the initial and final states, between which the inelastic scattering event takes place. The initial state consists of the incident plane wave $\exp(i\mathbf{k}.\mathbf{r})$ scattering elastically off the entire crystal, with the target Ni^{2+} ion in its ground state i. This is represented by the LEED state:

$$|\psi_{\mathbf{k}}^+, i\rangle = (1 + G_0 T)|\mathbf{k}, i\rangle, \qquad (24)$$

where G_0 is the free-electron Green function, and T is the elastic T-matrix for the crystal. After inelastic scattering the electron is detected in the plane wave $\exp(i\mathbf{k}'.\mathbf{r})$, and the target is in state i'. It can be shown that the final state is given by:

$$|\psi_{\mathbf{k}'}^-, i'\rangle = (1 + G_0^\dagger T^\dagger)|\mathbf{k}', i'\rangle, \qquad (25)$$

– a time-reversed LEED state in which we let $\exp(-i\mathbf{k}'.\mathbf{r})$ scatter off the surface and then take the complex conjugate. The transition amplitude for the inelastic scattering is then given by:

$$f(\mathbf{k}', i' \leftarrow \mathbf{k}, i) \propto \sum_{\Gamma} \sum_{(p_2 hl; p_2' h'l')} \langle \psi_{\mathbf{k}'}^-, i' | \bar{\Phi}_{i'(p_1')p_2'h'l'}^{\Gamma}\rangle \times$$
$$\langle \bar{\Phi}_{i'(p_1')p_2'h'l'}^{\Gamma} | t | \bar{\Phi}_{i(p_1)p_2hl}^{\Gamma} \rangle \langle \bar{\Phi}_{i(p_1)p_2hl}^{\Gamma} | \psi_{\mathbf{k}}^+, i\rangle. \qquad (26)$$

The initial and final states can be readily found by multiple scattering theory, if a muffin-tin form of potential is used. We are currently working on this multiple scattering formalism.

ACKNOWLEDGEMENTS

Much of this research has been done in collaboration with others, particularly Sean Clarke, Simon Crampin, Jeroen van Hoof, John Michiels, Maziar Nekovee, Cliff Noble, and Val and Phil Burke. I would especially like to acknowledge the insight which Phil Burke brings to theoretical and computational studies.

REFERENCES

1. P.G. Burke and K.A. Berrington, *Atomic and Molecular Processes: an R-Matrix Approach*, IOP Publishing, Bristol (1993).
2. J.E. Inglesfield, *J. Phys. C: Solid State Phys.* 14:3795 (1981).
3. M. Nekovee and J.E. Inglesfield, *Europhys. Lett.* 19:535 (1992).
4. S. Clarke, M. Nekovee, P.K. de Boer and J.E. Inglesfield, *J. Phys. Condens. Matt.* 10:777 (1998).
5. S. Crampin, M. Nekovee and J.E. Inglesfield, *Phys. Rev. B* 51:7318 (1995).
6. J.J.M. Michiels, J.E. Inglesfield, C.J. Noble, V.M. Burke and P.G. Burke, *Phys. Rev. Lett.* 78:2851 (1997).
7. A. Gorschlüter and H. Merz, *Phys. Rev. B* 49:17293 (1994).
8. J.E. Inglesfield and G.A. Benesh, *Phys. Rev. B* 31:6682 (1988).
9. S. Crampin, J.B.A.N. van Hoof, M. Nekovee and J.E. Inglesfield, *J. Phys. Condens. Matt.* 4:1475 (1992).
10. J.E. Inglesfield, *J. Phys. F: Metal Phys.* 11:L287 (1981).
11. P.G. Burke, A. Hibbert and W.D. Robb, *J. Phys. B: Atom. Mol. Phys.* 4:153 (1971).
12. C. Bloch, *Nucl. Phys.* 4:503 (1957).
13. P.J.A. Buttle, *Phys. Rev.* 160:712 (1967).
14. P.G. Burke and C.J. Joachain, *Theory of Electron-Atom Collisions*, Plenum, New York (1995).
15. J.E. Inglesfield, in *Cohesion and Structure of Surfaces*, North-Holland, Amsterdam (1995).
16. R.O. Jones and O. Gunnarsson, *Rev. Mod. Phys.* 61:689 (1989).
17. N.V. Smith, *Rep. Prog. Phys.* 51:1227 (1988).
18. K.R. Brownstein, *Phys. Rev. Letts.* 71:1427 (1993).
19. E. Dix and J.E. Inglesfield, *J. Phys. Condens. Matt.* 10:5923 (1998).
20. J.J.M. Michiels, J.E. Inglesfield, C.J. Noble, V.M. Burke and P.G. Burke, *J. Phys. Condens. Matt.* 9:L543 (1997).
21. R. Liu, D. Salamon, M. Klein, S.L. Cooper, W.C. Lee, S-W. Cheong and D.M. Ginsburg, *Phys. Rev. Lett.* 71:3709 (1993).
22. S. Sugano, Y. Tanabe and H. Kamimura, *Multiplets of Transition-Metal Ions in Crystals*, Academic Press, New York (1970).
23. G.C. Aers, T.B. Grimley, J.B. Pendry and K.L. Sebastian, *J. Phys. C: Solid State Phys.* 14:3995 (1981).

NEGATIVE ION RESONANCE OF MOLECULES ON SURFACES: FROM SPECTROSCOPY TO DYNAMICS

Lidija Šiller and Richard E. Palmer

Nanoscale Physics Research Laboratory,
School of Physics and Astronomy,
The University of Birmingham,
Birmingham B15 2TT, UK

1. INTRODUCTION

The formation of negative ion resonances (NIR), created via the temporary capture of an electron into an anti-bonding molecule orbital, has now been studied in many molecular systems in the gas phase[1] and on the surfaces.[2,3] After being created, the negative ion state may decay via electron emission to a vibrationally or electronically excited state (i.e. inelastic electron scattering) or even by dissociation of the molecule (i.e. dissociative electron attachment). In Fig. 1 the schematic diagram[4] illustrates these mechanisms of vibrational excitation and molecular dissociation via formation of a negative ion resonance in the gas phase (for a homonuclear diatomic molecule).

In order to obtain information on the energy, the lifetime and the symmetry of the resonant states of molecules on surfaces, electron scattering experiments i.e. high resolution electron energy loss spectroscopy (HREELS) are typically employed together with theoretical calculations. In HREELS a monochromatic electron beam is scattered from the surface region of the sample and the energy losses which the electrons may undergo, via e.g. rotational, vibrational or electronic excitation of adsorbed molecules, are monitored as a function of incident electron energy and scattering angle. Note also that there are additional possibilities for energy loss through excitation of the substrate, such as losses to phonons, plasmons, electron-hole pairs, excitons in the case of semiconductors, etc.

For molecules which are physisorbed, i.e. weakly bound via van der Waals forces to a surface, the observed resonances can usually be associated with those observed in free molecules. However, the angular distribution of scattered electrons can be different because of adsorption. The presence of the surface can change the symmetry of the molecular orbital[5], or the molecule can be oriented and/or laterally ordered on the surface, which creates novel pathways for multiple (elastic) electron scattering events[2,6] compared with the gas phase, so theoretical treatments are invaluable.

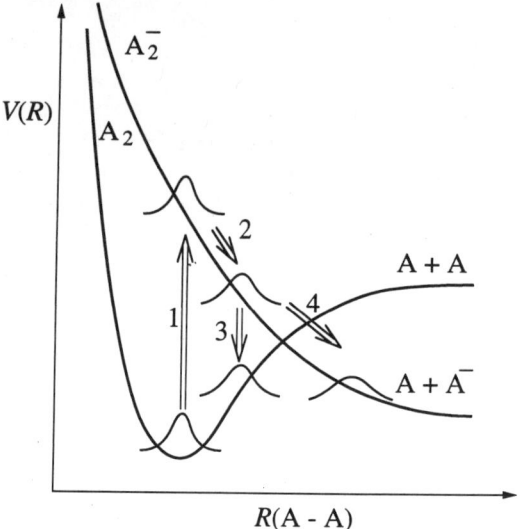

Figure 1. Schematic diagram to show the mechanism of vibrational excitation and molecular dissociation via the formation of a negative ion resonance. (1) Frank-Condon transition from the ground state of the neutral molecule (A_2) to the negative ion state; (2) propagation of the nuclear wavepacket over the potential energy surface of the negative ion, followed (3) by autoionisation leading to the vibrational excitation of A_2 or (4) by dissociation to the limit $A+A^-$.[4]

In order to illustrate the fundamental aspects of NIR phenomena at surface we will focus on two recent studies. The vibrational excitation of O_2 physisorbed on Ag(110) serves as a model system in surface dynamics.[7,8] The co-adsorption of K and O_2 on the graphite surface demonstrates the utility of resonance electron scattering as a probe of relatively sophisticated surface problems.[9] These examples are discussed in sections 2 and 3.

Many investigations in surface science deal with the thermal activation of elementary surface chemical processes, such as chemisorption, surface migration, surface reactions, desorption and diffusion. Another possible route to the activation of surface processes involves electronic excitation of surface species by means of photon, electron or ion bombardment. For example, the electronic transitions within the molecule can cause both ground state and excited neutral species to be desorbed, as well as positive and negative ions. In such cases it is possible to initiate higher energy processes which are thermally inaccessible at modest temperatures.

The investigations of stimulated desorption processes, in what might be called the "spectroscopic mode" - where the ion yield as a function of energy of the exciting particle is measured, provides (by monitoring spectral shapes and thresholds) information on both the basic excitations initiating the desorption processes and on the nature of the surface chemical bond. Thus two types of desorption spectroscopy experiments can be performed. First, to gain information about the nature of the surface bond, relatively well-defined desorption processes have to be exploited. As an example, we may mention the application of ESDIAD (Electron Stimulated Desorption Ion Angular Distributions) to quite a large number of molecule-surface systems, in order to determine the orientation of the surface bond.[10, 11] This technique is based on the fact that the desorption of both covalently and ionically bonded surface species proceeds via highly repulsive states (e.g. two hole or two hole one electron states), so that the initial repulsive interaction experienced by the desorbing ion is mainly along the direction of the ruptured bond.[10]

In the second type of stimulated desorption experiments, were the goal is to obtain information on the nature of the desorption process itself, well characterised adsorption systems have to be chosen. To illustrate this latter approach, the role of negative ion resonant states will be discussed in section 4, with special reference to a new mechanism of desorption in well characterised physisorption systems, specifically O_2 and CO physisorbed on graphite.[12]

In the last section of this review article, section 5, new method of investigating negative ion resonance dynamics, by measuring the depletion of the loss intensities in HREELS (rather than the desorption or vibrational measurements) will be suggested. Preliminary experimental evidence, in the case of physisorbed CH_3Cl on graphite surface, will be discussed which suggests the possibility of the dissociation of a negative ion resonance state into two *neutral* fragments.[13]

2. NEGATIVE ION RESONANCE SCATTERING BY PHYSISORBED O_2 ON Ag(110)

The adsorption of oxygen on the Ag(110) and Pt(111) surfaces has been the subject of considerable interest for a long time because of two features which these systems have in common. First, they act as prototypical systems for studies of precursor-mediated dissociation.[14] Specifically, on both surfaces oxygen adsorbs in three forms: physisorbed, molecularly chemisorbed and dissociatively chemisorbed, depending on the surface temperature.[15,16] Secondly, on the commercial front, the Ag(110) surface provides useful model of commercial catalysts for the ethane epoxidation reaction,[17] while the Pt(111) surface provides useful model of the commercial catalysts used in cars for the oxidation of CO.[18] However, there are of course differences between these systems. In particular, Ag(110) is a corrugated surface with two inequivalent high symmetry directions ([1$\underline{1}$0] and [001]),[19] where the Pt(111) has 6-fold symmetry.[20]

The two chemisorbed states, i.e. molecular chemisorbed and dissociative chemisorbed, have been studied extensively with a variety of experimental techniques; however, the physisorbed systems O_2/Ag(110)[7,8,21] and O_2/Pt(111),[22-25] have been explored in detailed only recently. Here we will review recent work on experimental and theoretical investigations[7,8] of the orientation of physisorbed O_2 molecules on Ag(110) as a function of coverage. Photoemission experiments[15] and a near-edge x-ray absorption fine structure (NEXAFS)[26] study suggest that the molecules in the first monolayer are "lying down" while those in second physisorbed layer of O_2 exhibit a substantially increased molecular tilt away from the surface. Recent molecular dynamics simulations suggest that the physisorbed O_2 molecules align themselves along the [001] direction, perpendicular to the rows and thus also perpendicular to the traditional orientation of the molecular chemisorbed species.[27]

Typical HREEL spectra of physisorbed O_2/Ag(110) are illustrated in fig. 2, for both monolayer and multilayer coverages at 20 K.[7] The incident electron beam energy is 7 eV. The energy loss at ~190 meV is approximately the same as that observed in the gas phase and in physisorbed O_2 on both Pt(111)[24] and graphite.[28] The intense vibrational overtone excitations are characteristic of the resonance excitation mechanism, while the rate of decay of the intensities of these features reflects the lifetime of the O_2 negative ion resonance. In fig. 3, the intensity of the ν=0-1 vibrational excitation is plotted as a function of the incident electron energy (after normalisation to the diffuse elastic intensity) for (a) monolayer and (b) multilayer coverages. In the monolayer the resonance is centred at ~ 7 eV while in the multilayer the resonance is shifted to higher energy, with a peak at ~ 9 eV.[7] In the gas phase, the $^4\Sigma_u^-$ resonance is observed at an energy of ~ 9.5 eV.[29] When O_2 is physisorbed on graphite, this resonance is observed at ~ 8.5 eV, the lowering in energy being attributed to the image potential.[28] Since graphite is a semimetal the energy shift on

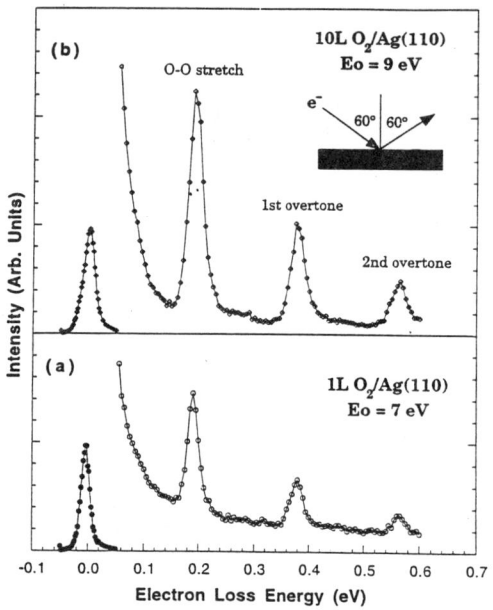

Figure 2. Typical electron-energy-loss spectra from physisorbed O_2/Ag(110) at 20K for (a) monolayer and (b) multilayer coverage.[7]

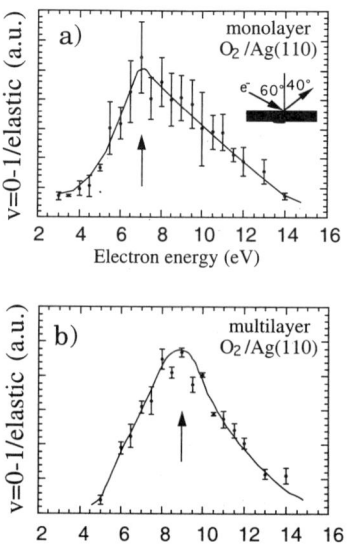

Figure 3. Resonance energy profiles obtained from physisorbed O_2 on Ag(110) at ~ 20K for (a) monolayer and (b) multilayer coverage. The solid curves are drawn as a guide to the eye.[7]

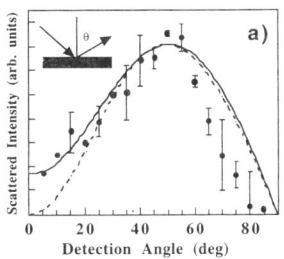

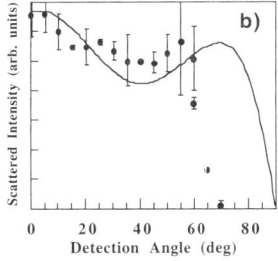

Figure 4. Angular distributions of electrons ejected by the $^4\Sigma_u^-$ resonance of O_2 physisorbed on Ag(110), obtained from the intensity of the n=0-1 vibrational excitation of the intramolecular mode observed in HREELS for (a) monolayer coverage and (b) multilayer coverage.[8]

the metallic substrate is expected to be bigger. Thus, in the case of O_2 physisorbed on Pt(111) the $^4\Sigma_u^-$ resonance is observed at ~7.25 eV.[24] Consequently, the resonant state at ~ 7 eV in the monolayer phase of physisorbed O_2/Ag(110), fig. 3. (a), is assigned to the $^4\Sigma_u^-$ state.[7] The same resonance appear at ~ 9 eV in the multilayer, fig. 3. (b), since the reduction in resonance energy by the image potential is lower when the molecule is further from the surface.

The angular distribution from the monolayer phase of O_2/Ag(110) is shown in fig. 4 (a). The incident electron energy is 7 eV (i.e. in the centre of the $^4\Sigma_u^-$ resonance state) and the incident angle 60° with respect to the surface normal.[8] An important feature of this angular distribution is the minimum along the surface normal. The angular distribution from the multilayer, fig. 4 (b),[8] is strongly modified; in particular, substantial signal is observed in the direction of the surface normal.

In the case of resonance electron scattering by free molecules the angular distributions are characteristic of the resonance symmetry, and usually associated with a very small number of spherical harmonics. On the surface, when molecule is oriented, the angular distribution reflects not only the symmetry of the resonant state but also its orientation.[30, 31] Using the selection rules for resonance scattering,[32] the minimum along the surface normal is an indication that the molecular axis of O_2/Ag(110) in the monolayer regime is approximately parallel to the surface, if one assumes scattering via partial waves of σ_u symmetry consistent with the $^4\Sigma_u^-$ resonance state.[8] The decay in the distribution towards the surface plane, fig. 4(a), is due to the image charge potential, which retards the component of velocity of emitted electrons normal to the surface.[33,34] The absence of the node along the surface normal in the angular distribution from the multilayer reflects a different orientation in this phase.[8] This agrees with the NEXAFS study,[26] which indicates that in the multilayer regime the O_2 molecules become progressively more oriented towards the surface normal. In addition to the experimental data, theoretical calculations,[8,33,34] which take into account the orientation of the molecular axis in both polar and azimuth angles as well as the refraction of electrons by the surface image potential, have been employed to predict the shape of the angular distribution of the resonantly scattered electrons. Good agreement has been achieved with the general conclusions from the NEXAFS study.[8] The overall conclusion is that the angular distributions obtained in resonance electron scattering experiments represents a useful probe of the orientation of molecules on the surfaces, especially in the case of weak adsorption.

The analysis of the decay of the vibrational overtone intensities (from figure 2) was also performed in order to extract the (relative) lifetime of the $^4\Sigma_u^-$ resonance state[7] in the physisorbed O_2 molecule on the Ag(110) surface. The theoretical analysis follows the displaced harmonic oscillator approach proposed by Gadzuk.[35] The calculated rate of decay of the overtone intensities for the monolayer phase is in agreement with the measured intensities if the resonance lifetime of the adsorbed molecule is reduced to 70% of the lifetime of the free molecule. The lifetime reduction for the multilayer coverage is even smaller; the resonance lifetime of the condensed O_2 molecule is 90% of the lifetime of the free molecule. It has been suggested that the interaction of the resonance state with the unoccupied electronic states of the substrate is principally responsible for the reduction in lifetime in the monolayer.[5,36,37] Specially interesting is the idea that this electron wave "interference" effect would actually lead to increased negative ion resonance lifetime for appropriate values of the molecule-surface separation.

3. NEGATIVE ION RESONANCE ELECTRON SCATTERING BY O_2 CO-ADSORBED WITH K ON GRAPHITE

The adsorption of alkali metals on surfaces is of particular interest because alkali metals act as promoters in catalytic reactions. In addition, the relative simplicity of the electronic structure of alkalis metals means they have become a model system for understanding the bonding and the structures of atoms to surfaces.[38,39] Experimental investigations of the structures and excitation spectra arising from alkali adsorption in the submonolayer regime have been important in the development of an understanding the nature of charge transfer between adsorbate and substrate. One key issue is the critical density at which an alkali layer becomes fully metallic.[40-42] In addition, the co-adsorption of alkali metals and small molecules on surfaces is of specific interest[43-45] in modelling "real" catalytic situations. From the fundamental perspective, the challenge is to understand the competing routes for charge transfer (or redistribution) from the alkali to (a) the surface versus (b) the coadsorbed molecule.

Here we review the co-adsorption of potassium with O_2 on the graphite surface, since it serves as a good example of the utility of negative ion resonance electron scattering in the determination of the nature of bonding in a (relatively) complex system. The co-adsorption experiments grew out of prior experiments concerned (separately) with the adsorption (phase diagram) of potassium on graphite[41,42,46] and the physisorption of oxygen on graphite where resonance scattering was extensively applied.[31,47,48] The experimental work[9,45] reviewed here has also triggered additional theoretical approaches[49] and further experimental investigations[50], and these we also will be reviewed at the end of this section.

First we summarise briefly what is known experimentally about K on graphite. At low potassium coverage and a temperature of T= 90 K a "dispersed" potassium phase, with a giant K-K spacing of up to 60 Å, has been identified through high resolution, low energy electron diffraction measurements.[41,42] This phase continuously compresses with increasing adatom density until a critical coverage Θ_c (which would correspond to a 7x7 K structure, with K-K spacing about 14 Å), in the solid phase at which point a close-packed 2x2 K structure nucleates. At low K coverage the potassium atom is ionised, and the combination of this K^+ ion with corresponding image (i.e. screening) charge in the graphite surface region creates a substantial dipole, which results in turn in a strong adsorbate-induced reduction of the work function. The charge redistribution also leads to two kinds of repulsive interaction in such a system. The first is the repulsive interaction via electric field between the potassium dipoles, while the second is (another) long-range K-K repulsive interaction which is derived from the filling the graphite π^* band by charge transfer from the alkali to the graphite.[41] These two repulsive forces operating at low K coverage and low temperature explain the dispersed hexagonal structure with maximal K-K separation. The charge transfer between the potassium and the graphite has been studied experimentally,[41,51,52] with HREELS, by monitoring the energy of the low frequency substrate surface plasmon mode which depends critically on the charge transfer from the potassium overlayer to the π^* band of the semimetallic graphite substrate.

The interaction between graphite and molecular oxygen has also been studied experimentally by a range of techniques.[31,46-48,53] On the clean graphite surface at low temperatures (T<47K) O_2 physisorbs with an O-O stretch frequency almost identical to the gas phase value, indicating that the perturbation of the O_2 electronic structure by the graphite surface is minimal, i.e. the interaction can be described via van der Waals forces. The phase diagram of the system is extremely rich,[46] exhibiting a variety of different molecular phases as a function of temperature and coverage for example, at a temperature of 25 K the physisorbed layer exhibits an orientational phase transition from the lying down (δ) phase to the standing up (ζ) phase with increasing coverage in submonolayer regime.

The co-adsorption experiments were performed using high resolution electron energy loss spectrometer which doubles as a high resolution low energy electron diffractometer.[54] Fig. 5. shows the HREELS spectra obtained when the "open" 7x7 phase of K on graphite (K-K separation ~ 14Å) is exposed to O_2.[9] The O_2 exposure is given in langmuirs (1L = 1x10^{-6} Torr sec). The HREEL spectra are obtained in the specular direction, where the dipole scattering mechanism dominates so that the low frequency surface plasmon mode is the dominant energy loss feature. Upon the adsorption of O_2 the plasmon peak shifts from ~ 320 meV to a value of ~130 meV. This reduction indicates that (some of) the electronic charge donated to the substrate by the alkali is withdrawn when O_2 is co-adsorbed. In addition, the diffractional peak character of the 7x7 phase of K disappear, since this phase is stabilised by the repulsive K-K interaction arising from charge transfer to the surface.[41] Fig. 6. presents the HREELS spectrum obtained with an electron beam energy of E= 4 eV and with an off specular scattering geometry. In this case two series of vibrational loss features are observed with the help of resonant electron scattering. In addition to the physisorbed O_2 vibrational excitation at ~ 190 meV (ν=0-1) and corresponding overtone at ~ 380 meV (ν=0-2) a second loss feature is evident at 140 meV (with an overtone at ~ 280 meV). The loss feature at 140 meV is absent from the EELS spectrum of physisorbed O_2 on graphite,[31,47,48] but has a frequency close to the vibrational frequency of the $^2\Pi_g$ ground state of the gas phase O_2^- negative ion, 133 meV.[55] This result suggests the local transfer of electron charge from the alkali to the co-adsorbed O_2 molecule, rather than to the substrate. The intensity of this "O_2^-" mode as a function of electron beam energy is shown in fig. 7, together with the cross section for the ν=0-1 excitation of physisorbed O_2 (i.e., the 190 meV feature in fig. 6). In the case of the physisorbed O_2 species, the well defined resonance peak observed at ~ 8 eV is assigned to the $^4\Sigma_u^-$ negative ion shape resonance (the "σ resonance") of the O_2 molecule.[31,47,48] In the case of the 140 meV feature, a resonance is again observed, but this time shifted down in energy to ~ 4 eV. This result is consistent with an increase in the O-O bond length as a result of the transfer of one electron into the antibonding π^* orbital of the O_2 molecule, leading to a decrease in the energy of the σ^* resonance.[24,49]

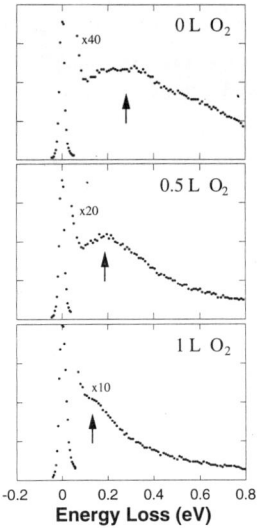

Figure 5. Electron energy loss spectra from co-adsorbed K and O_2 on graphite as a function of increasing O_2 exposure.[9]

The oxidation of graphite is catalysed by potassium and other alkali metals.[56,57] The formation of a K-O_2 complex is believed to be the first step in the oxidation process.[49] Clearly the bonding geometry of the complex formed by O_2 (or O) and the potassium atom are important in such a process. Recent electronic structure calculations suggest that O_2 chemisorbs "side on" to K on top of on graphite, and that the barrier for dissociation is much smaller than that of the gas phase or physisorbed molecule.[49] The vibrational frequency found in this calculated adsorption configuration is 151 meV, relatively close to

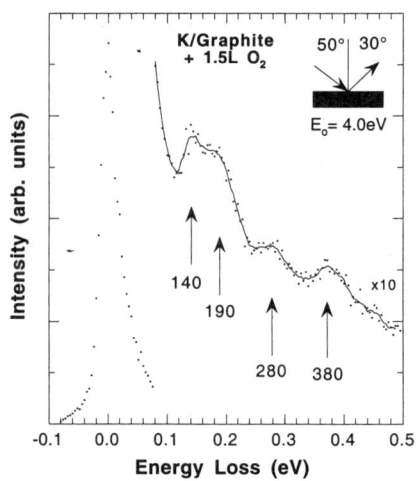

Figure 6. Electron energy loss spectrum from co-adsorbed K and O_2 (1L) on graphite at T<30 K. The incident electron energy is 4 eV and the off specular scattering geometry is shown (inset).[9]

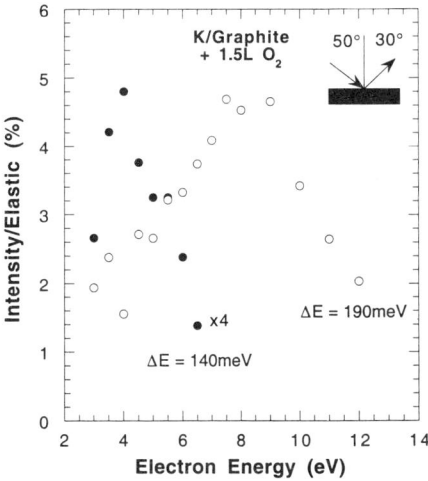

Figure 7. Resonance energy profile for O2 (1.5 L) co-adsorbed with potassium (0.1 ML) on graphite at T< 30K. The intensities of 140 meV and 190 meV loss features, normalised to the diffuse elastic intensity, are plotted as a function of incident electron energy. The solid curves are drawn as a guide to the eye.[9]

the experimentally observed frequency, 140 meV, found in the experiment. Thus the theoretical calculations are consistent with the conclusions of the experimental study.

Recent photoemission and X-ray absorption spectroscopy studies of oxygen coadsorbed with a (dense) (2x2) layer of K on graphite have shown three different adsorption phases for oxygen.[50] At lower oxygen doses a species akin to K_2O is identified. At higher oxygen doses, K_2O_2 and KO_2 are present on the surface up to saturation coverage. These data confirm the notion that the electron charge initially donated from the alkali to the graphite is gradually withdrawn from the substrate with increasing oxygen coverage, in agreement with the HREELS results for the dispersed K phase.[41]

4. ELECTRON ATTACHMENT TO OPTICALLY EXCITED MOLECULES: O_2 AND CO/ GRAPHITE

The topics of primary concern in the investigation of stimulated desorption processes, often termed "DIET" (desorption induced by electronic transitions) processes,[58] are the nature of chemical bonding at surfaces (in both the ground and excited states), surface dynamical processes involving charge and energy transfer, interactions among adsorbates and the conversion of electronic potential energy into nuclear motion.

The main cause of the fragmentation of a molecule in the gas phase is the electronic excitation of metastable states of the molecule. When molecules are adsorbed on a surface, such states may manifest themselves as precursor states in photon stimulated desorption (PSD) of their fragments (the so-called "direct" process). Because of the presence of the surface, PSD may also proceed via the production of intermediate secondary electrons in the substrate (the so-called "indirect" process), and in this case the electronic structure of the substrate is crucial. However, direct processes may also be modified on the surface, through the perturbation of the molecular orbitals by the substrate (e.g. due to chemisorption) or by co-adsorbed molecules, or as a result of the influence of the substrate on the dynamics of desorption.

Here we review recent investigations of the photon stimulated desorption of ions from physisorbed layers of CO and O_2 on graphite using vacuum ultraviolet (VUV) synchrotron radiation in the energy range of 13 to 30 eV.[12] In both systems there is a significant enhancement, in the monolayer regime, of the desorption of O^- ions via neutral resonant states which are created by direct molecular photoabsorption.

CO was physisorbed on graphite at a temperature ~30 K, and the O^- and C^+ ion yields were monitored as a function of photon energy. In Fig. 8. we plot the yield of photodesorption ions from a thin film of CO on graphite as a function of photon energy. This coverage corresponds to the low density commensurate monolayer phase.[59] A single peak is observed at a photon energy of 22 eV and no other ions (positive or negative) were observed at this coverage. C^+ ions were observed only for higher coverages. For an exposure of 6 L of CO, two peaks were identified at ~ 21.5 eV and ~ 27 eV.[59] Since both the O^- ion yield and C^+ ion yield for CO on graphite show a peak at ~ 22 eV, the mechanism of dissociation is assigned to dipole dissociation (DD).[59] Gas phase measurements in the photon energy range 20.9 eV to 22.5 eV reveal seven intense bands, attributed to DD via predissociation of $3d\sigma$ and $3p\sigma$ Rydberg states converging to the $D^2\Pi(5\sigma)^{-1}(1\pi)^1(6\sigma)^1$ state.[60] Fig. 9. shows the coverage dependence of the O^- and C^+ ion yields at the photon energy of 21.5 eV, where dipolar dissociation occurs. Note that the C^+ signal begins to appear at a coverage where the O^- ion yield reaches a maximum. This coverage corresponds (approximately) to the completion of one monolayer of CO on the surface. Thus the yield of O^- ions is enhanced in the monolayer regime while that of C^+ is suppressed.

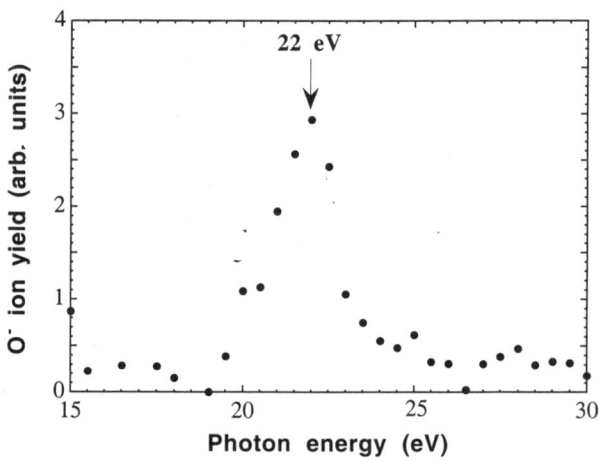

Figure 8. Photon stimulated desorption yield of O^- ions from submonolayer (3L) CO physisorbed on graphite as a function of photon energy.[12]

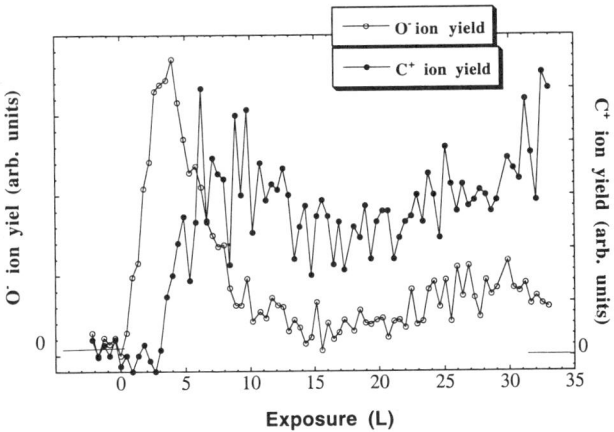

Figure 9. The yield of O⁻ and C⁺ ions photodesorbed from CO/graphite as a function of exposure for a photon energy of 22 eV.[12]

The photon stimulated desorption of ions from O_2 physisorbed on the graphite surface reveal the same type of behaviour, i.e., strong enhancement of the negative ion signal in the monolayer regime together with suppression of the corresponding positive ion.[12] Even more surprising, a resonant feature is observed at 15.3 eV in the desorbed O⁻ yield, which falls below the thermodynamic threshold for gas phase production of O⁻ (17.3 eV).[60] The mechanism invoked in both cases to explain this behaviour is the surface electron attachment to the physisorbed molecule "following" photoexcitation of the molecule into series of Rydberg states.[12] Additional experiments need to be done to explain unambiguously the state selectivity, e.g. in the case of O_2, why the resonant state observed in photoabsorption at 17 eV (in the gas phase) is quenched (it was suggested that the Rydberg states contributing to dipolar dissociation in O_2 at 17 eV have predominantly σ symmetry, while the 15 eV states have predominantly Π symmetry[61]).

A similar enhancement in the desorption yield of O⁻ ions from monolayers of O_2 physisorbed on graphite was also observed in electron stimulated desorption (ESD) experiments,[62] and was attributed to a similar substrate mediated mechanism. However, in this case it was argued that surface electron attachment may happen after the neutral atomic fragments have been formed, rather than via direct electron attachment to the photoexcited molecule.

5. EVIDENCE FOR DISSOCIATION OF NEGATIVE ION RESONANCE TO NEUTRAL FRAGMENTS: CH₃Cl/GRAPHITE

As illustrated by the previous section, negative ion resonance states are commonly invoked in the interpretation of surface dynamical processes, driven by both electrons and photons. However, in photon stimulated desorption experiments, were resonance states can be populated by photo-excited electrons, the desorbed fragments usually detected are neutrals (i.e. not ions).[63-65] Yet the formation of neutral fragments from negative ion resonant states has rarely, if ever, been reported in the gas phase. The mechanism by which the neutral fragments are thought to be formed is due to (surface) electron capture and subsequent re-emission, such that the molecule is left with sufficient vibrational energy in the ground electronic state to dissociate.[66] Experiments looking at electron impact on adsorbed molecules are usually set up to detect negative ion fragments produced by dissociative electron attachment (DEA).[67] In DEA the molecule remains in the electronically excited negative ion state until the fragments separate, one of which carries the negative charge. Here we review a recent study of the dissociation of physisorbed CH_3Cl on graphite, which provides first direct evidence of the decay of a negative ion resonance state into neutral products.[13]

The condensation of CH_3Cl on graphite at ~100 K, and the associate coverage calibration, have previously been characterised by photodesorption studies using synchrotron radiation, with reference to the low coverage orientational phase transition in the molecular film (i.e., the β-α phase transition).[68] Fig. 10 shows a series of HREELS spectra for a thin layer of CH_3Cl (6 monolayers), and demonstrates a decay in intensity of the C-Cl stretch[69] mode over a period of time.[13] The spectra are normalised to the elastic peak in order to account for any drift in the spectrometer transmission function with time. Since the spectra recorded one hour apart but without electron irradiation reveal no significant change in the intensity of C-Cl mode, the conclusion is reached that it is the e-beam irradiation which leads to the depletion of the C-Cl mode of CH_3Cl/graphite. In order to obtain the cross section for the depletion of this mode as a function of electron impact energy the rate of the decay of the C-Cl mode has been monitored for various e-beam energies. Fig. 11 is a plot of the intensity of the C-Cl mode versus time for three different impact energies. The data has been normalised to the intensity of the C-Cl vibration at the start of each experiment (t=0), since the coverage is the same for each impact energy. For each energy the data is fitted by a straight line consistent with zero order kinetics.[13] The slope of the line enables one at least, in principle, to obtain the absolute cross section if one knows the exact profile of the incident electron beam on the surface. Fig. 12 shows the

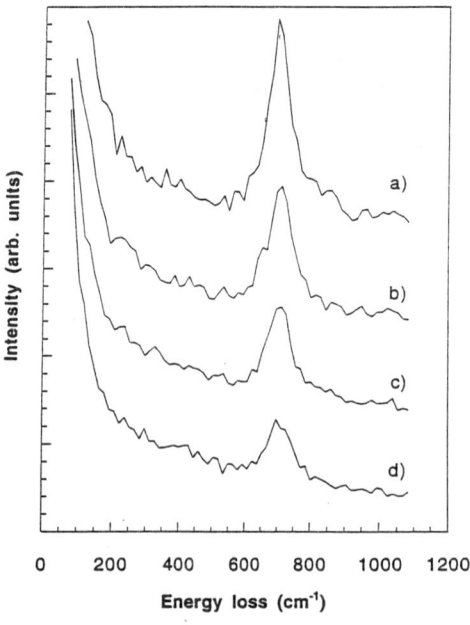

Figure 10. Series of HREELS spectra (a thin layer of CH_3Cl is condensed on graphite at 100 K) showing the decay in intensity of the C-Cl stretch mode after: a) 2 min, b) 10 min, c) 20 min, d) 30 min.[13]

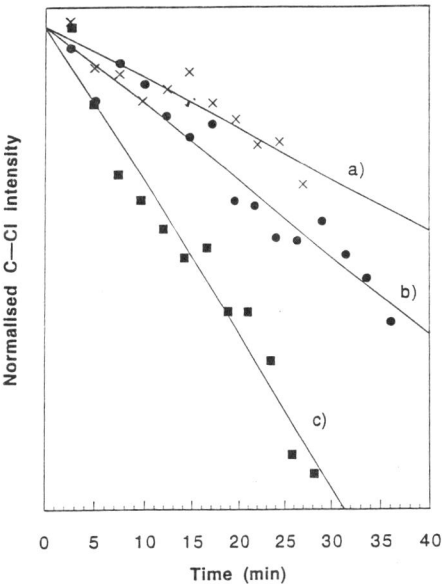

Figure 11. The normalised intensity of the c-Cl stretch mode as a function of time for different electron energies: a) 1.5 eV, b) 4 eV, c) 3 eV. A least squares straight line fit to each set of data is also shown.[13]

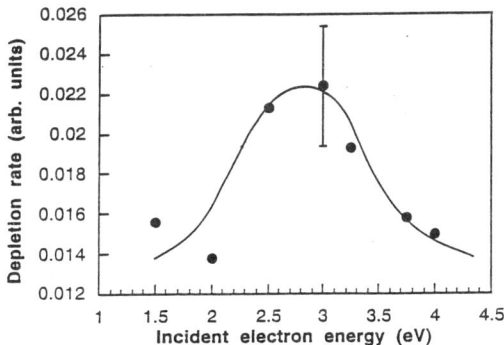

Figure 12. Plot of the rate of the normalised C-Cl stretch mode intensity as a function of incident electron energy. Each point shown represents the gradient of a line such as those shown in fig.11. A resonance is observed at 2.8 eV.[13]

slopes (i.e. rates of the decay) as a function of the electron impact energy over the range 1.5-4 eV, and reveals a clear resonance feature centred at 2.8 eV.[13] The approximate cross section for C-Cl decay at the peak of this curve is 10^{-15} cm^2.

In the gas phase, the Cl$^-$ ion yield due to dissociative electron attachment to CH$_3$Cl, shows a resonance at 7-8 eV together with a much weaker resonance below 1 eV.[70] Electron stimulated desorption experiments for CH$_3$Cl in the condensed phase have observed a Cl$^-$ ion yield only above ~ 6eV.[71,72] Charge trapping experiments in condensed CH$_3$Cl (which measure the formation of negative ions which remain on the surface rather than desorb) show a resonance near 0.5 eV.[71,72] Thus, in previous studies, there is no evidence for a resonance near 3 eV leading to the production of Cl$^-$ ions. However, measurements of electron scattering by gaseous CH$_3$Cl do show a resonance in the cross section for vibrational excitation at 3.6 eV.[69] This resonance energy would be shifted down slightly by the polarisation potential in the condensed layer.[2] Thus it is suggested that the resonance identified in the "depletion" measurements[13] corresponds to this gas phase state ($^2A_1$), which then leads not only to vibrational excitation of the molecule but also to C-Cl bond dissociation to create neutral fragments without production of Cl⁻ ions.

The experiments reviewed[13] also demonstrate a new type of measurement, using high resolution energy loss spectroscopy, which allows one to obtain cross sections for the "depletion" of a specific vibrational mode as a function of time, incident electron energy and angle. In the future it would be of interest to obtain such depletion measurements for the individual modes of polyatomic molecules adsorbed on surfaces.

6. SUMMARY

In this article we have reviewed a number of examples of our recent research in the field of negative ion resonance phenomena at surfaces, with emphasis in the fundamental aspects of the work.

The studies of resonance scattering by physisorbed O_2 on Ag (110) demonstrate the role of molecular alignment on the surface and the role of theory in understanding the changes in resonance character compared with the gas phase. The investigations of a more complex system, the coadsorption of K and O_2 on the graphite surface, demonstrates the utility of resonance electron scattering in solving real surface problems. In this case, resonance scattering obtained not just from physisorbed O_2 but also from a chemisorbed O_2 species. A new mechanism for surface dynamics, i.e. surface electron attachment to a photoexcited molecule (the Rydberg state) is proposed in the case of photo desorption of negative ions from CO and O_2 monolayers on graphite. Finally, a new kind of measurements, of the "depletion" of the vibrational modes of CH_3Cl condensed on the graphite surface, provides possibly the first direct evidence of the dissociation of a molecular negative ion state into two neutral fragments.

ACKNOWLEDGEMENTS

The experiments reviewed in this article were genuinely supported by the EPSRC and EU. L.Š. is grateful to the Royal Society for the award of a Dorothy Hodgkin Research Fellowship.

REFERENCES

1. G.J. Schultz, *Rev. Mod. Phys.* 45:423 (1973).
2. R.E. Palmer and P.J. Rous, *Rev. Mod. Phys.* 64:383 (1992).
3. R.E. Palmer, *Prog. Surf. Sci.* 41:51 (1993).
4. R.E. Palmer, *Comments At. Mol. Phys.* 30:77 (1994).
5. P.J. Rous *Phys. Rev. Lett.* 74:1835 (1995).
6. E.T. Jensen, R.E. Palmer and P.J. Rous, *Phys. Rev. Lett.* 64:1301 (1990).
7. K.B.K. Tang, P.J. Rous and R.E. Palmer, *Phys. Rev. B* 52:12395 (1995).
8. K.B.K. Tang, J. Villette, D. Teillet-Billy, J.P. Gauyacq and R.E. Palmer, *Surf. Sci.* 368:43 (1996).
9. K.M. Hock, J.C. Barnard, R.E. Palmer and H. Ishida, *Phys. Rev. Lett.* 71: 641 (1993).
10. R.D. Ramsier and J.T. Yates, Jr., *Surf. Sci. Reports* 12:243 (1991).
11. T.E. Madey, *Surf. Sci.* 300:824 (1994).
12. L. Šiller, S.L. Bennett, M.A. MacDonald, R.A. Bennett, R.E. Palmer and J.S. Foord, *Phys. Rev. Lett.* 76:1960 (1996).
13. J. Wilkes, R.E. Palmer and C.L.A. Lamont, to be published.
14. A. Zangwill, *Physics at surfaces*, Cambridge University Press, 1990.
15. K.C. Prince, G. Paolucci and A.M. Bradshaw, *Surf. Sci.* 175:101 (1986).

16. J.L. Gland, B.A. Sexton and G.B. Fisher, *Surf. Sci.* 95:587 (1980).
17. D.J. Sajkowski and M. Boudart, *Catl. Rev. Sci. Eng.* 29:325 (1987).
18. A. Naumovets, private communications.
19. L. Vattuone, M. Rocca, C. Borgano and U. Valbusa, *J. Chem. Phys.* 101:713 (1994).
20. M.A. VanHove, J.Cerda, P. Sautet, M.L. Bocquet and M. Salmeron, Prog. Surf. Sci. 54:315 (1997).
21. K.B.K. Tang and R.E. Palmer, *Phys. Rev. B* 53:1099 (1996).
22. L. Šiller, R.E. Palmer and J.F. Wendelken, *J. Chem. Phys.* 99:7175 (1993).
23. L. Šiller, K.M. Hock, R.E. Palmer and J.F. Wendelken, *Surf. Sci.* 287/288:165 (1993).
24. L. Šiller, J.F. Wendelken, K.M. Hock and R.E. Palmer, *Chem. Phys. Lett.* 210:15 (1993).
25. J.C. Barnard, K.M. Hock, L. Šiller, M.R.C. Hunt, J.F. Wendelken and R.E. Palmer, *Surf. Sci.* 291: 139 (1993).
26. R.J. Guest, B. Hernnas, P. Bennuch, O. Bjorneholm, A. Nilsson, R.E. Palmer and N. Martensson, *Surf. Sci.* 278:239 (1992).
27. F. Bartolucci, R. Franchy, J.C. Barnard and R.E. Palmer, *Phys. Rev. Lett.* 80:5224 (1998).
28. P.J. Rous, E.T. Jensen and R.E. Palmer, *Phys. Rev. Lett.* 63:2496 (1989).
29. S.F. Wong, M.J.W. Boness and G.J. Schultz, *Phys. Rev. Lett.* 31:969 (1973).
30. J.W. Davenport, W. Ho and J.R. Schrieffer, *Phys. Rev. B* 17:3115 (1978).
31. E.T. Jensen, R.E. Palmer and P.J. Rous, *Phys. Rev. Lett.* 64:1301 (1990).
32. P.J. Rous, R.E. Palmer and E.T. Jensen, *Phys. Rev. B* 41:4793 (1990).
33. D. Teillet-Billy and J.P. Gauyacq, *Nucl. Instrum. Methods B* 58:393 (1991).
34. D. Teillet-Billy and J.P. Gauyacq, *Nucl. Instrum. Methods B* 101:88 (1995).
35. J.W. Gadzuk, *J. Chem. Phys.* 79:3982 (1983).
36. P.J. Rous, *Surf. Sci.* 326:67 (1995).
37. P.J. Rous and D.M. Hartley, *Chem. Rev. Lett.* 236:299 (1995).
38. See, for example, C. Stampfl, M. Scheffler, H. Over, J. Burchardt, M. Nielsen, D.L. Adams and W. Moritz, *Phys. Rev. Lett.* 69:1532 (1992).
39. See, for example, O. Pankratov and M. Scheffler, *Phys. Rev. Lett.* 70:351 (1993).
40. M.S. Dresselhaus and G. Dresselhaus, *Adv. Phys.* 30:139 (1981).
41. Z.Y. Li, K.M. Hock and R.E. Palmer, *Phys. Rev. Lett.* 67:1562 (1991).
42. Z.Y. Li, K.M. Hock, R.E. Palmer and J.F. Annett, *J. Phys. C* 3:S103 (1991).
43. H.P. Bonzel, *Surf. Sci. Rep.* 8:43 (1987).
44. See, for example, T. Gerber, R. Grossecher, A. Morgante, A. Bottcher and G. Ertl, Phys. Rev. Lett. 70:1331 (1993).
45. C. Janiak, R. Hoffman, P. Sjovall and B. Kasemo, *Langmuir* 9:3427 (1993).
46. M.F. Toney and S.C. Fain, Jr., *Phys. Rev. B* 36:1248 (1987).
47. E.T. Jensen, R.E. Palmer and P.J. Rous, *Chem. Phys. Lett.* 169:204 (1990).
48. E.T. Jensen, R.E. Palmer and P.J. Rous, *Surf. Sci.* 237:153 (1990).
49. D. Lamoen and B.N.J. Persson, *J. Chem. Phys.* 108:3332 (1998).
50. C. Puglia, P. Bennich, J. Hasselstrom, P.A. Bruwiler, A. Nilsson, A.J. Maxwell, N. Martensson and P. Rudolf, *Surf. Sci.* 383:149 (1997).
51. K.M. Hock and R.E. Palmer, *Surf. Sci.* 284:349 (1993).
52. J.C. Barnard, K.M. Hock and R.E. Palmer, *Surf. Sci.* 287/288:178 (1993).
53. H.S. Youn and G.B. Bess, *Phys. Rev. Lett.* 64:443 (1990).
54. E.T. Jensen, R.E. Palmer and R.F. Willis, *Rev. Sci. Instrum.* 60:2408 (1988).
55. K.P. Huber and G. Herzberg, Molecular Spectra and Molecular Structure, Vol. 4.,Van Nostrand Reinhold, New York (1979).
56. R.A. de Paola, F.M. Hoffmann, D. Heskett, E.W. Plummer, *J. Chem. Phys.* 87:1361 (1987).
57. P. Sjovall, B. Hellsing, K. Keck, B. Kasemo, *J. Vac. Sci. Technol. A* 5:1065 (1987).
58. P. Avouris and R.E. Walkup, *Annu. Rev. Phys. Chem.* 40:173 (1989).
59. L. Šiller, P. Laitenberger, R.E. Palmer, S.L. Bennett, J.S. Foord and M.A. MacDonald, *Nucl.Inst. & Methods in physics research section B - Beam Interactions with Materials* 101:73 (1995).
60. H. Ortel, H. Schenk and H. Baumgartel, *Chem. Phys.* 46:251 (1980).
61. P.M. Dehmer and W.A. Chupka, *J. Chem. Phys.* 62:4525 (1975).
62. K.B. Tang, R.Azria, Y. LeCoat, M. Tronc and R.E. Palmer, *Z. fur Phys. D* 38:41 (1996).
63. Q.Y. Yang, W.N. Schwarz, P.J. Lasky, S.C. Hood, N.L. Loo, R.M. Osgood, Phys. Rev. Lett. 72: 3068 (1994).
64. X.L. Zhou, X.Y. Zhu and J.M. White, Surf. Sci. Rep. 13:73 (1991).
65. D.G. Bush and W. Ho, Phys. Rev. Lett. 77:1338 (1996).
66. J.W. Gadzuk, Phys. Rev. B 44:13466 (1991).
67. G.A. Kimmel and T.M. Orlando, Phys. Rev. Lett. 77:3983 (1996).
68. J. Wilkes, C.L.A. Lamont, L. Siller, J.-M. Coquel and R.E. Palmer, *Surf. Sci.* 390:237 (1997).
69. X.Shi, V.K. Shen, G.A. Gallup and P.D. Burrow, *J. Chem. Phys.* 104:1855 (1996).
70. P.D. Burrow and D.M. Pearl, *Nucl. Instrum. Meth.* B 101:219 (1995).
71. P. Ayotte, J. Gamache, A.D. Bass, I.I. Fabrikant and L. Sanche, *J. Chem. Phys.* 106:749 (1997).
72. L. Sanche, A.D. Bass, P. Ayotte and I.I. Fabrikant, *Phys. Rev. Lett.* 75:3568 (1995).

BERTHA – 4-COMPONENT RELATIVISTIC MOLECULAR QUANTUM MECHANICS

I. P. Grant

Oxford University Mathematical Institute
24/29 St. Giles'
Oxford OX1 3LB

INTRODUCTION

The need to include relativistic effects in calculations of atomic structure and processes has been appreciated for many years. Although Dirac (1928) only considered the hydrogen atom, for which relativistic effects are tiny, his formalism showed that relativistic effects grow rapidly with atomic number Z, and Swirles (1935) was the first to formulate a relativistic self-consistent field formalism for many-electron atoms based on the Dirac hamiltonian. Little was done to apply this development until serious calculations on many-electron atoms became possible in the 1960s with the introduction of electronic computers, coinciding with a compact reformulation of the relativistic self-consistent field problem using Racah algebra (Grant 1961). Relativistic correlation effects were first studied in the 1970s, (Desclaux 1975, Grant et al., 1976) and relativistic electronic structure calculations can now be made for any open shell atom in the Periodic Table, limited only by the available computing power.

Relativity profoundly affects the properties of matter. Although relativity cannot be switched off, it is a comforting fallacy to think that because an electron has a binding energy of only a few electron volts it exhibits no relativistic behaviour, and that perhaps relativistic effects can therefore be ignored. The analogy with a comet in a low energy solar orbit is instructive. The body accelerates in the solar gravitational field, reaching its maximum speed at perihelion, when it is closest to the sun. Similarly, a valence electron which probes the high potential regions near an atomic nucleus may move there with speeds approaching c, the velocity of light, even though its total energy is small. Quantitative radial density plots for hydrogen-like atoms were first made by Val Burke and myself (Burke et al., 1967) in the case of hydrogen-like mercury ($Z = 80$). Such density plots are all more compact than their nonrelativistic counterparts, and the orbital binding energies are correspondingly increased. Whereas Schrödinger states of hydrogenic atoms can be labelled by quantum numbers n, l, m_l, m_s, where n is the principal quantum number, $l\hbar$ and $m_l\hbar$ are the orbital angular momentum of the electron and its projection on the quantization axis, and m_s is the spin (up or down), Dirac states of a hydrogenic atom must be labelled by n, j, l, m_j, where now $j\hbar, m_j\hbar$

are the total (orbital +spin) angular momentum and its projection, in which j can take values either $j = l + 1/2$ or $j = l - 1/2$. This is conveniently expressed in terms of a quantum number $\kappa = (2j+1)(l-j)$, taking non-zero integer values. The fact that, in a hydrogenic atom, the orbital energies, $\varepsilon_{n,|\kappa|,m_j}$ are degenerate with respect to j rather than to l, is usually referred to as "spin-orbit splitting", which grows roughly like Z^4 as Z increases.

Things become more complicated in many-electron atoms because of the interaction between electrons. Although the degeneracy of hydrogenic levels $\varepsilon_{n,|\kappa|,m_j}$ with respect to j is lifted in a non-Coulomb field, the tightly bound electrons display the same strong relativistic contraction and increased binding. However, high angular momentum electrons have a lower probability of probing the high potential regions near the nucleus, and screening by the compact penetrating orbitals opposes the dynamical effects of relativity. Thus the relative binding of electron shells is altered, and this may be sufficient to change the order in which electron shells are filled in the heavier elements (Figure 1). Many of the subtle effects which can be attributed to relativity have been documented in detail, especially by Pyykkö (1978, 1988).

The qualitative insights reviewed by Pyykkö are largely based on atomic Dirac-Hartree-Fock calculations, supplemented by results from 1-centre molecular Dirac-Hartree-Fock schemes of limited validity, or from even cruder relativistic Hueckel models. More recent work is based on semi-relativistic schemes involving effective core potentials or density functionals in which it is very hard to quantify the errors of approximation. Relativistic calculations at a similar level to those of atomic physics are few and far between. As in nonrelativistic molecular structure, *ab initio* relativistic calculations based on the Dirac Hamiltonian must use basis set methods such as were pioneered by Hall (1951) and Roothaan (1951), in which each atomic or molecular orbital is expanded in a set of simple three-dimensional functions centred on each atomic nucleus. The first reasonable attempt to do this for the Dirac-Hartree-Fock atomic problem was made by Kim (1967). He obtained a solution for Be, but encountered many computational problems that were not resolved for many years. Many further attempts were made in the 1970s which revealed all sorts of horrors: for example, the energies predicted in many one electron calculations converged to values well below their proper position, a phenomenon dubbed "variational collapse". Moreover many papers by Sucher, for example (Sucher 1980), argued that the Dirac-Coulomb Hamiltonian has no bound states, implying that solutions of the Dirac-Hartree-Fock equations do not exist. His claim that the only way to avoid this disaster, and another "disaster" dubbed "continuum dissolution" (Brown and Ravenhall 1951), was to surround all operators with projectors to exclude "negative energy states" has persuaded many workers that working directly with the Dirac operator is "fundamentally flawed". This perception has dominated relativistic quantum chemistry since 1980.

The BERTHA program is based on a wholesale rejection of this view, as argued in depth by Grant (1996) and Quiney et al. (1998a). The first point is that all variational approaches for computing bound states assume that the trial wavefunctions must satisfy appropriate boundary conditions, which are part of the definition of the domain of Hilbert space in which calculations are to be performed. In the case of the Dirac one-electron atomic Hamiltonian, the crucial boundary condition is at the singular point at the origin. Analysis shows that the four components of the trial spinor are everywhere strongly coupled by the Dirac operator. Provided the basis functions used to expand the trial solution are compatible with this coupling, there is an effective lower bound supporting the bound states somewhere in the energy interval $-mc^2 < E < mc^2$. Variational calculations for bound states will thus converge in much the same way as

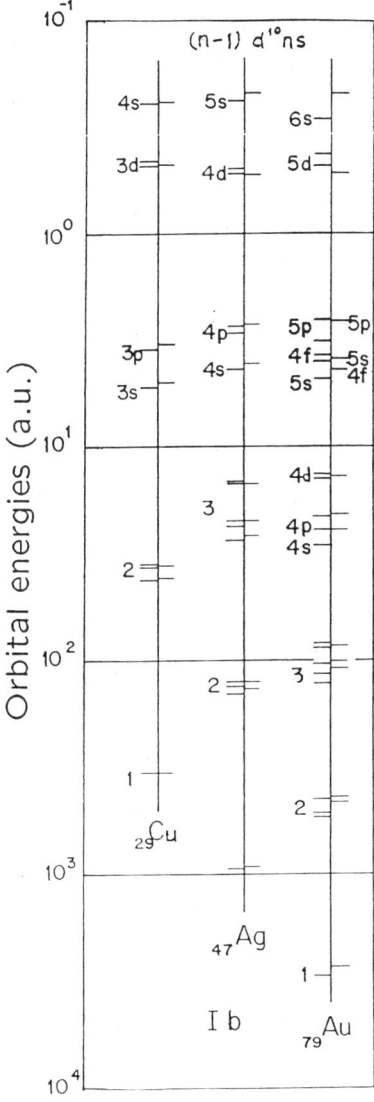

Figure 1. Ground state configuration average orbital eigenvalues for the coinage metals, copper, silver and gold, illustrate the competition between the stabilizing effect of relativity and increased screening of the nuclear charge. Relativistic eigenvalues are indicated by tick marks to the left, nonrelativistic ones to the right, of each vertical line. The increased binding due to relativity is most marked for s and $p_{1/2}$ electrons, decreasing as l increases, whilst screening destabilizes shells of higher angular momentum that scarcely probe the high potential region near the nucleus. The electron shells will fill in a different order in gold, the heaviest element of the group. DHF eigenvalues from Desclaux (1973), HF from Fischer (1973).

equivalent nonrelativistic calculations (Grant, 1996, §22.6.2). The solutions with energies in the positive energy continuum ($E > mc^2$) or the negative energy continuum ($E < -mc^2$) can be interpreted as square integrable wavepackets. The energy of a continuum eigensolution depends on the number of basis functions employed, and does not converge to any limit as the basis set expands towards completeness. Instead, the energies of continuum pseudo-states span an increasing range of the energy spectrum $|E| > mc^2$. Although for many purposes, one is only interested in bound states and states of the positive energy continuum, the negative energy states are needed for mathematical completeness. Indeed one of the difficulties of projecting out negative energy states in the manner recommended by Sucher (1980) is that smooth perturbations of the potential require *the entire spectrum*, including the negative energy states, to represent the bound states of the perturbed system properly (Quiney et al. 1985). Finally, scrutiny of those papers in which "variational collapse" has been observed confirms the view that they employ basis functions which fail to satisfy physically sensible boundary conditions at the atomic nucleus.

Once we have constructed a complete set of one-particle solutions of the Dirac operator for a given potential, normally for an atomic mean field, the way is open to use the usual methods of second quantization embodied in QED (quantum electrodynamics) to set up the many-particle problem. The rules for this are given in all standard textbooks and there has never been any dispute that all the disasters catalogued by Sucher and others are avoided when these rules are followed. This is implicit in all our work with BERTHA.

ASPECTS OF BERTHA

Whilst relativity affects the physics of atoms and molecules throughout the Periodic Table, its effects are most pronounced for high atomic numbers. High-Z systems have correspondingly large numbers of electrons, so that it is imperative to design algorithms which build in as much as possible of the physically important properties of Dirac spinors at the outset. Basis functions which have appropriate symmetries and satisfy correctly the boundary conditions are essential to minimize the amount of necessary computation. The G-spinor basis functions, which possess these properties, have proved to be extremely effective in a wide range of atomic and molecular calculations.

Dirac's equation for a central potential

As in nonrelativistic quantum mechanics, molecular orbitals are constructed as linear combinations of spinor basis functions centred on the atomic nuclei. The form of Dirac spinors near the nuclei is most easily expressed in spherical coordinates, so that the primitive basis spinors must have the appropriate component structure.

Dirac's equation for the stationary state, $\psi_k(\boldsymbol{r}, t) = \psi_k(\boldsymbol{r}) \exp -iE_k t$, of an electron moving in a classical external electric field can be written

$$\hat{h}_D \psi_k(\boldsymbol{r}) \equiv \{c\boldsymbol{\alpha} \cdot \boldsymbol{p} + \beta mc^2 + V(\boldsymbol{r})\}\psi_k(\boldsymbol{r}) = E_k \psi_k(\boldsymbol{r}), \tag{1}$$

where $\boldsymbol{\alpha} \equiv (\alpha_x, \alpha_y, \alpha_z)$ and β are the standard 4×4 Dirac matrices, c is the velocity of light and $V(\boldsymbol{r})$ is the potential of the external field. These matrices can be expressed in terms of the familiar Pauli matrices

$$\alpha_q = \begin{bmatrix} 0 & \sigma_q \\ \sigma_q & 0 \end{bmatrix}, \quad \beta = \begin{bmatrix} I & 0 \\ 0 & I \end{bmatrix},$$

where
$$\sigma_x = \begin{bmatrix} 0 & 1 \\ 1 & 0 \end{bmatrix}, \quad \sigma_y = \begin{bmatrix} 0 & -i \\ i & 0 \end{bmatrix}, \quad \sigma_z = \begin{bmatrix} 1 & 0 \\ 0 & -1 \end{bmatrix}.$$

and I is the unit 2×2 matrix. It will be convenient also to define $\sigma_0 = I$ in what follows.

In a central field $V(\mathbf{r})$ depends only on the radial coordinate r, and will usually have the form
$$V(\mathbf{r}) = Z(r)/r$$
where $Z(r)$ may take account of the fact that nuclei have a finite size, in which case,
$$Z(r) \sim z_1 r + O(r^2), \quad r \to 0,$$
and it may also represent some *atomic mean field* behaviour for large r. Solutions to this problem in the usual spherical coordinates (r, θ, φ) have the form
$$\psi_{E_k,\kappa,m}(\mathbf{r}) = \frac{1}{r} \begin{bmatrix} P_{E_k,\kappa}(r) \chi_{\kappa,m}(\theta,\varphi) \\ i Q_{E_k,\kappa}(r) \chi_{-\kappa,m}(\theta,\varphi) \end{bmatrix}. \tag{2}$$

The quantum numbers κ and m are related to the total angular momentum quantum number of the state, $j\hbar$, and of its projection $m\hbar$. The angular 2-spinor amplitudes are given by
$$\chi_{\kappa,m}(\theta,\varphi) = \sum_\sigma Y_{l,m-\sigma}(\theta,\varphi) \phi_\sigma \langle l, m-\sigma, 1/2, \sigma \,|\, l, 1/2, j, m \rangle, \tag{3}$$
where
$$\phi_{1/2} = \begin{bmatrix} 1 \\ 0 \end{bmatrix}, \quad \phi_{-1/2} = \begin{bmatrix} 0 \\ 1 \end{bmatrix},$$
and $Y_{l,m}(\theta,\varphi)$ is a spherical harmonic with the usual Condon and Shortley phase convention. The $\chi_{\kappa,m}(\theta,\varphi)$ are eigenfunctions of the spin $\mathbf{s}^2, s_z$, with eigenvalues $s(s+1)\hbar^2$ (with $s = 1/2$) and $s_z = \pm 1/2$ respectively, as well as of $\mathbf{l}^2$ with eigenvalue $l(l+1)\hbar^2$ and of $\mathbf{j}^2$ with eigenvalue $j(j+1)\hbar^2$. The two possibilities $\mathbf{l} = \mathbf{j} \pm \mathbf{s}$ are conveniently identified by a single quantum number
$$\kappa = (2j+1)(l-j).$$

Thus four component wave functions of the form (2) are eigenfunctions of $\mathbf{j}^2, j_z$, and we can identify the two possible signs of κ by the parity, $(-1)^l$, associated with the "large" component $P_{E_k,\kappa,m}(r)$, which has the corresponding Schrödinger amplitude as its formal nonrelativistic limit as $c \to \infty$ (in which light signals travel instantaneously).

The character of the radial components depends upon the eigenvalue E_k; when $mc^2 > E_k > -mc^2$, the solutions have bound state character. They must be square integrable near $r = 0$ and as $r \to \infty$. The Dirac operator couples the two amplitudes, and the behaviour near $r = 0$ depends upon the form of $Z(r)$ as $r \to 0$. A detailed analysis of this behaviour is given in (Grant 1996, §22.5.3), and is important to motivate the construction of the G-spinor basis used in BERTHA.

The angular density distribution, $A_{\kappa,m}(\theta,\varphi) = \chi_{\kappa,m}(\theta,\varphi)^\dagger \chi_{\kappa,m}(\theta,\varphi)$, was shown by Hartree (1929) to be independent of the sign of κ, so that it is possible to define a radial density distribution $\rho(r) = P^2_{E_k,\kappa}(r) + Q^2_{E_k,\kappa}(r)$ as discussed in the introduction.

Finally, we note that spinor functions of the form (2) are natural building blocks for the construction of relativistic molecular symmetry orbitals for finite double point groups (Meyer et al. 1996). Meyer et al.'s TSYM program provides the means of projecting out relativistic symmetry orbitals for some 45 different finite double point groups, taking time reversal invariance into account.

G-spinors

BERTHA uses molecular orbitals of the form

$$\psi_k(\mathbf{r}_A) = \begin{bmatrix} \sum_{\mu=1}^{N_B} c_{\mu k}^L M[L, \mu, \mathbf{r}_A] \\ \sum_{\mu=1}^{N_B} c_{\mu k}^S M[S, \mu, \mathbf{r}_A] \end{bmatrix}, \quad (4)$$

where each 2-component basis element $M[T, \mu, \mathbf{r}_A]$, in which T denotes the label L or S, is summed over N_B multi-indices $\mu := \{A, \kappa, j, m_j, \lambda_\mu\}$ defining the centre A on which the element is centred, its angular quantum numbers, and its Gaussian exponent λ, with linear expansion coefficients $c_{\mu k}^T$ and $\mathbf{r}_A = \mathbf{r} - A$. The individual elements have the central field structure

$$M[L, \mu, \mathbf{r}_A] = \frac{1}{r_A} f_\mu^L(r_A) \chi_{\kappa_\mu, m_\mu}(\theta_A, \varphi_A) \quad (5)$$

$$M[S, \mu, \mathbf{r}_A] = \frac{1}{r_A} f_\mu^S(r_A) \chi_{-\kappa_\mu, m_\mu}(\theta_A, \varphi_A), \quad (6)$$

where the radial basis functions take the Gaussian form (hence the term G-spinor)

$$f_\mu^L(r_{A_\mu}) = N_\mu^L r_{A_\mu}^{l_\mu+1} \exp(-\lambda_\mu r_{A_\mu}^2),$$

and

$$f_\mu^S(r_{A_\mu}) = N_\mu^S [(\kappa_\mu + l_\mu + 1) - 2\lambda_\mu r_{A_\mu}^2] r_{A_\mu}^{l_\mu} \exp(-\lambda_\mu r_{A_\mu}^2),$$

and N_μ^L and N_μ^S are normalization constants. The more complex form of the "small" component function arises from the so-called *strict kinetic balance* condition

$$M[S, \mu, \mathbf{r}_A] \propto \boldsymbol{\sigma} \cdot \mathbf{p}\, M[L, \mu, \mathbf{r}_A]$$

This definition of $M[S, \mu, \mathbf{r}_A]$ in terms of $M[L, \mu, \mathbf{r}_A]$ is particularly useful when nucleus A is modelled as a finite size object. Our construction then ensures the correct asymptotic behaviour of the G-spinor as $r \to 0$ as well as ensuring that the matrix of the Dirac Hamiltonian in this basis gives the *exact* nonrelativistic kinetic energy matrix in the limit $c \to \infty$ (Grant 1996, §22.6.2). The situation with point charge nuclei is quite different: see Grant (1996).

Coupling of the electron and photon fields in relativistic QED

It is essential for the physical interpretation of relativistic wave equations that there exist a conserved charge-current four-vector, $j^q(x)$, where the index q labels Minkowski time, $t = x^0/c$, and space, $\mathbf{r} = (x^1, x^2, x^3)$, coordinates. In nonrelativistic notation, the charge density and electron current vector components are respectively

$$\rho(\mathbf{r}, t) := j^0(x)/c = -e\psi^\dagger(x)\psi(x), \quad (7)$$

and

$$j^q(\mathbf{r}, t) := -ec\psi^\dagger(x)\alpha_q \psi(x), \quad q = 1, 2, 3, \quad (8)$$

and must satisfy the usual continuity equation

$$\frac{\partial \rho}{\partial t} + \operatorname{div} \mathbf{j} = 0.$$

If $\psi(x)$ is expanded in G-spinors as in (4) each charge-current component will be a linear combination of terms of the form

$$\mathcal{R}_{\mu\nu q}^{T,T'}(\boldsymbol{r}) = M[T, \mu, \boldsymbol{r}_A]^\dagger \sigma_q M[T', \nu, \boldsymbol{r}_B], \tag{9}$$

where $\boldsymbol{r}_A = \boldsymbol{r} - \boldsymbol{A}$, $\boldsymbol{r}_B = \boldsymbol{r} - \boldsymbol{B}$, in which centre $\boldsymbol{A}$ and centre $\boldsymbol{B}$ may be the same or different points in space. The properties of Gaussian products of this sort allow us to combine the exponential terms to give an expression of the form (Quiney et al. 1997)

$$\mathcal{R}_{\mu\nu q}^{T,T'}(\boldsymbol{r}) = N_{\mu\nu}^{TT'} K_{\mu\nu} \sum_{s,t,u} E_q[\boldsymbol{A}, \boldsymbol{B}; p, \boldsymbol{P}; T, \kappa_\mu, m_\mu; T', \kappa_\nu, m\nu; s, t, u] H(p, \boldsymbol{r}_P; s, t, u). \tag{10}$$

Here

$$H(p, \boldsymbol{r}_P; s, t, u) = \left(\frac{\partial}{\partial x_P}\right)^s \left(\frac{\partial}{\partial y_P}\right)^t \left(\frac{\partial}{\partial z_P}\right)^u \exp\left(-p r_P^2\right)$$

is an Hermite Gaussian function on a new centre

$$\boldsymbol{P} = (\lambda_\mu \boldsymbol{A} + \lambda_\nu \boldsymbol{B})/p$$

where $p = \lambda_\mu + \lambda_\nu$, and $\boldsymbol{r}_P = \boldsymbol{r} - \boldsymbol{P}$, $N_{\mu\nu}^{TT'} = N_\mu^T N_\nu^{T'}$, $K_{\mu\nu} = \exp\{-\lambda_\mu \lambda_\nu (\boldsymbol{A} - \boldsymbol{B})^2/p\}$. Information about the geometry of the molecule and the structure of the component basis functions is encoded in the coefficients $E_q[\boldsymbol{A}, \boldsymbol{B}; p, \boldsymbol{P}; T, \kappa_\mu, m_\mu; T', \kappa_\nu, m_\nu; s, t, u]$, and the only reference left to the four-component structure is now the dependence of the E_q-coefficients on the indices T and T'. The Hermite-Gaussian functions $H(p, \boldsymbol{r}_P; s, t, u)$ are, of course independent of the spinor structure. Those familiar with nonrelativistic quantum chemistry will recognise (10) as the relativistic generalization of the McMurchie-Davidson algorithm for computing molecular interaction integrals (McMurchie et al. 1978). The speed and efficiency of BERTHA depend crucially on the algorithms we have constructed for computing the E_q-coefficients on demand; a full account of the technical details will be given elsewhere.

It remains to explain how interactions in QED are treated in this formalism. The interaction of the (second) quantized Dirac electron-positron field with the quantized Maxwell photon field can be represented by an interaction Hamiltonian

$$\hat{H}_{int} = -\int j^\mu(x) A_\mu(x) d^3 x \tag{11}$$

where $A_\mu(x)$ is the four-potential at the space-time point x. In nonrelativistic notation, which is perhaps more familiar to most readers, we have

$$j^\mu(x) A_\mu(x) = \rho(\boldsymbol{r}, t)\phi(\boldsymbol{r}, t) - \boldsymbol{j}(\boldsymbol{r}, t).\boldsymbol{A}(\boldsymbol{r}, t).$$

The four-potential will satisfy Maxwell's equations, in particular

$$\Box A^\mu(x) = J^\mu(x), \tag{12}$$

ignoring gauge terms, where $J^\mu(x) = j^\mu(x) + j_{ext}^\mu(x)$, the last term representing any classical charge-current distributions due, for example, to the nuclei. The QED S-matrix formalism, briefly outlined in Grant (1996), §22.3.4, enables us to extract physical information about atoms and molecules using (11) as a perturbation. Many of the terms that appear in the QED perturbation series can be interpreted as due to an effective interaction between two electron charge-current distributions of the form

$$\iint j^\mu(x) D_{\mu\nu}(x - y) j^\nu(y) dx\, dy, \tag{13}$$

where $D_{\mu\nu}(x-y)$ is a causal propagator for the photon field. In effect, $A_\mu(x) = \int D_{\mu\nu}(x-y) j^\nu(y) dy$, is the four-potential at x generated by the charge-current distribution $j^\nu(y)$. In its most general form, this interaction includes coupling to negative energy (positron) states; most of these terms are too small to be observable. However some, for example those which can be interpreted as electron self-energy or as polarization of the electron vacuum, give observable contributions. They are formally described by mathematically divergent integrals which become finite after renormalization. We ignore such contributions here.

The four-potential is only unique up to a gauge transformation, and it is most convenient for electronic structure calculations to choose the Coulomb gauge to calculate electron-electron interaction integrals. This separates the sum over components μ, ν into two disjoint parts, one giving the (static) Coulomb interaction, the other the retarded current-current (transverse photon) interaction. The former gives Coulomb interaction integrals of the familiar form

$$(ac|g_{12}|bd) = \iint \frac{\rho_{ac}(\boldsymbol{r}_1) \rho_{bd}(\boldsymbol{r}_2)}{R_{12}} d\boldsymbol{r}_1 d\boldsymbol{r}_2, \tag{14}$$

coupling overlap density distributions of the form

$$\rho_{ac}(\boldsymbol{r}) = \varphi_a^\dagger(\boldsymbol{r}) \varphi_c(\boldsymbol{r})$$

at points with relative separation $\boldsymbol{R}_{12} = \boldsymbol{r}_1 - \boldsymbol{r}_2$. The transverse photon interaction similarly involves the 3-current overlap distributions $\boldsymbol{j}_{ac}(\boldsymbol{r})$ and $\boldsymbol{j}_{bd}(\boldsymbol{r})$, and in general depends upon the energy differences $\omega_{ac} = E_a - E_c$ and $\omega_{bd} = E_b - E_d$. For bound state interactions, it is usually adequate (to within a few percent of the small transverse photon correction) to assume that ω_{ac}, ω_{bd} are negligible. The Breit interaction, which is the limit as $\omega_{ac} \to 0, \omega_{bd} \to 0$ gives interaction integrals of the form

$$(ac|b_{12}|bd) =$$
$$-\frac{1}{2} \iint \left(\frac{\boldsymbol{j}_{ac}(\boldsymbol{r}_1) \cdot \boldsymbol{j}_{bd}(\boldsymbol{r}_2)}{R_{12}} + \frac{\boldsymbol{j}_{ac}(\boldsymbol{r}_1) \cdot \boldsymbol{R}_{12}\, \boldsymbol{j}_{bd}(\boldsymbol{r}_2) \cdot \boldsymbol{R}_{12}}{R_{12}^3} \right) d\boldsymbol{r}_1 d\boldsymbol{r}_2, \tag{15}$$

where

$$\boldsymbol{j}_{ac}(\boldsymbol{r}) = \varphi_a^\dagger(\boldsymbol{r})\, c\boldsymbol{\alpha}\, \varphi_c(\boldsymbol{r})$$

is an overlap current density. The nonrelativistic limit of (15) is part of the Breit-Pauli Hamiltonian, giving rise to the two-body spin-orbit, spin-spin, spin-other-orbit, and contact interactions as listed in standard texts such as Bethe and Salpeter (1957). Evaluation of such integrals using the Gaussian product expansion, (10), is straightforward. Expressions for the molecular integrals involving both the Coulomb and Breit interactions are given in Quiney et al. (1997).

Further developments: electron correlation

Historically, many-body theories of electron correlation originated as adaptations of textbook QED perturbation theory (Lindgren and Morrison 1982). In relativistic QED, the contributions order by order are described by means of Feynman diagrams (Greiner and Reinhardt 1992). The order of a QED diagram is equal to the number of interaction vertices, each with an incoming and outgoing electron line and a single photon line. In nonrelativistic many-body theories, the most elementary interaction between electrons involves a diagram which would be second order in QED, with a Coulomb interaction coupling a pair of incoming and outgoing electron lines. Thus nonrelativistic many-body theory will contain a well-defined subset of the equivalent relativistic QED diagrams. Of those diagrams that need to be taken into account in electronic structure calculations, those omitted mainly involve interactions with the negative energy continuum, especially those which can be interpreted in terms of electron self-energy and vacuum polarization contributions which may involve difficult problems of renormalization. As the most important diagrams to take into account in a relativistic electronic structure calculation of electron correlation are precisely those which contribute in the nonrelativistic case, a good deal of the machinery required can be taken over to BERTHA with little change. Although problems involving heavy atom centres will inevitably require greater computer resources, the outlook for such calculations is highly promising.

THE BERTHA PROGRAM

BERTHA is still being developed, although certain parts have now stabilized, and are unlikely to change much from now on. At present its major components are

- An atomic Dirac-Hartree-Fock program using G-spinor basis functions, but exploiting 1-centre machinery from the earlier SWIRLES code (Quiney 1990). This is a very inexpensive way to generate DHF solutions for atomic cores (normally only a few minutes computer time, even for a big atom) as the first stage of a molecular electronic structure calculation.

- A molecular structure code using the same G-spinor basis. The use of converged atomic core wavefunctions as the initial guess for the molecular structure calculation helps to obtain smooth convergence to the final DHF solution.

- For molecules with point group symmetry, the TSYM program of Meyer et al. (1996) can be invoked to classify molecular symmetry orbitals according to the relevant double group. We have not so far attempted to use these symmetry orbitals to simplify the SCF calculation.

- The full Breit interaction can be included in both atomic and molecular calculations.

- So far, we have only implemented correlation calculations at the second order of MBPT. Higher order terms will be included if manpower permits. A prototype MCSCF program is in the testing stage.

- Calculation of some electromagnetic properties, for example moments of the molecular charge-current distribution, nuclear hyperfine interactions, and NMR shielding have been implemented so far. Similar techniques can be used to study parity-violating interactions in atoms and molecules (Quiney et al. 1998b), particularly in chiral molecules (Quiney et al. 1998a).

So far we have published only a limited number of results as part of the program for testing BERTHA. Some of these results have appeared in a more extended version of this article (Quiney et al. 1998a), along with a discussion of some of the underlying policies. In particular, we describe relativistic corrections to the Zeeman effect in hydrogen-like Ne, the calculation of NMR shielding tensors in small molecules such as H_2O and NH_3, parity violation energy differences at the DHF level in the chiral molecule CHBrClF, and a study of the feasability of doing calculations for molecules with many centres illustrated by germanocene, Ge(cp)$_2$, in which the heavy atom is sandwiched between two planar 5-membered CH rings with 21 atoms in all. Clearly, such calculations are likely to require the evaluation of large numbers of multi-centre integrals. The more complicated structure of the spinor small components means that these are, at least superficially, extremely expensive, which has proved a big deterrent to the use of four-component molecular calculations. We have largely overcome this in BERTHA by observing that the small components are generally localized close to the nuclei. As there are no exchange interactions between non-overlapping distributions, it is only necessary to compute long range electrostatic interactions between the aggregated localized distributions, so eliminating most of the really expensive multi-centre integrals over small ocmponents. Moreover, it is possible to devise a test for eliminating batches of integrals whose contributions are negligible without having to calculate them.

BERTHA is economical on memory, and the adoption of a strategy which only demands the calculation of batches of molecular integrals when they are to be used immediately minimizes the amount of disk storage. The ability to use atomic structure methods to do preliminary calculations for atomic cores using the same G-spinors as in the molecular calculation, means that there is no need to use programs such as GRASP to calculate atomic wavefunctions numerically and then to have to fit them to Gaussian basis sets. BERTHA therefore avoids many of the problems that beset other four component schemes, and it is entirely feasible to do many calculations on modest PCs or workstations.

A simple calculation, which demonstrates the program's capabilities for a simple molecule, is provided by water (Quiney et al. 1998c). It is usually assumed that relativistic four-component calculations are so expensive that they should only be contemplated for very heavy systems, and then only as a means of benchmarking other calculations, for example those using effective core potentials. We have found not only that BERTHA is relatively cheap to run, but that there are also positive advantages to performing relativistic calculations even for molecules as light as water. We have compared DHF (relativistic) and HF (nonrelativistic) calculations, the Breit contributions to the total energy of the molecule in its ground state, and the second order correlation energy for several different basis sets taken from the literature. The nonrelativistic calculations, obtained by multiplying the velocity of light by a large factor, reproduce published data extremely well. The relativistic corrections are small in light molecules, and it has usually been assumed that although the absolute value of the correction may be significant, its variation with geometry is too small to be noticed. However, (Császár et al. 1998) found recently that the addition of one electron mass-velocity and Darwin terms from the Breit-Pauli Hamiltonian as first order perturbations to a high quality nonrelativistic potential energy surface improved the agreement of calculated vibrational band origins and rotational term values for $H_2{}^{16}O$ more than Born-Oppenheimer diagonal corrections. The effect was particularly marked for vibrational bending modes and for rotational terms with high $K_a{}^*$. However, the average relativistic correction was still quite large. A comparison of Császár et al.'s relativistic correction surface over the more restricted range of bond angles and bond lengths that we considered shows remarkably good agreement. BERTHA thus shows promise as a tool for investigation of relativistic effects in light molecules as well as for the heavier systems which have hitherto been the focus of attention.

---

*This is of particular importance for the detailed interpretation of the infrared spectrum of hot water in the Sun (Polyansky et al. 1998).

DISCUSSION

The BERTHA program is largely the work of my colleague Harry Quiney, assisted by Haakon Skaane, who were responsible together for most of the novel ideas and procedures embodied in the code. Development of BERTHA did not start until 1995, although many of the insights that we exploited have accumulated over the last forty years. Some of these, which were first presented in Burke et al. (1967), were first investigated in response to a question about how to visualize relativistic effects raised by the late Professor Charles Coulson at an Oxford seminar in 1966. However, the realization that basis set methods could be used not only for molecular structure calculations but for setting up calculations in a framework consistent with bound state QED dates back to the mid 1980s (Quiney 1990), and the charge-current formalism presented by Quiney et al. (1997) is even more recent. Development of BERTHA was greatly accelerated by writing large parts of the program in the algebraic language Maple, which could be used to produce the bodies of Fortran subroutines automatically. The speed with which Maple code could be changed enabled us to experiment with a large number of alternative computational strategies easily. We are indebted to Dr. Tony Scott for his help and advice at that stage of the project.

It has been a pleasure to present this work at a conference to honour Phil Burke's distinguished scientific career, and to highlight the collaboration with Val Burke which produced some fundamental insights into relativistic effects over 30 years ago.

EPSRC are thanked for support of computing facilities for this project and for the award of an Advanced Fellowship to Harry Quiney from 1992 to 1997, and for a postdoctoral award which supported Dr Tony Scott for 3 years starting 1993. Haakon Skaane was supported by a pre-doctoral award from the Norwegian Research Council.

I am also indebted to Prof J. Tennyson and Dr. O. L. Polyansky for illuminating discussions after the conference on relativistic effects in the infra-red spectroscopy of water.

REFERENCES

Bethe, H. A. and Salpeter, E. E., 1957, *Quantum Mechanics of One and Two Electron Systems*, Springer, New York.
Brown, G. E. and Ravenhall, D. G. 1951, *Proc. Roy. Soc. A* **208**, 552.
Burke, V. M. and Grant, I. P., 1967, *Proc. Phys. Soc.* **90**, 297.
Császár, A. G., Kain, S. J., Polyansky, O. L., Zobov, N. F. and Tennyson, J., 1998, *Chem. Phys. Letts.* **293**, 317.
Desclaux, J. P., 1973, *At. Data Nucl. Data Tables* **12**, 311.
Dirac, P. A. M., 1928, *Proc. Roy. Soc. A* **117**, 610.
Fischer, C. F., 1973, *At. Data Nucl. Data Tables* **12**, 87.
Grant, I. P., 1961, *Proc. Roy. Soc. A* **262**, 555.
Grant, I. P., Mayers, D. F. and Pyper, N. C., 1976, *J. Phys. B* **9** 2777.
Grant, I. P., 1996, *Atomic, Molecular & Optical Physics Handbook*, ed. G. W. F. Drake, Chapter 22. AIP Press, Woodbury, New York.
Greiner, W. and Reinhardt, J., 1992, *Quantum Electrodynamics*. Springer-Verlag, Berlin.
Hall, G. G., 1951, *Proc. Roy. Soc. A* **205**, 541.
Hartree, D. R., 1929, *Proc. Camb. Phil. Soc.* **25**, 225.
Kim, Y.-K., 1967, *Phys. Rev.* **154**, 17; *Phys. Rev.* **159**, 190.
Lindgren, I. and Morrison, J., 1982, *Atomic Many-Body Theory*. Springer-Verlag, Berlin.
McMurchie, L. E. and Davidson, E. R., 1978, *J. Comput. Phys.* **26**, 218.
Meyer, J., Sepp, W. D., Fricke, B., and Rosén, A., 1996, *Comput. Phys. Commun.* **96**, 263.
Polyansky, O. L. and Tennyson, J., 1998, *Contemp. Phys.* **39**, 283.
Pyykkö, P. 1978, *Adv. Quantum Chem.* **11** 353.
Pyykkö, P. 1988, *Chem. Rev.* **88**, 563.

Quiney, H. M., 1990, Relativistic atomic structure calculations I: Basic theory and the finite basis set approximation, pp. 159–184, II: Computational aspects of the finite basis set method. pp. 185–200, *in*: "Supercomputational Science", ed. R. G. Evans and S. Wilson, Plenum, New York.
Quiney, H. M., Grant, I. P. and Wilson, S., 1985, *J. Phys. B* **18**, 577.
Quiney, H. M., Skaane, H. and Grant, I. P., 1997a, *J. Phys. B* **30**, L829.
Quiney, H. M., Skaane, H. and Grant, I. P., 1998a, Adv. Quantum Chem. **32**, 1.
Quiney, H. M., Skaane, H. and Grant, I. P., 1998b, *J. Phys. B* **31**, L85.
Quiney, H. M., Skaane, H. and Grant, I. P., 1998c, *Chem. Phys. Letts.* **290**, 473.
Roothaan, C. C. J., 1951, *Rev. Mod. Phys.* **23**, 69.
Sucher, J., 1980, *Phys. Rev. A* **22**, 348.
Swirles, B., *Proc. Roy. Soc. A* **152**, 625.

JET APPLICATIONS OF ATOMIC COLLISIONS

H. P. Summers, R. W. P. McWhirter[†], H. Anderson, C. F. Maggi[‡], M. G. O'Mullane

Dept. of Physics and Applied Physics, University of Strathclyde, Glasgow G4 0NG, UK
[†] Rutherford Appleton Laboratory, Chilton, Didcot, Oxon. OX11 0QX, UK
[‡] JET Joint Undertaking, Abingdon, Oxon. OX14 3EA, UK

INTRODUCTION

Historically JET represents the largest and one of the most recent in a long series of large toroidally shaped magnetic confinement devices starting, in the U.K., with ZETA in the midddle fifties. As the fusion programme has evolved over these years with its peaks and troughs, so also has the confidence in the spectroscopy and atomic modelling used to interpret the radiating properties of the fusion plasma. These early days must have been very exciting. It is clear the spectra from ZETA, as exemplified in Fig. 1, were very novel, rich in spectrum lines and that there was, at least initially, little idea how to model the emission. The assumption of Saha-Boltzmann populations was suggested as a starting point by Sir Harrie Massey, although now we realise this is very much in error. However by 1958, Thonemann (private communication) was working in his notes with equations for ionisation balance very like those in use today and was even concerned about relaxation and populations of metastables. It was of course in 1958, following much newspaper speculation, that the paper 'Production of High Temperatures and Nuclear Reactions in a Gas Discharge' was published. It was believed that we had made it - especially after Sir John Cockcroft's response at a press conference that he was 90% certain that some at least of the neutrons were thermal (Hendry and Lawson, 1993). The public interest was enormous and prizes were awarded. ZETA of course became a political machine, like JET, and the overly high expectations and as yet unfulfilled promises of limitless cheap power, which has dogged fusion research since, started there.

There was some disagreement within the ZETA team about the temperature of the plasma. The neutron yield, which later turned out to be non-isotropic and non-thermal, and spectral line widths, which turned out to be turbulent, indicated temperatures of ~ 200 eV. However, the spectral emission of trace gasses in the plasma rose and then decayed and there were two interpretations for this, firstly that the gas was diffusing rapidly out of the plasma again (Burton and Wilson, 1961) and that the 200eV tem-

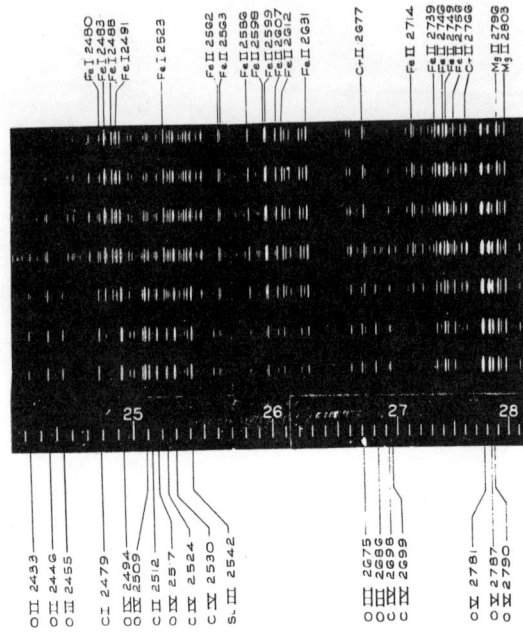

Figure 1. A portion of a plate spectrum from ZETA 1A in the quartz ultraviolet. Note the rich spectrum of the low ionisation stages of carbon, oxygen and neon as well as the neutral and singly ionised metals chromium, iron and nickel.

perature was sound, and secondly that the trace gas spectrum was 'burning through' by ionising to higher stages and that the temperature was ~ 50 eV. The latter position proved more nearly correct and the great success slipped away. However, it was already clear that this was a dynamic plasma from the atomic physics point of view, many processes contributed to observed emission and cross-sections for ionising reactions etc. were essential for interpretation. A little later at Harwell and Belfast collisional-radiative modelling was born (Bates, Kingston and McWhirter, 1962). Interestingly, in 1965, a calculation was performed on the excitation cross-sections of lithium-like nitrogen using a new five state close-coupling method (Burke, Tait and Lewis, 1966) so that the $2s-2p$, $2s-3p$ line ratio could be used as as an electron temperature diagnostic. We quote from that paper: 'In conclusion, we feel it worth mentioning here that preliminary calculation (McWhirter, private communication) of the effect of using Coulomb-Born cross-sections rather than the Bethe-Seaton values for Q(2s-2p) and Q(2s-3p) in an estimate of the temperature in a typical run on ZETA changed the electron temperature from $\sim$20 eV to $\sim$80eV'. Some strongly held views became rooted in these years, for example, the plasma physicist opinion that spectroscopists with their atomic modelling based diagnosis never get in right! Also about this time, Seaton (1964) encapsulated the ionisation and excitation state of elements in the solar corona as a balance of collisional ionisation and excitation by electrons and spontaneous capture (radiative recombination) and spontaneous emission. This was the 'coronal picture'. It had to be adjusted when Burgess showed the importance of resonance capture contributions, that is, dielectronic recombination (Burgess, 1964), but the picture was enthusiastically taken up by the fusion plasma modelling community as very relevant to their plasmas and easy to apply. Simple approximate formulae for

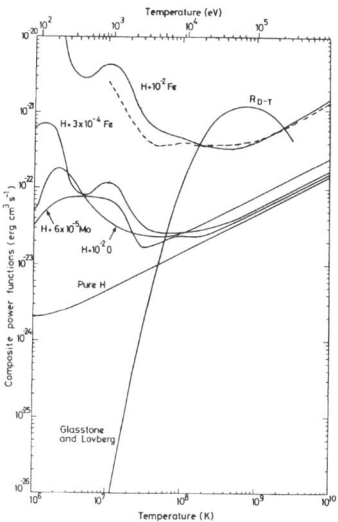

Figure 2. Composite radiation power loss functions for various impurity proportions in combination with hydrogen Vs electron temperature (McWhirter and Summers; 1984). The dashed line is a composite power function for iron from McWhirter(1959).

the basic processes in the coronal picture, developed by Seaton(1964), Burgess (1965), Van Regemorter (1962), Lotz (1967) and others added to the model's usefulness. In particular the anxiety about radiative losses preventing fusion could be examined. Fig. 2 shows the competition between the deuterium-tritium fusion rate and impurity radiative losses (McWhirter and Summers, 1984). The quite close agreement for iron with a much older and simpler calculation (McWhirter, 1959) is a little embarrrassing. It sometimes seems that the quality of atomic physics in general plasma transport models stuck at this point and has scarcely evolved since.

However, the fusion plasma is not quite like the solar corona. It is more dense, ion/atom collisions play a role, it is usually penetrated by particle beams, heated by electron and ion cyclotron radiation and of course has some residual but crucial contact with containing walls. Also, total radiated power at reasonably high temperatures is the least sensitive quantity both to plasma conditions and quality of atomic coefficients. Fig. 3, which compares the equilibrium radiated power function for argon, is more revealing, particularly in the low temperature regime < 10 eV. It is in the experimental spectroscopic diagnosis of fusion plasmas that atomic modelling has its greatest impact. In this area it has made large strides, underpinned by the scope and quality of collision calculations by ever more powerful methods on ever larger computers. In the following sections, the development, scope and future directions of atomic collisions and atomic modelling for JET will be described. JET is representative of similar activities at other fusion laboratories.

ATOMIC COLLISIONS IN PLASMA TRANSPORT MODELLING

Plasma transport modelling is a generic term spanning the range from quite basic zero-dimensional, single species, calculations to computationally intensive two or three

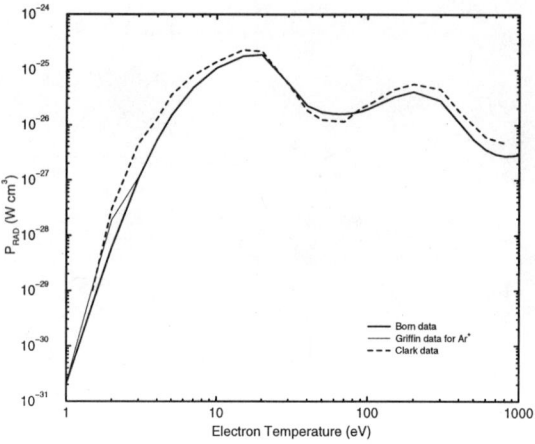

Figure 3. Radiative power function for argon. Curve 1 uses JET parametrised formulae with the parameters built up from Cowan/O'Mullane atomic structure calculations and Born approximation collisional rates. Curve 2 is from Clarke et al. (1995) and is based mainly on distorted wave calculations for the collisional rates. Curve 3 is as for curve 1 except that the power loss coefficient for Ar^+ uses collisional rates from a sophisticated 40-state R-matrix calculation by Griffin et al. (1997)

dimensional multi-species calculations. It is helpful to distinguish the dominant plasma species (deuterons and deuterium usually) and the electrons from the sub-dominant impurities which can be treated as trace species. In a simple approximation, the impurity transport is modelled as occurring in a fixed background plasma. The distributions of electron temperature and electron density in the plasma are determined in a prior step by solving the deuterium, deuteron and electron equations alone or by measurement. This type of approach is typical of modelling for the core plasma where a poloidally symmetric picture (more properly a closed flux surface averaged picture) is used to establish the radial distribution of impurity ionisation stage number densities and the electron temperature and density radial distributions are obtained from LIDAR measurements. For an impurity element A of nuclear charge z_0 with charge stages A^{+z} and stage number densities $N^{(z)}$, the number density continuity equations are

$$\frac{\partial N^{(z)}}{\partial t} + \frac{\partial \Gamma^{(z)}}{\partial r} = S^{(z-1 \to z)} N_e N^{(z-1)} - (\alpha^{(z \to z-1)} + S^{(z \to z+1)}) N_e N^{(z)} + \alpha^{(z+1 \to z)} N_e N^{(z+1)} \quad (1)$$

for $z = 0, ..., z_0$, where the flux, $\Gamma^{(z)}$, is written as

$$\Gamma^{(z)} = -D \frac{\partial N^{(z)}}{\partial r} + V N^{(z)} \quad (2)$$

D is the anomalous diffusion coefficient and V gives an inward pinch velocity. Both may have a prescribed behaviour as a function of r. Radiant losses are then obtained from the solution. Thus the power function for the impurity, P_{tot}^A is given by

$$P_{tot}^A = (\sum_{z=0}^{z_0} P^{(z)} N^{(z)}) / N_{tot}^{(z_0)} \quad (3)$$

where $P^{(z)}$ is the power coefficient for A^{+z} and $N_{tot}^{(z_0)}$ the number density of element A summed over all ionisation stages. Evidently non-diffusive stationary ionisation balance

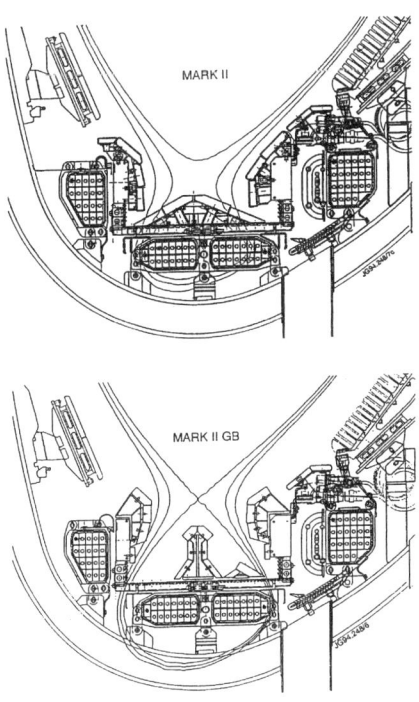

Figure 4. Schematic of a poloidal section of JET showing recent divertor patterns. The Mark II GB (gas box) was installed remotely in 1998 following vessel activation by a series of deuterium/tritium experiments.(Maggi, 1996)

is a simplification of this model with the left hand side of Eqns. 1 set equal to zero. The atomic collisions enter only in the source/sink terms on the right hand side of Eqn. 1 and in the power coefficients. The model is used in practice to determine the diffusion coefficient and pinch velocity by following the transient ionisation of various medium and heavy trace impurities laser ablated into the plasma. It is found that, depending on the type of pulse, there may be zones of different transport character, including apparent transport barriers. Thus the diffusion coefficient is given a restricted radial dependence in the model. The ionisation stage number densities are tracked experimentally by measuring the variation with time of radial line of sight integrated selected main resonance line emission. Absolute excitation cross-section data are not critical for this, it is enough to know that the coronal model for excitation is approximately valid. The model is also used to infer absolute impurity concentrations by matching to measurements of bolometric power and/or soft x-ray power. Such a model is of course only relevant to the confined plasma where ion number densities, electron temperature etc. are constant over closed flux surfaces. Boundary conditions at the plasma edge can only be a very simplified gross representation of the cross-field transport between the poloidally asymmetric edge/scrape-off-layer plasma and the symmetric confined plasma.

Detailed transport modelling is now strongly focussed on the edge, scrape-off-layer and divertor plasma. JET, in the nineties, was altered internally to a divertor machine. Divertor design is still evolving and only a few months ago, a new divertor structure (the MK2 gas box) was installed remotely by robotic systems. Fig. 4 shows schematically, in poloidal section, the lay-out of the divertor region in JET. The geometry here is two dimensional (since toroidal symmetry is assumed) in which both the very strong parallel (to the magnetic field) and slower cross-field transport must be modelled. The plasma, that is the deuterons, electrons and the impurity ions, are described in a fluid approximation by a set of transport equations on a computational grid established from the magnetic equilibrium of the plasma discharge. These equations are written in the form of conservation equations for particles momentum and energy following Braginskii (1965), from the first three moments of the Vlasov equation and are reduced to 2-D equations describing the transport along ($\|$) and across ($\perp$) the magnetic field lines. Neutral deuterium, usually treated separately by Monte Carlo codes, contributes to the sources and losses in the particle conservation equations for electrons and ions, to the momentum sources and losses for ions and to the energy losses for electrons. The impurity ions mainly contribute to radiant losses and the electron energy losses. Thus all the fluids are coupled. The typical source/loss terms, in which atomic collision data enter, are illustrated in the continuity, parallel momentum and energy equations for the deuterons,

$$\frac{\partial(N_i v_{i,\|})}{\partial x_\|} + \frac{\partial(N_i v_{i,\perp})}{\partial x_\perp} = N_e N_0 S_i^{(0\to1)} - N_e N_i \alpha_i^{(1\to0)} \quad (4)$$

$$\frac{\partial(p_i + N_i m_i v_{i,\|}^2)}{\partial x_\|} + \frac{\partial(m_i v_{i,\|} N_i v_{i,\perp})}{\partial x_\perp} - eN_i E_\| = -R_{i,\|} + S_{i,\|}^M \quad (5)$$

$$\frac{\partial[(\frac{5}{2}kT_i + \frac{1}{2}m_i v_i^2)N_i v_{i,\|} + q_{i,\|}]}{\partial x_\|} + \frac{\partial[(\frac{5}{2}kT_i + \frac{1}{2}m_i v_i^2)N_i v_{i,\perp} + q_{i,\perp}]}{\partial x_\perp}$$
$$-eN_i \vec{E}\cdot\vec{v}_i = Q_\Delta + S_i^E \quad (6)$$

and in the energy equation for the electrons

$$\frac{\partial(\frac{5}{2}kT_e N_e v_{e,\|} + q_{e,\|})}{\partial x_\|} + \frac{\partial(\frac{5}{2}kT_e N_e v_{e,\perp} + q_{e,\perp})}{\partial x_\perp} + eN_e \vec{E}\cdot\vec{v}_e$$
$$= -Q_\Delta - \vec{R}_e \cdot \vec{u} - S_e^E \quad (7)$$

where the suffices i, e and 0 denote deuterons, electrons and deuterium respectively. The effective ionisation and recombination coefficients for deuterium, $S_i^{(0\to1)}$ and $\alpha_i^{(1\to0)}$, in Eqn. 4 appear again in the equivalent equation for the electrons. For the impurity ions, there is such an equation for each ionisation stage with source/sink coefficients as for Eqn. 1, but including charge exchange effective recombination from deuterium (cf. Eqn. 8). The momentum transfer between deuterons and electrons $R_{i,\|}$ on the right-hand-side of Eqn. 5 is partly collisional (electron-ion friction) but includes the thermal force due to the electron temperature gradient. The same term appears oppositely signed in the equivalent electron momentum equation. The second term $S_{i,\|}^M$ is the important ion momentum loss due to charge exchange collisions with neutral deuterium and does not appear in the equivalent electron momentum equation. Q_Δ in Eqn. 6 is the electron-ion equipartition term expressing the heat acquired by ions in collisions with electrons. S_i^E is the ion energy source/sink due to charge exchange with neutral deuterium. In Eqn. 7, the heating of the electrons by the friction with ions, $\vec{R}_e \cdot \vec{u}$, occurs, depending on the relative velocity of the deuteron and electron fluids, $\vec{u}$. The electron energy source/sink, S_e^E, is a key term which includes ionisation, excitation and recombination energy transfered in collisions not only with deuterium but also impurity ions.

The most complete approach at JET is to solve simultaneously these complete 2-D multifluid, multispecies equations using the code EDGE2D (Simonini, 1994). The transport of the deuterium neutrals is treated by the 2-D Monte Carlo code NIMBUS (Cupini et al., 1984).

A second approach, which is convenient and faster for later spectral analysis, uses the simpler Onion Skin Model, rather than solving the full 2-D transport equations above. Since parallel transport for ions and electrons occurs on a much faster timescale than perpendicular transport, it is possible to describe the motion parallel to the field lines neglecting perpendicular transport. This means that each flux tube in the scrape-off-layer can be modelled individually, starting from appropriate plasma boundary conditions for each flux tube. The perpendicular profiles are defined at one poloidal location and then mapped along the scrape-off-layer assuming classical transport. It is believed that the most reliable experimental profiles are measured at the target plates and so the transport equations are integrated from the target. The major problem when mapping the plasma solution around the scrape-off-layer is posed by the source/loss terms in the transport equations. Initially a simplified explicit analytic estimate of the source/loss terms is put in and a first estimate of the background plasma computed by the Onion Skin Model. This approximate background plasma is then fed into NIMBUS, for the neutral deuterium, which returns a better estimate of the source/loss terms. A converged solution is sought by iterating between the Onion Skin Model and NIMBUS. Impurity radiation losses do of course influence the solution and the analytic expression continues to be used for them. The impurities transport itself is then modelled in a trace impurity approach using a second Monte Carlo code called DIVIMP (Stangeby and Elder, 1992). As for the hydrogenic plasma, the perpendicular transport for the impurities is anomalous. It is treated by specifying a constant cross-field diffusion coefficient $D_\perp$ which is the same for all stages. The parallel transport for the impurities is considered to be classical. Taking s as the coordinate along the flux tube, the continuity equation for an impurity charge state z is

$$\frac{d(N^{(z)}v^{(z)})}{ds} = N_e S^{(z-1 \to z)} N^{(z-1)} - (N_e \alpha^{(z \to z-1)} + N_e S^{(z \to z+1)} + N_H q^{CX,(z \to z-1)}) N^{(z)} + (N_e \alpha^{(z+1 \to z)} + N_H q^{CX,(z+1 \to z)}) N^{(z+1)} \quad (8)$$

Note the source and loss terms involving charge exchange with neutral hydrogen. Time constants for ionisation, recombination and charge exchange recombination, $\tau_{ion}^{(z)}$, $\tau_{rec}^{(z)}$ and $\tau_{CX}^{(z)}$, for the impurity ions evaluated at the local electron temperature and density, are formed from these and used in the Monte Carlo formulation. The momentum balance equation is

$$m^{(z)} v^{(z)} \frac{d(v^{(z)})}{ds} = -\frac{1}{N^{(z)}} \frac{d(p^{(z)})}{ds} + ezE + \frac{m^{(z)}(v - v^{(z)})}{\tau_S} + \alpha \frac{dT_e}{ds} + \beta \frac{dT_i}{ds} + F_\eta \quad (9)$$

where successive terms on the right hand side are the impurity pressure force, the electrostatic force due to the ambipolar electric field, E, the frictional force between impurity ion and hydrogenic plasma flow, the thermal and the diffusion force. The last term, the viscous force, is neglected in the model. The constants, τ_S, α and β are given by Spitzer (1962) and Neuhauser et al.(1984). The energy balance equation is replaced by thermalisation of the impurity ions with the background deuterons, occurring in a timescale τ_T, the Spitzer energy transfer collision time.

Fundamental and Derived Atomic Data.

Of most relevance for atomic collision calculations are the various charge exchange source/sink terms and the inelastic stage to stage effective ionisation and recombination coefficients such as those of the form $\alpha^{(z \to z-1)}$ and $S^{(z \to z+1)}$ in Eqns. 1 and the equivalent quantities for deuterium, α_i and S_i, in Eqn. 4. These are properly collisional-radiative coefficients and in general depend on electron density N_e as well as electron temperature T_e. For complex species, full collisional-radiative coefficients may not be available and so simplifications are often made such as $S^{(z \to z+1)}(T_e, N_e) \simeq S^{(z \to z+1)}(T_e, N_e = 0)$. This is not a bad assumption for medium and highly ionised ions at JET densities ($N_e = 10^{13} - 2*10^{14} cm^{-3}$) but quite incorrect for deuterium and neutral impurities and inaccurate for few times ionised ions. In high density, low temperature divertor plasmas, three-body recombination is the dominant contribution to the effective recombination of deuterons to form deuterium. Generally, the true collisional-radiative recombination coefficient is used for deuterons in models. For impurities, it is often assumed that $\alpha^{(z \to z-1)}(T_e, N_e) \simeq \alpha_{rad}^{(z \to z-1)}(T_e, N_e = 0) + f(N_e)\alpha_{diel}^{(z \to z-1)}(T_e, N_e = 0)$ with $f(N_e)$ a non-specific density correction and three-body recombination ignored. These approximations are much less sound and lead to more severe consequential errors. It is also to be noted that the effective coefficients refer to the population of the whole ionisation stage which is identified with the ground population. Basic treatments ignore the presence of populated metastables. True collisional-radiative coefficients in the stage to stage picture exist for hydrogen, helium and first and second period elements and in the more advanced generalised collisional-radiative picture (metastable resolved) for key light elements in the first period which implies a metastable resolution of Eqns. 1 and 4 for their utilisation. The added complexity of the generalised picture is not included in most purely theoretical plasma transport modelling but only in the more detailed diagnostic spectroscopic modelling. Modern generalised coefficient data, of which increasing amounts are in production and which are much more accurate, are condensed onto the stage to stage picture for the theoretical transport modellers at this point in time.

Detailed calculations and measurements for the $S_{i,\parallel}^M$ and and S_i^E momentum transfer and energy transfer rates involving charge transfer reactions between deuterium and deuterons exist and are used. The key final term is the electron energy loss S_e^E. It is more usual in spectroscopy to be concerned with the radiant energy loss (which includes Bremsstrahlung and free bound recombination radiation). S_e^E however includes free electron kinetic energy changes from potential energy losses and gains in non-radiative excitation, de-excitation, ionisation and recombination. In collisional-radiative theory however there is an exact result

$$S_e^E = N_e \sum_{z_0} (\sum_{z=0}^{z_0} (P^{(z)} N^{(z)} - r^{(z)} S'^{z \to z+1} N^{(z)} + I^{(z)} \alpha^{(z+1 \to z)} N^{(z+1)})) \qquad (10)$$

where the α's and S's are the true collisional-radiative coefficients. $I^{(z)}$ denotes the ionisation energy. Note that since in ionisation balance $N^{(z+1)}/N^{(z)} = S^{(z \to z+1)}/\alpha^{(z+1 \to z)}$ the electron energy loss is equal to the radiant energy loss in ionisation equilibrium. We see that there is no need to independently evaluate S_e^E but it can be obtained from the usual tabulations of $P(z)$, $S^{(z \to z+1)}$ and $\alpha^{(z+1 \to z)}$. These same coefficients enter the time constants used for Monte Carlo decision nodes. The various coefficients are evaluated from manipulations of the collisional-radiative excited population equations in which electron impact Maxwell averaged rate coefficients enter directly.

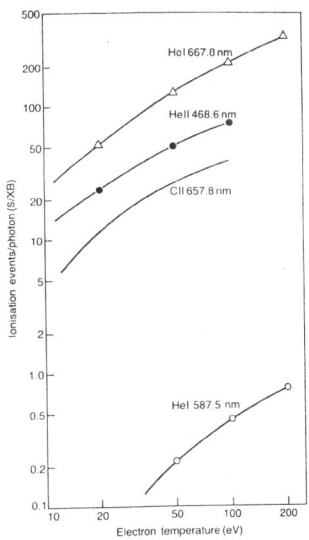

Figure 5. Ionisation events per photon quantities Vs electron temperature for some spectrum lines of helium and carbon ions based on the coronal picture. (Stamp et al., 1989)

ATOMIC COLLISIONS IN SPECTRAL DIAGNOSTICS

Although the complexity of plasma tranport modelling in the fusion plasma has often led to economising on the quality and completeness of atomic data used, the same is certainly not true in spectral analysis. The influence of high quality collision data on spectroscopic diagnosic analysis is now very extensive, spanning many different parts of the plasma environment and touching many different species and spectral ranges. It is usual to distinguish three zones, namely, the high temperature core plasma, the edge and scrape-off-layer plasma and the divertor plasma. The zones may be observed passively or the diagnostic method may involve active switching on and off of a localised perturbation of the plasma. Laser induced fluorescent emission and neutral beam induced emission are common active techniques although the former is not used at JET. Ion cyclotron heating of minority species can be quite a localised perturber and so falls into the active technique category. The study of the emission associated with beams, that is from the beams themselves or from the beam penetrated plasma has become very rewarding at JET. Active beam methods now span from low energy (sub eV) supersonic gas jets in one extreme, through essentially edge probes using low energy ($\sim$ 2-10keV/amu helium and lithium beams) to high energy, high power hydrogen isotope and helium beams ($\sim$ 40-70keV/amu). We wish to outline how diagnostic inferences are drawn in different zones using both passive and active techniques at JET and how particular atomic collision data are required to enable this.

SPECTRAL ANALYSIS OF THE DIVERTOR PLASMA

Measuring Impurity Influx.

Plasma contact with limiting surfaces (either the walls of the vessel or specially constructed buffers) is the primary unavoidable source of impurities which enter the plasma. Nonetheless the release of impurities from the sources, which is mostly by sputtering, can be controlled to some extent by reducing the energy and energy flux density of plasma ions impacting the limiting surfaces. Also, the penetration of released impurities into the confined plasma can be modified by plasma flows, connection lengths etc. Modern divertor designs, such as those illustrated in Fig. 4 seek to realise such controls. The impurity influx from divertor target plates is clearly a key diagnostic measurement. There is a useful result which relates the influx, Γ, of an impurity element from a localised surface to the integrated emission, $I_{i \to j}$, in a spectrum line from the impurity along a line of sight directed at the surface. That is

$$\Gamma = [(S_1^{(z \to z+1)}/(q_{1 \to j}^{(z)})(A_i/A_{i \to j})]I_{i \to j} \tag{11}$$

where the spectrum line is emitted by the z-times ionised ion of the element, with $S_1^{(z \to z+1)}$ the ionisation rate coefficient from the ground state of the ion, $q_{1 \to j}^{(z)}$ the excitation rate ceofficient from the ground state of the ion to the ith excited level, A_i is the reciprocal radiative lifetime of the level i and $A_{i \to j}$ the radiative transition probability. Since $A_i/A_{i \to j}$ is a branching ratio often written as 'B' and the excitation coefficient as 'X', the quantity in square brackets is usually called the 'S/XB' ratio. It is a theoretical quantity determined from atomic collision physics and assumes the coronal picture for excitation. In principle, if this assumption is valid, any spectrum line of any ionisation stage of the impurity which ionises close to the surface can be used for the measurement. Fig. 5 shows S/XB ratios for the light impurities helium and carbon. JET studies have been restricted to the light impurities beryllium and carbon as limiter and divertor target materials. However, light gases such as nitrogen and neon are from time to time introduced into the divertor and are then observed as inflowing species recycling from the divertor target regions.

The S/XB analysis is over-simplified. At JET, there is a strong focus on the use of visible lines for influx studies since absolute intensity calibration, multiple views directed at the divertor etc. are easier to implement. Such spectrum lines are emitted from higher principal quantum shells of low ionisation stages of light impurity ions. At the typical electron temperatures and electron densities of the JET divertor, the upper emitting levels may be populated indirectly and their lifetimes modified by re-distributive collisions. Also, ionisation may occur in a stepwise manner. These 'collisional-radiative' effects mean that the S/XB ratios must be replaced by more complex functions of T_e and N_e which involve the effective ionisation rate coefficient and effective excitation coefficient. The electron temperature has a second important effect for complex ions which have low-lying metastable states. Such metastables can be significantly populated in plasmas with populations comparable to ground states. For inflowing ions, which find themselves in a highly ionising environment, there is a competition between the rate of populating the metastables and collisional ionisation destroying the ground state. The competition is electron temperature dependent but can result in metastable populations being substantially out of excitation balance at the local electron temperature where they emit. Thus influx measurements must address metastables and ground states of an ionisation stage as though they are separate weakly linked species. The situation is illustrated in Fig. 6 where the population distributions of the metastables of Mo^{+0}, Mo^{+1} and Mo^{+2} are mapped at different times following ground state Mo^{+0} entering a plasma of different electron temperatures. Modelling these complexities requires large and precise fundamental data inputs. These include collision cross-sections spanning the complete manifold of levels up to the $n = 4$ shell of selected light impurity neutrals and ions, from low energy up through the intermediate energy regime; collision data for the neutral stage of complex heavy metals which are

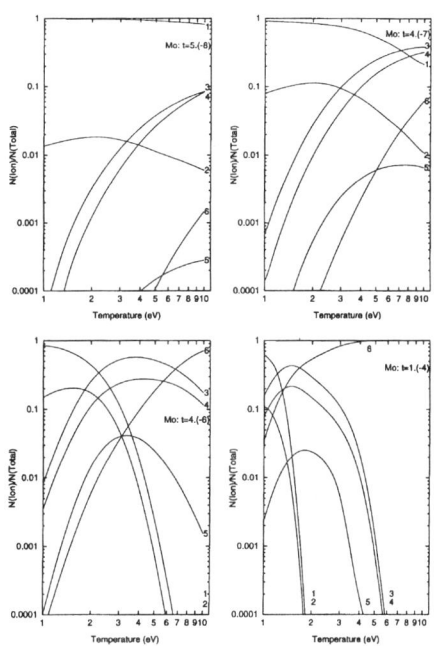

Figure 6. Population fractions for Mo at the times $t = 5 * 10^{-8}s$, $t = 4 * 10^{-7}s$, $t = 4 * 10^{-6}s$ and $t = 1 * 10^{-4}s$. Index 1, $Mo\ ^7S$; 2, $Mo\ ^5S$; 3, $Mo^+\ ^6S$; 4, $Mo^+\ ^6D$; 5, $Mo^+\ ^4D$; 6, Mo^{+2} unresolved. (Badnell et al., 1996)

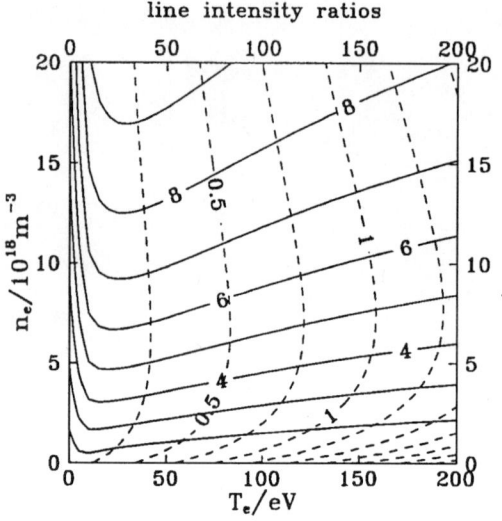

Figure 7. Some theoretical HeI line ratios as a function of electron temperature and electron density. Dashed line: T_e ratio $I(\lambda_1)/I(\lambda_3)$, solid line: N_e ratio $I(\lambda_2)/I(\lambda_1)$. $\lambda_1 = 2\,^1P - 3\,^1S = 7281\text{Å}$, $\lambda_2 = 2\,^1P - 3\,^1D = 6678\text{Å}$ and $\lambda_3 = 2\,^3P - 3\,^3S = 7065\text{Å}$. (Brix and Schweer, 1997)

divertor target candidates; electron impact ionisation cross-section data for metastables as well as ground states which are consistent with the excitation data. Demonstration calculations, for example on Be^{+0} (Bartschat et al., 1991) and Cr^{+0} (Reid et al., 1992; Hibbert et al., 1988) have piloted the way for others to follow.

Measuring Electron Temperature and Density.

The elaborated influx spectroscopic analysis described above does require knowledge of local electron temperature and density. Although such information is routinely and accurately available for the high temperature confined plasma using laser scattering, this is not so in the divertor. Measurements by multiple Langmuir probes embedded in the divertor target plates allow a reconstruction but with significant uncertainty and probably more substantial mis-interpretation in the event of non-Maxwellian distortions of the electron distribution function. Classic passive line ratio spectroscopy for electron temperature and density of the type described in the introduction is difficult to use as an *ab initio* diagnostic because of the geometrical complexity of the observational lines-of-sight, although it remains valuable in retrospective verification. Active edge/divertor neutral beam probes can provide important alternatives. Medium energy neutral lithium and neutral helium beams are suited to the edge/divertor plasma, but the helium gas puff probe has received particular attention in recent years. The helium is injected through a nozzle at thermal ($\sim .1\text{eV}$) energies. The technique was pioneered at TEXTOR (Brix et al.,1997; Schweer et al, 1992) and copied in most other laboratories. An experimental advantage is that suitable spectral lines for diagnostic line ratios lie in the visible. In the low temperature regime, it is electron collisions which excite and ionise helium and through such processes helium is an effective spectrum line radiator above 3-4eV. Because it has singlet and triplet spin systems and a high lying $1s2s\,^3S$ metastable, it yields spectral line ratios diagnostic of both electron temperature and

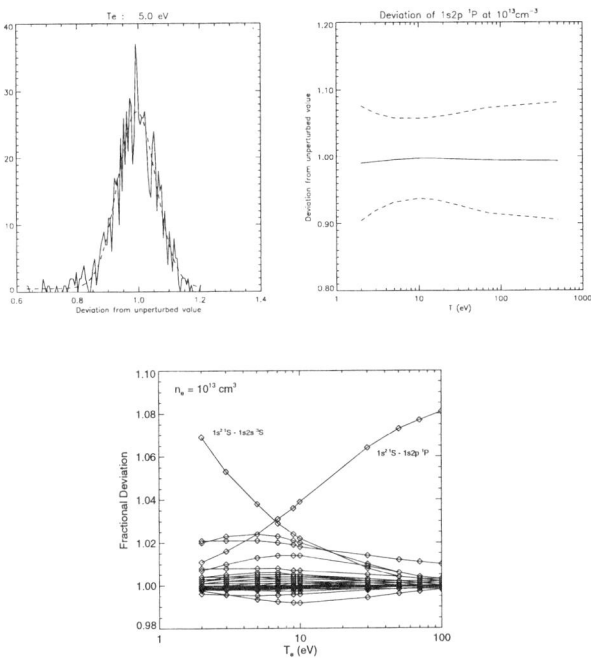

Figure 8. Sensitivity analysis of the $1s2p\ ^1P$ population of He^0 in a plasma. (a) Counts Vs deviation of population from unperturbed value from random sampling at $T_e = 5.0eV$ and $N_e = 10^{13}\ cm^{-3}$. (b) Centroid and half-width from count distribution plots Vs electron temperature. (c) Individual collisional excitation process contributions to fractional deviation of the population Vs electron temperature at $N_e = 10^{13}\ cm^{-3}$.

electron density, as shown in Fig. 7. Full generalised collisional-radiative calculations underpin the theoretical prediction of the ratios for the dynamically ionising helium. It is to be noted that the diagnostic extends to temperatures well above the helium ionisation potential and that there is marked sensitivity to particular collisional rate coefficients such as the ionisation out of the triplet metastable. The highest possible precision of atomic collision data has been required here and its creation and appraisal (de Heer et al., 1996) led through many-state R-matrix (Berrington & Burke, 1996), the problems of the intermediate energy region, R-matrix with pseudo-states (Bartschat et al., 1996) and Converged Close Coupling (Fursa et al., 1997). It has required the assistance of many researchers and is on going.

Analysis of sensitivity to cross-section error

Faced with the realities of calibration in experimental measurements and averaging effects in plasma modelling, moderate precision in atomic data is often adequate. Nonetheless, certain data are required to high precisions, perhaps better than 10-15%. It is important to know which data these are and also to be able to give an assessment of consequential errors in the derived data used for diagnostics from errors in the fundamental data. Most fundamental cross-section data which enters plasma models or spectroscopic analysis are from multi-electron multi-configuration theoretical calculations and so are without error estimates. Experience, gained from a limited number of benchmark studies, has led to broad expectations of the potential accuracy of particular methods. It is however to be noted that these comparisons are often of theoretical

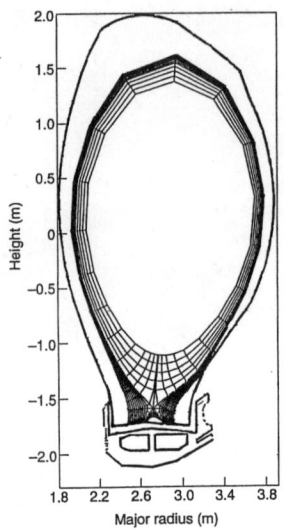

Figure 9. Rings and knots used in the dicretisation of the edge and divertor region of JET for numerical plasma transport simulation. (Maggi, 1996)

method versus theoretical method and also that the potential accuracy of calculation of a given class is probably not achieved in practice because of supplementary imposed simplifications. At JET, it is our current practice to assign percentage uncertainty limits to sub-sets of the collisional (and other) data collections which enter our collisional-radiative calculations. In the case of an atom such as neutral helium above, the work of de Heer (1996) has allowed detailed assignments while for most other ions, only much larger sub-sets can be used. These 'uncertainty blocks' associated with the atomic data for an ion are used to determine the uncertainty in the population of any chosen level of the ion in a plasma. Two measures have proved useful. Firstly, the deviations of the selected population through varying each fundamental process individually within its bounds enables us to identify the precision limiting processes and so target fresh fundamental calculations. Secondly, multiple sampling of the uncertainities in the fundamental processes by random number generation within assumed Gaussian shapes allows the statistics of the resultant error in the derived population to be examined. Convergence is reasonably rapid and we deduce from the distribution of deviations a half-width which is our expected error in the population. Both types of result are illustrated in Fig. 7 for the $1s2p\ ^1P$ population of He^0. The calculations are based on the 'uncertainty blocks' in the JET data for helium from the work of de Heer. At JET, the ability to perform such analysis is being provided as part of the spectral diagnostic modelling and it is hoped that this will provide a focus and incentive to fundamental data producers.

Comprehensive light impurity simulation

Once high quality collision and other atomic data for key species of the type used in the previous sub-sections are sufficiently complete and available, self-consistent and comprehensive confrontation between experiment and models of the experiment can be contemplated. This is attempted at JET for important species. The generalised

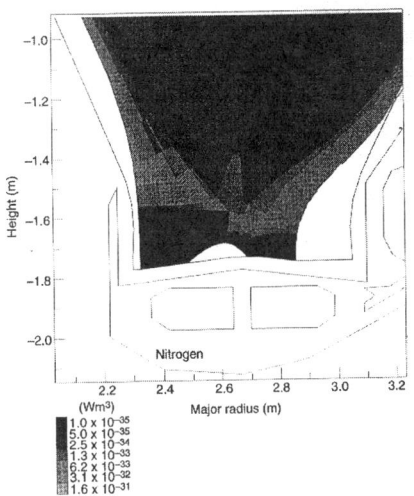

Figure 10. Simulation of the spatial variation of the radiated power loss function for nitrogen in a JET discharge. Note the enhanced radiation away from the strike zones indicating a detached plasma (Maggi, 1996)

collisional-radiative calculations for every ion of an impurity element deliver not only the effective ionisation, recombination, radiated power and electron energy loss coefficients used in the transport modelling but also effective emissivity coefficients of all spectrum lines emitted by these ions, whatever their dynamic state. The deuteron, deuterium and impurity ion distributions in the reconstructed magnetic field topology of the JET pulse are modelled first. Then in a post-processing step, all observed signals along spectrometer and bolometer lines of sight are simulated. An example is shown here from the JET MK1A divertor campaign with graphite target plates when experiments were carried out with combined nitrogen seeding and deuterium fuelling to investigate the establishment and characteristics of a detached radiating divertor plasma. Resulting spectra, observed in the VUV were rich in line emission from NII - NV. Additional diagnostic measurements included bolometric power along lines of sight directed at the divertor. Fig. 9 shows the numerical model grid derived from the reconstructed magnetic field. Fig. 10 shows the simulated spatial variation of the nitrogen radiated power loss function. Evidently all diagnostic measurements can be addressed in this way and the effect of plasma control parameters such as neutral gas injection assessed quantitatively in detail. Many years of effort has devoted to building the collections of fundamental and derived data to support this enterprise. These years have seen a steady progession in quality of data in use. Recently, a step increase in completeness and quality was made when new derived data for helium and carbon were put in place in the JET databases. This will be matched in the very near future with similar data for nitrogen, oxygen and neon. The speed as which such improvements occur depends largely on the output of fundamental collision data by the main producers and the ease with which they can be prepared in the JET formats.

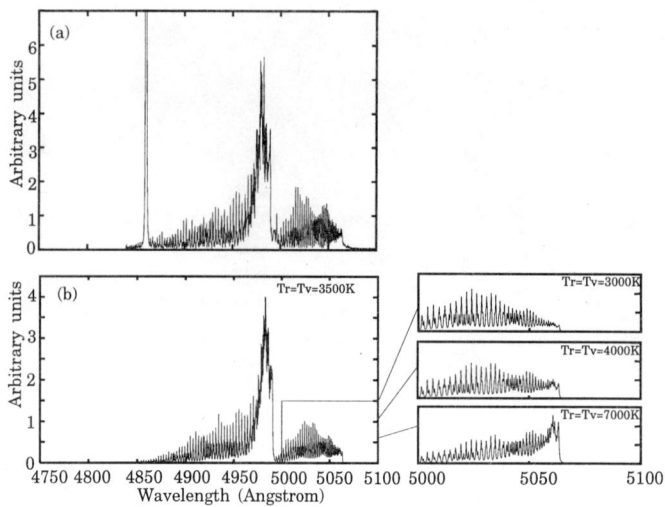

Figure 11. (a) Observed spectrum of the 4850-5100Å region (JET Pulse #35687 at 14.5s) showing BeD ($A\ ^2\Pi - X\ ^2\Sigma$). (b) Simulated 0-0, 1-1 and 2-2 bands at T_r=3500K and T_v = 3500K. The insert shows the variation of the P branch with temperature. (Duxbury et al., 1998)

Measurements on molecules

Although almost all modelling and spectral analysis of divertor plasma has been of atom and ion distributions and emission based on a thermal electron, optically thin plasma, there is clear spectral evidence in JET of the presence of molecular species, of optical thickness in deuterium lines, of volume recombination and a strong probability of non-Maxwellian electron distributions. Supplementary studies at JET are carried out in parallel with the main themes to evaluate some of these effects and to assess their impact on the main radiating properties and modelling of the plasma. The studies on the role of molecules in the divertor at JET are unique since JET alone uses beryllium as a divertor target. We observe occurrence of molecular diatomic band emission in the visible from the $BeD(A\ ^2\Pi - X\ ^2\Sigma)$. Fig. 11 shows modelled relative emission from the superposition of sequence bands from this electronic transition with varying rotational and vibrational temperature in comparison with the observed spectrum. These results suggest a useful and quite sensitive diagnostic. Modelling of the absolute intensity of diatomic emission is a much more complex task. Observations do indicate that BeD emission occurs where it is released from the surface and that this is only where the plasma flow down the scrape-off-layer strikes the target plates. CD molecular band emission by contrast indicates that CD occurs, at least partly, in a catabolic chain from higher deuterides within the plasma. Queen's University, Belfast and JET have been collaborating on the generation of vibronic excitation cross-sections for CD using the R-matrix method. It is planned that over the next year or so, this will lead to initial vibronic population models for CD in the fusion plasma. It is evident that in this area, JET has a significant overlap of interest with the technical plasmas community.

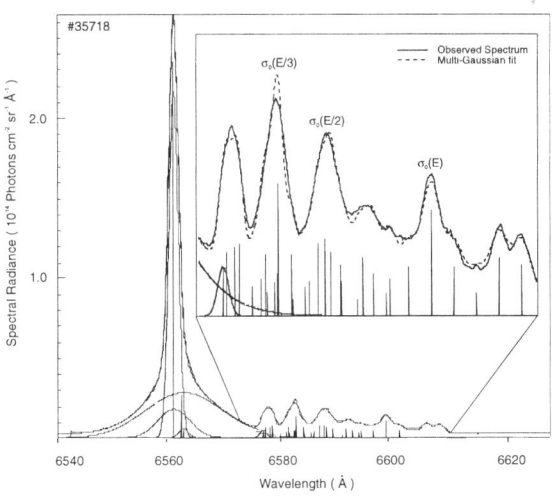

Figure 12. Observed motional Stark D_α spectral feature. The emission feature was recorded during the JET pulse 35718 using track 4 of the multichord visible spectroscopy system. The tangential bank was active with a primary energy of 140keV. (cf. von Hellermann (1993) for a specification of the geometry). The constrained multigaussian spectral fit to the interval is also shown. Particular attention should be brought to the annotation of the full, half and one-third σ_0 Stark components. Due to the existence of three fractional energy components in the beam this gives rise to three Stark multiplets. Each Stark multiplet is Doppler shifted according to the velocity of the relevant beam neutrals and as a result the overal picture is an overlap of each Stark feature. A somewhat more complicated situation arises when the radial bank is also on as this results in the overlap of 6 Stark multiplets.

SPECTRAL ANALYSIS OF BEAM PENETRATED PLASMA

JET is a neutral beam heated plasma. The two very large neutral beam assemblies on opposite sides of the torus are capable of delivering ~ 20 MW of neutral beam power. In normal operation, deuterium is the beam species with the individual atom energies ~ 70keV/amu. The atomic interactions of the plasma with the beam atoms has spawned an enormously fruitful spectroscopic diagnostic development. Not only do the beam atoms radiate characteristically but they also act as charge exchange donors to ions in the plasma. In the confined plasma, light impurity ions are fully ionised, so the charge exchange reaction with the beam atoms leads to hydrogen-like emission from the impurities at positions in the plasma where such radiation does not occur passively. Also both the beam atom emission and the charge exchange emission can be detected spectroscopically in the visible. Fig. 12 shows the $D(n = 3-2)$ beam emission spectrum as a complex Stark multiplet Doppler shifted from the passive $D(n = 3 - 2)$ emission from the plasma edge. The atomic modelling effort at JET which has been required to exploit the diagnostic potential of the deuterium beams is fully as large as that for passive thermal electron driven emission in the divertor. It is also to be noted that the primary collision data for beam modelling is of ion/atom processes. Most neutral beam studies at JET have been directed at the fast deuterium heating beams. JET does have a neutral lithium edge diagnostic beam and there has been some use of neutral helium in the primary injectors. Extensive preparations have been made and are continuing for more extended use of neutral helium beams as a probe both in the fast heating beam regime ~ 50 keV/amu and in the low/medium energy regime ~ 5-10keV/amu.

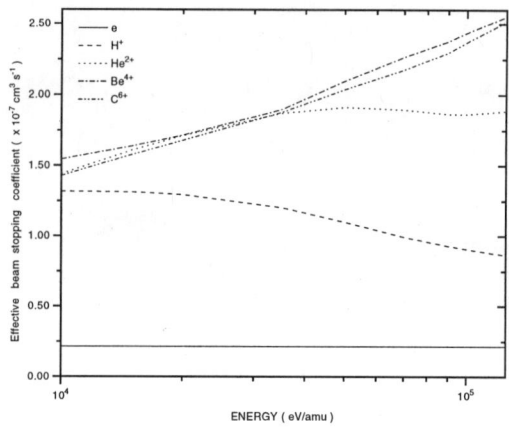

Figure 13. Calculated beam stopping coefficients for a neutral deuerium beam entering various pure impurity plasmas. The beam stopping coefficient is written in terms of the electron density. The impurities are assumed to be fully ionised with number densities required for charge neutrality. $N_e = 3.0 * 10^{13} \ cm^{-3}$; $T_e = T_i = 2.0 * 10^3 \ eV$.

Calculating & Measuring Beam Attenuation.

Deuterium atoms in beams penetrate the confining magnetic field of the plasma as neutrals until their point of ionisation. The rate of beam energy deposition at any point along the beam path is obtained therefore from the ionisation rate, R, or from the effective ionisation rate coefficient, $S_B^{(e)}$ with $R = N_e N_{D,beam} S_B^{(e)}$ if the electron density N_e is known. $S_B^{(e)}$ is usually known as the stopping coefficient. The neutral deuterium density in the beam, $N_{D,beam}$ at any point of the beam path can be obtained by using $N_e S_B^{(e)}$ to calculate the attenuation from where the beam enters the plasma. $S_B^{(e)}$ is written with reference to the electron density in the plasma, but it is primarily collisions with thermal deuteron and thermal impurity ions (fully ionised) which cause the ionisation. So $S_B^{(e)}$ depends on the impurity concentrations. Unfortunately, the concentrations, N_{D+}/N_e, and $\{N_{Z_i}/N_e : i = 1, ...\}$ in the plasma are not initially known except in so far as their mean Z_{eff} can be inferred from bremsstrahlung measurements. However, a part of the ionisation of the beam atoms takes place via a charge transfer reaction to the impurity ions in the plasma which then emit measurable spectral line radiation. Thus, the charge exchange spectroscopic line-of-sight intensity of radiation, $I_{Z_i, n \to n'}$ in a spectrum line $n \to n'$, which is localised at the intersection of the beam/ spectrometer viewing line intersection,L, may be used to infer N_{Z_i}, from $I_{Z_i, n \to n'} = \int_L N_{D,beam} N_{Z_i} q_{Z_i, n \to n'} dl$, if the effective emission coefficient $q_{Z_i, n \to n'}$ is known. We are therefore led to a circular iterative analysis based on plasma measurements and spectroscopy. The beam stopping coefficient depends on density. This is because there are stepwise losses through excited states of the beam deuterium atoms at fusion plasma densities. At the precision required of beam stopping coefficients for calculating attenuation to the centre of the plasma ($\leq 10\%$) such effects matter. Thus the beam stopping coefficient is an effective coefficient including the influence of the excited states. It is again a collisional-radiative coefficient. The secondary collisions which cause this stepwise ionisation are principally ion collisions (deuterons and impurity nuclei) and so the effective stopping coefficient is itself a function of the impurity concentrations. The *ab initio* re-evaluation of the stopping coefficient for varying impurity mixtures in the

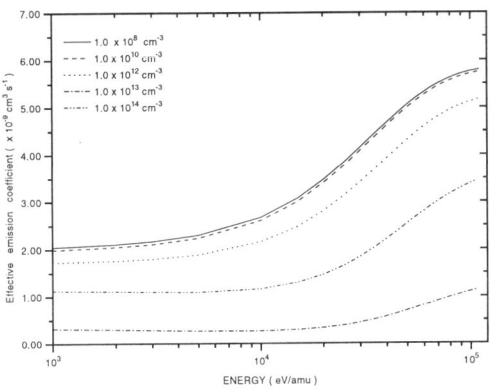

Figure 14. Calculated beam emission coefficient for $D(n = 3 - 2)$ for a neutral deuterium beam entering a pure deuteron plasma Vs beam energy for various electron densities. Note the collisional-radiative nature of the coefficient. The beam emision coefficient is written in terms of the electron density. $T_e = T_i = 2.0 * 10^3 \, eV$.

iterative analysis is quite impractical. At JET, we have evaluated the beam stopping for pure impurity plasmas, that is with only a single impurity and with the electron number density being that which comes from charge balance. The behaviours of such coefficients are shown in Fig. 13. Then a linear combination of coefficients for a mixed impurity plasma can be produced. The error of this step is assessed by comparison with exact calculation for a mixed plasma to be below 5%. Thus a tabulation of the derived stopping coefficient is made for every light impurity $H, He, Li, Be, B, C, N, O, F$ and Ne. The effective coefficients are also functions of N_i and E_{beam} primarily but also of T_i. We define reference conditions for the plasma parameters and beam energy relevant to JET and establish the stopping coefficient at a two-dimensional grid in E_{beam}/N_i at the reference condition of T_i. Then a one-dimensional scan in T_i is made at the reference values of E_{beam} and N_i. Such efforts allow extraction of the stopping coefficient from the tabulations sufficiently rapidly for inter-pulse analysis.

As is seen from Fig. 12, the $D(n = 3 - 2)$ emission from the beam atoms is observed. The collisional-radiative calculations for the beam stopping provide also the effective emission coefficients which allow the reduction of the measured $D(n = 3 - 2)$ emission to the ground state number density. That is the beam attenuation can be measured. The theoretical beam emission coefficient is illustrated in Fig. 14. At JET, it is calculated and tabulated in the same manner as the beam stopping coefficient. Consistency between the calculated and measured attenuation is a strong test of the atomic data, atomic modelling and iterative plasma analysis loop. At this time, the difference is $\leq 20\%$ suggesting that the concentrated attention on the atomic data for deuterium beams is bearing fruit.

Measuring Impurity Concentrations

Charge transfer from fast deuterium atoms to impurity bare nuclei gives consequential cascade radiation over a very wide spectral range including the visible. At JET, charge exchange spectroscopy is performed in the visible using multichord observations along the beam line approximately in the horizontal plane of the torus. The primary state selective charge exchange cross-sections from $D(n = 1)$ are largest to

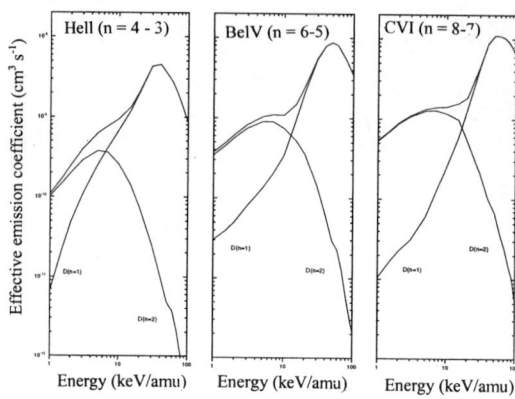

Figure 15. Effective emission coefficients for charge exchange lines of leading impurities in the JET plasma. The graphs contrast the contributions to the emission coeffficients from capture from the D(n=1) ground state and the D(n=2) excited state. The population ratio $D(n=2)/D(n=1)$ is calculated in a very many level beam attenuation code and the coefficient is normalised to the D(n=1) populations. $T_e = T_i = 8*10^3$ eV; $N_e = 3*10^{13}$ cm^{-3}; $Z_{eff} = 2$. For JET heating beams at 70keV/amu, the D(n=1) donor dominates. (Hoekstra et al, 1998)

one or two dominant receiver states but decrease relatively slowly ($\sim n^{-3}$) to higher sub-dominant n-shells for fast beams. It is transitions from these sub-dominant levels which are observable in the visible. Charge exchange cross-sections also cut-off quite sharply at high beam energies when the beam atom speed exceeds the electron orbital speed in the initial donor state. In addition, charge exchange may occur from excited deuterium donor states. Such charge exchange cross-sections are much larger than $D(n=1)$ donor cross-sections but the population of excited donor states in the beam is a small fraction of the ground state population. The dominant receiver states from the excited donor are the sub-dominant levels from the ground state donor, but the high energy cut-off for the excited donor charge exchange cross-sections is much lower than for the ground state donor. These various effects lead to a complex competition between different processes which requires extensive fundamental collision cross-section data for its elucidation. Also, it means that apparently absolute cross-section independent measurement such as the width and shifts of charge exchange spectrum lines are in fact cross-section dependent. The effective emission coefficients for the visible charge exchange lines must be calculated in the collisional-radiative picture. There is substantial redistribution, mainly amongst angular states of the same principal quantum shell, caused by ion impact. The re-distribution is not complete for JET conditions but it is only the unresolved principal quantum shell transition arrays which are observed. Fig. 15 illustrates the contributions from ground state and excited state capture to the effective emission coefficients for charge exchange lines measured routinely at JET (Hoestra et al., 1998). Fundamental atomic data used in the collisional-radiative calculations of effective emission coefficients for JET includes nl selective charge exchange cross-sections from $D(n=1)$ and $D(n=2)$ over very wide beam energy ranges for light impurity nuclei up to neon, re-distributive collision data for hydrogen-like impurites up to neon for impact of bare nuclei of all species up to neon. The assembly and calculation of such data for JET has taken many years supported by scientists at F.O.M, Amsterdam; KVI, Groningen; Dept of Physics, Queens's University, Belfast; Hahn-Meitner Institute, Berlin and La Rolla, Missouri.

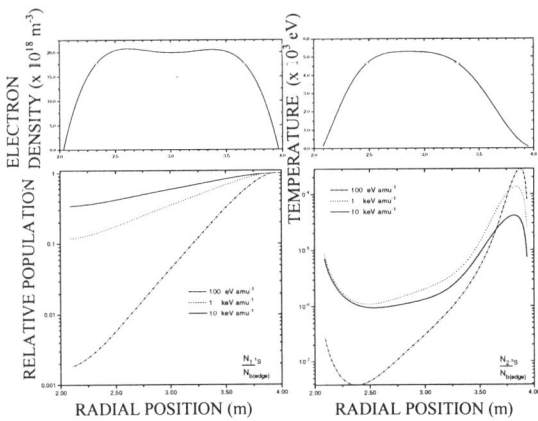

Figure 16. Calculated populations for the 1 ^1S ground state and 2 ^3S metastable state for a neutral helium beam penetrating into the JET plasma. Injection is from the outermost radius (at ~4.0m with only the 1 ^1S ground state populated. The electron temperature and electron density profiles are given in the upper figures as a function of radial coordinate from the tokamak central vertical axis.

Neutral Helium Beams

The beam diagnostic studies at JET has been extended over the last seven years to calculations and modelling for neutral helium beams. This was prompted by the expectation of significant use of the ^3He isotope in the heating beams at JET. There has also been a shared interest in helium beam attenuation and emission at a number of fusion laboratories in Europe. This is to exploit its potential as both a central and edge diagnostic at lower energies ~ 5-10keV/amu. The presence of the triplet metastable gives the promise of additional diagnostic information as well as differential sensitivity to ion and electron impact collisions. Fig. 16 shows the calculated ground and triplet metastables populations along a helium beam line for JET conditions. The triplet metastable is set to zero at the beam entry point into the plasma, so its population is collisionally generated within the plasma. Beam emission from $n = 2 - 3$ transitions on both the singlet and triplet sides of helium is observed in the visible. Charge exchange spectroscopy is also possible with neutral helium beams with markedly different cross-sections for charge transfer depending on whether the ground state or the triplet metastable is the donor. Beam stopping and beam emission must again be calculated in the collisional-radiative picture. At the higher beam energies, ion impact is more effective than electron impact in exciting the beam atoms. Ion impact cross-section data for exciting and ionising neutral helium from its ground state is much more difficult to calculate than for neutral deuterium. This is because of the two correlated 1s electrons. JET has expended considerable effort in seeking to obtain all the ion and electron impact data required to model helium beams to the same standard as deuterium beams. This effort is continuing. Collisional-radiative modelling for helium must be elaborated over that for deuterium. Systematic production of the derived generalised collisional-radiative ionisation and cross-coupling coefficients (which replace the single stopping coeficient for deuterium beams), the beam emission ceofficients (driven by both ground and metastable) and charge exchange emission coefficients (for both ground and metastable donors) is underway but further enhancement of the database is required. In the next years, diagnostic analysis with the low/medium energy helium beam will be the subject of intensive study within the European association of fusion laboratories.

CONCLUSIONS

Since the beginning of the commissioning and experimental programmes on the JET tokamak, the JET Project has provided continuous support for the provision of spectroscopy-based diagnostics of the plasma and impurity state. This support has included the establishment of the infrastructure of atomic modelling at a level of sophistication sufficient for the foreseeable future. Also it has been possible to set up long-term collaborations with interested University groups with expertise in the measurement or calculation of fundamental processes which must underpin such fusion plasma modelling. JET has been fortunate in being able to collaborate with the group of Professor P. G. Burke, Department of Applied Mathematics and Theoretical Physics, Queen's University, Belfast on fundamental electron impact processes for fusion for nearly fourteen years. The work stemming from this collaboration has touched many of the areas of application of atomic data at JET described in the earlier sections. There have been considerable successes. A raft of spectroscopic/atomic collision-based diagnostics have been set up, such as influx spectroscopy, which are now viewed as essential and routine in the fusion community. It has been possible also to share this atomic data with all the fusion laboratories in Europe and coordinate on its use in analysis. Likewise, the fundamental collision community generally in Europe has willingly supported these efforts.

At an international level, the focus on key atomic data seems to have been less good. The demands by the fusion community to the fundamental producer community may have been too general and then the mechanisms for use of the data ill-prepared. It does seem that if fundamental atomic data for fusion is to get into use quickly and efficiently, then detailed agreements have to be reached which cover the pathways from source calculation to application and the structure of the data to be sent down these pathways. This must be done in advance of the substantive calculations. Only then are the data fully relevant and can they be properly maintained and updated. In this aspect, work remains for the future.

REFERENCES

Bates, D. R., Kingston, A.E., and McWhirter, R. W. P., 1962, *Proc. Roy. Soc.* 37:468.
Badnell, N. R. et al., 1996, *J. Phys. B:At. Mol. Opt. Phys.* 29:3683.
Bartschat, K. et al., 1996, *J. Phys. B:At. Mol. Opt. Phys.* 29:115.
Bartschat, K. et al., 1991, *JET Joint Undertaking Report*, JET-R(91)09.
Braginskii, S. L. 1965, *Reviews of Plasma Physics* 1:205.
Brix, M. and Schweer, B. 1997 *24th Eur. Phys. Soc. Conf. - Control. Fusion and Plasma Physics* P4.116:545.
Burke, P. J., Tait, J. H. and Lewis, B. A., 1966, *Proc. Roy. Soc.* 87:209.
Burke, P. J. and Berrington, K. A., 1993, *Atomic and Molecular Processes: and R-matrix approach*, IOP Publishing, London.
Burgess, A., 1964, *Astrophys. J.* 139:776.
Burgess, A., 1965, *Astrophys. J.* 141:1588.
Burton, W. M. and Wilson, R. 1961 *Proc. Phys. Soc.* 78:1416.
Clark, R., Abdallah, J. and Post, D., 1995 *J. Nucl. Materials* 120-122:1028.
Cupini, E., de Matteis, A. and Simonini, R., 1984, *NET Report*, EUR XII.
Duxbury, G., Stamp, M. F. and Summers, H. P., 1998, *Plasma Phys. Control. Fusion* 40:361.
Fursa, D. V. et al., 1997, *J. Phys. B:At. Mol. and Opt. Phys.* 30:3459.

Griffin, D. C.et al., 1997, *J. Phys. B:At. Mol. Opt. Phys.* 30:3543.
de Heer, F. J. et al.., 1996, *Atomic and Plasma Mater. Inter. Data for Fusion* 6:.
von Hellermann, M.. and Summers, H. P., 1993, *Atomic and Plasma-Material Interaction Processes in Controlled Thermonuclear Fusion* p135, Ed. Janev, R. K. and Drawin, H. W., Elsevier, Amsterdam.
Hendry, J. and Lawson, J. D., 1993, *AEA Technology Report*, AHO 1.
Hibbert, A. et al., 1988, *Nucl. Instrum. Methods* B31:276.
Hoekstra, R. et al., 1998, *Plasma Phys. Control. Fusion* 40:1541.
Lotz, W., 1967, *Astrophys. J. Supple.* 128:207.
McWhirter, R. W. P., 1959, *UKAEA Report* AERE-R2980.
McWhirter, R. W. P. and Summers, H. P., 1984, *Applied Atomic Collision Physics:Plasmas*, Ed. Barnett, C. F. and Harrison, M. F. A., Academic Press, Orlando.
Maggi, C. F., 1996, *Ph. D. Thesis: Measurement and Interpretation of Spectral Emission from JET Divertor Plasmas*, University of Strathclyde.
Neuhauser, J. et al. 1984, *Nucl. Fusion* 24:39.
Van Regemorter, H., 1962, *Astrophys. J.* 136:906.
Reid, R.H.G., Bartschat, K. and Burke, P. G., 1992, *J. Phys. B:At. Mol. Opt. Phys.* 25:3175.
Schweer, B. et al., 1992, *J. Nucl. Mater.* 196-198:174.
Seaton, M. J., 1964, *Planet. Space. Sci.* 12:55.
Simonini, R. et al., 1994, *Contrib. Plasma Phys.* 34:368.
Spitzer, L., 1962, *Physics of Fully Ionised Gases*, Interscience Publishers, New York.
Stangeby, P. C. and Elder, J. D., 1992, *J. Nucl. Mater.* 96-98:258.
Stamp, M. F. et al., 1989, *J. Nucl. Mater.* 162-164:404.

RADIATION PRESSURE AND ELEMENT DIFFUSION IN STELLAR INTERIORS

M.J. Seaton

Department of Physics and Astronomy
University College London
Gower Stree, London WC1E 6BT

INTRODUCTION

In recent years my work has been mainly concerned with the study of radiation-pressure forces which lead to diffusive separations of chemical elements in stellar interiors, and which makes extensive use of atomic data obtained by the Opacity Project (see The Opacity Project Team, 1995). I have given talks about the work at a number of meetings attended mainly by astronomers. At this meeting the audience will be mainly physicists and I will therefore put greater emphasis on some of the basic physics involved.

In a star one has an outward flux of radiant energy, $\mathcal{F}_\nu$ energy per unit area, unit time and unit frequency. The corresponding flux of momentum is $(1/c)\mathcal{F}_\nu$. Let an atom of some element k have a cross section $\sigma_\nu^{\mathrm{mta}}(k)$ for absorption of radiation with momentum transfer to the atom. The momentum absorbed per atom per unit time is then $(1/c)\int \sigma_\nu^{\mathrm{mta}}(k)\mathcal{F}_\nu\,d\nu$ which is just the force acting on the atom. Whereas the radiation-pressure forces act outwards, the star is held together by gravitational forces acting inwards. The gravitational forces are expressed as gravitational accelerations g_{grav} and it is usual to express the radiation-pressure forces as radiative accelerations $g_{\mathrm{rad}}(k)$,

$$M(k)g_{\mathrm{rad}}(k) = (1/c)\int \sigma_\nu^{\mathrm{mta}}(k)\mathcal{F}_\nu\,d\nu. \tag{1}$$

In some circumstances and for some elements k one can have $g_{\mathrm{rad}}(k) > g_{\mathrm{grav}}$ which can lead to those elements moving outwards — provided, of course, that the movements are not halted by convection or by circulating currents in the star. The process is important in explaining the strange abundances of the elements which are observed in the atmospheres of the chemically peculiar (CP) stars (Michaud, 1970 — a recent review is given by Smith, 1996).

We are faced with the following problems: (1) determine the cross sections $\sigma_\nu^{\mathrm{mta}}(k)$; (2) determine the radiative flux $\mathcal{F}_\nu$; (3) solve the equations of diffusion theory to obtain the abundance of element k as a function of time and of depth within the star.

LEVEL POPULATIONS

Element k can have many energy levels i and ionisation stages j. The positions of the levels are calculated using RMTRX STGB. The probability of (i,j) being populated is $P(i,j)$ ($\sum_{i,j} P(i,j) = 1$), which is calculated using methods described by Hummer and Mihalas (1988). The cross sections to be used in (1) are

$$\sigma_\nu^{\text{mta}}(k) = \sum_{i,j} \sigma_\nu^{\text{mta}}(i,j,k) P(i,j,k) \qquad (2)$$

where $\sigma_\nu^{\text{mta}}(i,j,k)$ is the cross section for level (i,j,k).

ATOMIC PROCESSESS

There are just two processes of any importance for the calculation of g_{rad}: absorption of radiation in spectrum lines (bound-bound, bb); and photo-ionisation (bound-free, bf).

Bound-Bound

Let absorption of a photon give a transition from an initial state i to a final state f. The cross-section is

$$\sigma_\nu^{\text{line}}(i,f) = (\pi e^2/mc)\Phi(f,i)\phi_\nu \qquad (3)$$

where $\Phi(f,i)$ is the oscillator strength, calculated using RMTRX STGBB, and ϕ_ν is a line-profile factor. Damping processes give a Lorentz profile,

$$\phi_\nu = \Gamma/[(2\pi(\nu - \nu_0))^2 + (\Gamma/2)^2] \qquad (4)$$

where $h\nu_0 = E_f - E_i$. For radiation damping, $\Gamma = A_i + A_f$ where A_i, A_f are probabilities for radiative decays of the two states. Collisional damping (pressure-broadening) is mainly determined by collisions with electrons. We can then express Γ in terms of an electron-impact rate-coefficient involving an electron-impact cross-section Q_D,

$$\Gamma = N_e \int v Q_D f(v)\, dv \qquad (5)$$

where N_e is the electron density, v the electron velocity and $f(v)$ the Maxwell velocity-distribution function. In a very crude first approximation one might expect Q_D to be equal to the sum of the cross sections for collisional de-excitation of the two states. An exact quantum-mechanical theory has been given by Baranger (1958), which provides an expression for Γ in terms of the scattering matrices for elastic scattering by the two states, i and f. The Baranger theory is implemented in RMTRX STGD, which I have used to obtain values of Γ for some 42 transitions (Seaton, 1988). However, for the calculation of stellar opacities and radiative accelerations one must consider of the order of 10^8 spectrum lines. Methods for estimating electron-impact excitation cross-sections, involving an empirical factor $\bar{g}$, have been given by Burgess (1961), van Regemorter (1962) and Seaton(1962) and adapted by Griem (1968) for the calculation of Γ. I have used the results of the STGD calculations to obtain empirical expressions of $\bar{g}$-type which can be used for all transitions. Some improved STGD calculations by Val Burke (1992), with better convergence in the R-matrix expansions, give improved agreement with the empirical formula.

Finally, the line profiles have to be convolved with a thermal Doppler profile.

Bound-Free

It may be assumed that, on average, any radiation emitted following the bb process will be isotropic, and hence that all momentum absorbed will be transferred to the atom. For the bf process the momentum absorbed is shared between the product ion and the ejected electron. Angular distributions of photo-electrons are often described in terms of the function $[1 + \beta P_2(\cos(\theta))]$ (see Burke, 1975)where θ is the angle between the direction of the photon and that of the ejected electron. That formula gives the same distribution for a forward direction ($\theta < \pi/2$) and a backward direction ($\pi - \theta$), and hence gives no net momentum to the electrons. The formula is, however, inadequate for our purpose, since we must consider a theory correct to order $1/c$. The problem has an interesting history. In the 1920's experiments had shown that electrons produced by absorption of X-rays were peaked in the forward direction but the old quantum-theory was not able to give a satisfactory explanation. Sommerfeld, in 1930, showed that agreement with experiment was obtained when one included effects of interference between dipole and quadrupole photo-ionisation amplitudes (giving a formula correct to order $1/c$). His results, which were for hydrogenic 1s states, have been used in much previous work on radiative accelerations. More recently results for hydrogenic 1s, 2s, 2p, 3s, 3p and 3d states have been obtained by myself (Seaton, 1995) and by Massacrier (1996).

The hydrogenic theory is used to estimate the fraction of absorbed momentum given to the ejected electrons, but care must be used. The basic bf cross-sections, calculated using RMTRX STGBF, contain many resonance structures. The resonance bf process is more akin to a bb process, and it is can be assumed that in the vicinities of resonances all absorbed momentum is given to the product ion.

THE RADIATIVE FLUX

Radiant energy is produced in the inner core of a star, by thermo-nuclear processes, and escapes at the stellar surface. In stellar interiors, regions well below the surface, conditions are very close those for a black-body enclosure at a local temperature T, but differ in that there is some net outward flux $\mathcal{F}_\nu$. Let $I_\nu(\hat{n})\,d\omega$ be the intensity of radiation in a direction $\hat{n}$ in a solid angle $d\omega$. The mean intensity, $J_\nu = (1/4\pi) \int I_\nu \, d\omega$, is very nearly equal to the intensity in a black-body, $I_\nu = B_\nu(T)$ where $B_\nu(T)$ is the Planck function. The outward flux is

$$\mathcal{F}_\nu = \int I_\nu(\hat{n}) \cos(\theta)\, d\omega \qquad (6)$$

where θ is the angle between $\hat{n}$ and the outward direction: $\mathcal{F}_\nu$ is non-zero because there is a small difference between the intensities in the outwards and inwards directions.

In interiors the equation of radiative transfer can be solved in a diffusion approximation to give

$$\mathcal{F}_\nu = \frac{\sigma_R}{\sigma_\nu} g_\nu \mathcal{F} \qquad (7)$$

where: σ_ν is the mean cross-section per atom for absorption or scattering of radiation; g_ν is a the weighting function

$$g_\nu = (\partial B_\nu/\partial T)/(\partial B/\partial T) = (15/4\pi^4) u^4 \exp(-u)[1 - \exp(-u)]^{-2} \qquad (8)$$

where $B = \int B_\nu \, d\nu$, $u = h\nu/(k_B T)$ and k_B is the Boltzmann constant; and σ_R is the Rosseland-mean cross-section defined by

$$\frac{1}{\sigma_R} = \int \frac{1}{\sigma_\nu} g_\nu \, d\nu \qquad (9)$$

which ensures that $\mathcal{F} = \int \mathcal{F}_\nu \, d\nu$. It can be assumed that the total flux $\mathcal{F}$ is a known quantity, usually expressed in terms of the effective temperature T_{eff} of a star.

Let chemical element m have an abundance $A(m)$ by number fraction, $\sum_m A(m) = 1$. The cross-section to be used in (7) and (9) is then

$$\sigma_\nu = \sum_m \sigma_\nu(m) A(m) \qquad (10)$$

where $\sigma_\nu(m)$ is the mean cross-section for element m. The cross-section $\sigma_\nu(m)$ differs from the cross section $\sigma_\nu^{\text{mta}}(m)$ in that the former includes contributions from scattering processes (of which the most important is scattering by free electrons) and free-free processes; and in that the latter includes corrections to bf cross-sections to allow for momentum transfer to ejected electrons.

CALCULATIONS OF RADIATIVE ACCELERATIONS

The dominant contributions to g_{rad} usually come from the bb processes, although the bf processes are also important. However, the flux $\mathcal{F}_\nu$ is reduced in the vicinities of the spectrum lines and of resonances for the bf processes. From equations (1) and (7) we obtain

$$g_{\rm rad}(k) = \frac{\sigma_{\rm R}}{M(k)c}\gamma(k)\mathcal{F} \tag{11}$$

where $\gamma(k)$ is the dimensionless quantity

$$\gamma(k) = \int \frac{\sigma_\nu^{\rm mta}(k)}{\sigma_\nu} g_\nu \, d\nu. \tag{12}$$

The weighting function g_ν is small for large values of $u = h\nu/(k_{\rm B}T)$. The integrals (9) and (12) for the calculation of $\sigma_{\rm R}$ and γ are evaluated in the range $0 \leq u \leq 20$. In order to have a reasonably complete resolution of all features in that range it would, in many cases, be necessary to include at least 10^7 frequency points. Fortunately, calculations of adequate accuracy can be made with an appropriate sampling of the points. In calculating $\sigma_{\rm R}$ it is found that results accurate to 1 or 2 per cent can be obtained with 10^4 points (Rogers and Iglesias, 1992; Seaton et al., 1994). Calculations of $g_{\rm rad}(k)$ are more sensitive and 10^5 points are required in order to obtain a similar accuracy.

Consider the contribution to $g_{\rm rad}(k)$ in the vicinity of a spectrum line. In that vicinity the cross-section $\sigma_\nu^{\rm mta}(k)$ will be large, but σ_ν in (12) also has a contribution from $\sigma_\nu(k)$ which will also be large. The value of γ is therefore sensitive to the abundance $A(k)$ of k occurring in (10). The value of $g_{\rm rad}(k)$ decreases as $A(k)$ increases. The effect is know as line-saturation. The results are sensitive to the expressions used for the line profiles (a very narrow lines becomes becomes saturated very quickly).

The diffusion processes which result from radiative accelerations can lead to marked changes in abundances. It is convenient to start with some standard set of abundances $A(m)$ which would be expected in the absence of diffusion and then to vary the abundance of one element at a time. Let the modified abundance of k be

$$A'(k) = \chi A(k) \tag{13}$$

and modify the abundances $A(m)$ of all other elements m so as to conserve the condition $\sum_\ell A'(\ell) = 1$. The quantity χ, which I will use throughout the following discussion, is the fractional change in the abundance of the selected element k.

Using Opacity Project atomic data, values of $g_{\rm rad}(k)$ have been calculated for the elements C, N, O, Ne, Na, Mg, Al, Si, S, Ar, Cr, Mn, Fe and Ni for a wide range of temperatures and densities and values of χ (Seaton, 1997) and the results obtained have been archived at the Centre de Données de Strasbourg (CDS).

DIFFUSION

Any momentum gained by an ion in a plasma is eventually lost due to collisions with other plasma particles. We define v to be the mean diffusion velocity.

The theory of diffusion is described in the monumental work by Chapman and Cowling (1952) and, of course, in many other books and papers. For the case of interest here, Chapman and Cowling give the diffusion velocity to be

$$v = D \times \left\{ \frac{M(k)}{k_B T}(g_{\text{rad}} - g_{\text{grav}}) - \frac{\partial \ln(c)}{\partial r} \right\} \tag{14}$$

where D is the diffusion coefficient, r is the stellar radial co-ordinate and $c = N(k)/N$ is the concentration of k, $N(k)$ being the number of atoms k per unit volume and N the total number of atoms.

In stars hydrogen is the most abundant element and in interiors the hydrogen will be almost completely ionised. The main loss of momentum will therefore be in collisions with protons. Aller and Chapman (1970) give the following expression for D assuming $A(k)$ to be small, that the proton number density is equal to the electron density, and that only collisions with protons are taken into account:

$$D = (3/4)(k_B T)^{5/2}(2/\pi M_p)^{1/2}/(e^4 N_e z^2 \ln(1 + x^2)) \tag{15}$$

where M_p is the proton mass, z is the charge on the diffusing ion and $x^2 = 4(k_B T)^3/(\pi N_e e^4 z^2)$. Numerous later refinements have been made (see Paquette et al., 1986) but we shall let (15) suffice for our present discussion. We note that D is proportional to $1/z^2$ and hence that the less highly ionised systems diffuse the most rapidly. The diffusion coefficient for neutrals ($z = 0$) is about about 100 times larger than that for singly-ionised ions (Gonzalez et al., 1995). It is necessary to calculate values of g_{rad} separately for each ionisation stage and to use some appropriate averaging to obtain the rates of diffusion for each chemical elements: details are given by Seaton (1997) which includes references to earlier work.

An interesting effect has been discussed by Michaud et al. (1979) and further by Gonzalez et al (1995): an accelerated ion can change its ionisation stage before being stopped by collisions with the protons. The most important process appears to be collisional ionisation and is particularly important for accelerated neutrals, which diffuse very rapidly. Further work would be of interest. For excited states, collisional ionisation may proceed via collisional excitation to higher excited states. The process may be similar to that involved in collisional-radiative recombination, as discussed by Bates et al. (1962).

THE CONTINUITY EQUATION

In order to simplify some subsequent formulae I will assume a plane-parallel model (in actual calculations stars are, of course, assumed to have spherical symmetry). The continuity equation for conservation of particles k is

$$\frac{\partial N(k)}{\partial t} + \frac{\partial vN(k)}{\partial r} = 0 \qquad (16)$$

where r is the stellar radial co-ordinate and t the time. Putting $N(k) = A(k)\chi N$ and introducing the column number-density

$$x = \int_r^\infty N\,\mathrm{d}r \qquad (17)$$

the continuity equation becomes

$$\frac{\partial \chi}{\partial t} = \frac{\partial F}{\partial x} \qquad (18)$$

where

$$F = vN\chi. \qquad (19)$$

I will refer to F as the particle flux but note that the true flux is $vN(k) = A(k) \times F$.

For any function $Y(x,t)$ I will use the notations

$$\dot{Y} = \partial Y/\partial T \quad \text{and} \quad Y' = \partial Y/\partial x. \qquad (20)$$

The continuity equation is then

$$\dot{\chi} = F'. \qquad (21)$$

Using (14) we obtain

$$F = A(x,\chi) + C(x)\chi' \qquad (22)$$

where:

$$A = ND(k_\mathrm{B}T/M(k))y; \qquad (23)$$

$$y = \chi(g_\mathrm{rad} - g_\mathrm{grav}); \qquad (24)$$

and

$$C = N^2 D. \qquad (25)$$

Equation (21) looks simple enough, but attempts to solve it encounter three problems: (1) the need to impose suitable boundary conditions; (2) the fact that some apparently obvious numerical techniques become hopelessly unstable; and (3) the fact that A is a fairly complicated function of χ.

Problem (3) can be overcome with a suitable linearisation. The quantity A depends on t only because χ depends on t and we can therefore put $\dot{A} = B\dot{\chi}$

where $B = \partial A/\partial \chi$ giving $\dot{F} = B\dot{\chi} + C\dot{\chi}'$. From (21), $\dot{\chi} = F'$ and $\dot{\chi}' = F''$ giving an equation for F,

$$\dot{F} = B(x,\chi)F' + C(x)F''. \tag{26}$$

The quantity B depends on χ but in one integration step, from t to $t + \delta t$, we neglect that dependence. After that step has been made, at $t + \delta t$ we have a new function F which is used to obtain a new function χ on solving (22), and hence to obtain a new B to be used for the next time-step. It must, of course, be checked that the values used for δt are such that the linearisation does not lead to significant error.

The term CF'' in (26) arises from the term $\partial \ln(c)/\partial r$ in (14) and will be referred to as the *concentration gradient* (CG) term. In many circumstances the CG term is small, $|CF''| \ll |BF'|$, but there are circumstances in which its inclusion is essential. If we neglect the CG term the equation to be solved is $\dot{F} = BF'$, which is a simple wave equation in which B can be interpreted as the velocity for movement of features in F. The general solution is

$$F = \Theta\left(t + \int (1/B)\,dx\right) \tag{27}$$

where $\Theta(\xi)$ is any differentiable function of ξ. Given $F(x,t)$, one can therefore calculate $F(x, t + \delta t)$ using simple interpolations.

There is no real difficulty in imposing lower boundary conditions since, when we go to deep enough layers, there is no significant change in χ during the entire time-scale for evolution of a star. Much more serious problems are encountered concerning the outer boundary condition. We define the Rosseland-mean optical depth

$$\tau_\mathrm{R} = \int_r^\infty N\sigma_\mathrm{R}\,dr. \tag{28}$$

The theory as presented here is valid only for larger values of τ_R, say $\tau_\mathrm{R} \geq 1$. The expression (7) used for $\mathcal{F}_\nu$ will not be valid for smaller τ_R. Furthermore, the diffusion theory as presented here becomes invalid if we attempt to go to the lower densities high in a stellar atmosphere (the diffusion coefficient D is proportional to the reciprocal of the density and we eventually obtain nonsensical diffusion velocities large compared with thermal velocities). In the outermost layers one should use a theory which is more akin to the theories used in studies of stellar winds.

My current interests are mainly in the study of diffusion of iron-group elements, which have large values of g_rad. For those elements one can find values of τ_0 of τ_R such that at τ_o we have $B > 0$ and very small values of the CG term. In such cases the solutions at τ_0 depend only on the behaviour of F for $\tau_\mathrm{R} \geq \tau_0$ and no further specification of outer boundary conditions is then required. In some cases one has $\tau_0 = 1$. I leave it to other workers to match such solutions to those for the stellar atmospheres.

We can distinguish three regions:

- **Regions A** have $B > 0$ and $|BF'| \gg |CF''|$. We obtain a first approximation using (27) and can then make a perturbation correction to include the CG term. Features in F move outwards.

- **Regions B** have $B < 0$ and $|BF'| \gg |CF''|$. Solutions are obtained as for Regions A. Features in F move inwards.

- **Regions C** are those for which B is small, and come between regions A or B. The equation (26) can be solved using stable Crank-Nicholson methods (see Press et al.. 1992) but it is essential to have known boundary conditions, obtained from the adjacent Regions A or B.

SOME ILLUSTRATIVE EXAMPLES

At this point the reader may feel that the theory is all becoming a bit involved. I hope to be able to provide some clarification by giving a few graphical results. I will take $\log(T)$ as a basic depth variable, where T is the temperature in a stellar interior, and give results for a stellar model with effective temperature $T_{\text{eff}} = 13000$ K and surface gravity $\log(g) = 3.87$.

Figure 1 shows the ratio $g_{\text{rad}}/g_{\text{grav}}$ for manganese in that model. Results are given for $\chi = 0.1, 1.0, 10$ and 100 and it is seen that g_{rad} decreases as χ increases, which is due to saturation effects. For $\chi = 1$ (the solar-system abundance) we obtain $g_{\text{rad}} > g_{\text{grav}}$ except in the very deepest layers, leading to outward diffusion of Mn atoms. Particularly large values of g_{rad} occur in the vicinity of $\log(T) = 5.2$, which is just the regions of the "Z-bump" in opacities, produced by iron-group elements: it was first noted by Rogers and Iglesias (1992) and confirmed in the Opacity Project work. There are 3 minima in the values of g_{rad}: one at $\log(T) \simeq 4.8$ where the dominant ionisation stage is Ar-like; one at $\log(T) \simeq 6$ where the Ne-like stage is dominant; and one at $\log(T) \simeq 6.9$ corresponding to the He-like stage. Such minima play a important rôles in the diffusion process. They will be referred to as the Ar, Ne and He barriers.

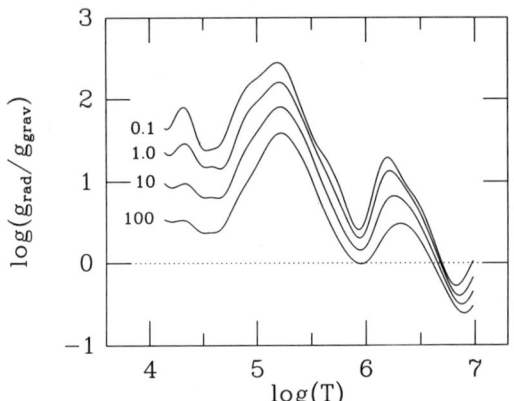

Figure 1. The ratio of radiative to gravitational acceleration for manganese in a star with $T_{\rm eff} = 13000$ K and surface gravity $\log(g) = 3.87$. Curves for concentrations $\chi = 0.1$, 1.0, 10 and 100 where $\chi = 1.0$ corresponds to standard solar-system abundances. The ratio is plotted against $\log(T)$ where T is the temperature in the stellar interior.

Figure 2 shows $y = \chi[g_{\rm rad} - g_{\rm grav}]$ as a function of χ at a point in the model with $\log(T) = 4.9$. At $\chi = 0$ we have $y = 0$, due to the external factor of χ. For the smaller values of χ we have $g_{\rm rad} > g_{\rm grav}$ and, again because of the external factor, y increases with increasing χ. However, due to saturation effects, $g_{\rm rad}$ decreases with increasing χ and eventually a maximum in y is reached. If we neglect the CG term, $C\chi'$ in (22), we have $F = A$ where A is proportional to y, and the maximum $y_{\rm mf}$ in y then corresponds to the maximum flux, $F_{\rm mx}$. At the maximum we have $\chi = \chi_{\rm mf}$. For $\chi < \chi_{\rm mf}$ we have $B > 0$ (recalling that $B = \partial A/\partial \chi$), and we are in a Region A. For $\chi > \chi_{\rm mf}$, on the other hand, we are in a region B. At $\chi \simeq \chi_{\rm mf}$ we have $B \simeq 0$ and we are in a region C.

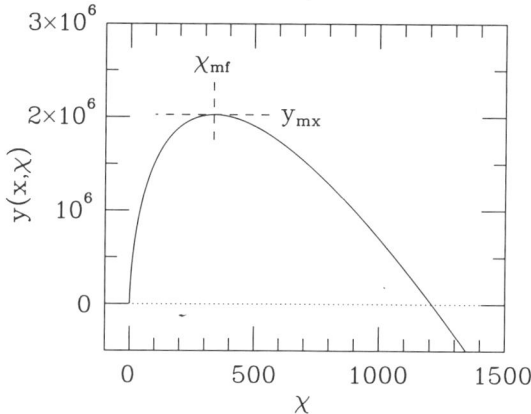

Figure 2. The quantity $y = \chi[g_{\rm rad} - g_{\rm grav}]$ for manganese as a function of χ at a point in the stellar model of Figure 1 with a temperature of $\log(T) = 4.9$.

Figure 3 shows the flux F for manganese at times $t = 0, 1, 2, 3, 4, 5, 10$ and 100 and, as a thick line, the flux $F_{\rm mx}$. We start at $t = 0$ with $\chi = 1$ and obtain a flux $F(t = 0)$ which has a peak at $\log(T) \simeq 5.2$. That feature moves outwards but for the whole diffusion process the flux for manganese remains smaller that $F_{\rm mx}$. The movement of the peak slows down in the vicinity of the Ar barrier at $\log(T) \simeq 4.8$ but has passed the barrier after times $t \simeq 4$; the fluxes for $t > 4$ are shown by dashed lines. The CG term is included but has only a minor effect, in leading to a slow reduction in the height of the peak flux. It is seen that there is practically no movement in F in the deeper layers, as the Ne barrier is approached.

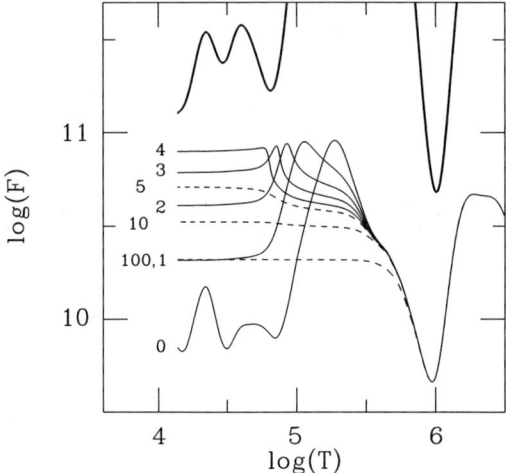

Figure 3. Fluxes for manganese in the stellar model of Figure 1. The fluxes are in arbitrary units. The top curves (heavy lines) give values of F_{mx}, the maximum flux which can occur when theconcentration-gradient (CG) term is neglected. The other curves show fluxes for times $t = 0$, 1, 2, 3, 4, 5 10 and 100, in units of 10^6 years. A maximum flux in the outer layers occurs at $t \simeq 4$. Curves for $t > 4$ are shown with dashed lines. Note that the Ar barrier occurs at $\log(T) \simeq 4.8$ and the Ne barrier at $\log(T) \simeq 6$.

Figure 4 shows values of χ for manganese, at times as for Figure 3. In the outer layers χ first increases, reaches a maximum value, and then decreases as the peak in F moves out through the surface. The abundance enhancements in the outer layers, and losses due to escape at the surface, are at the expense of depletions in the region of $\log(T) \simeq 5.3$.

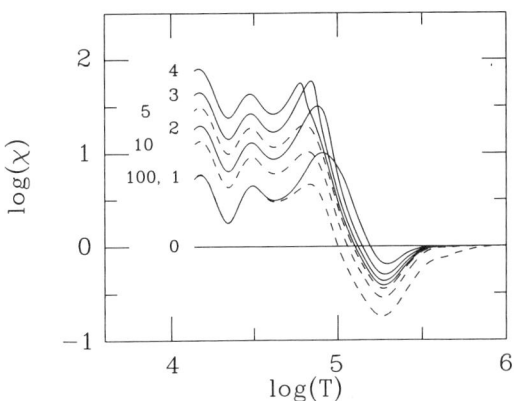

Figure 4. Concentration of manganese in the stellar moel. Curves as for Figure 3.

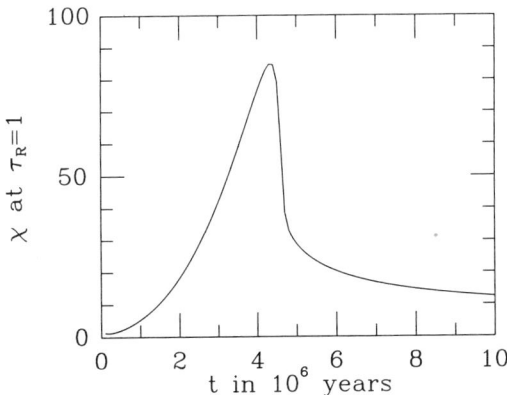

Figure 5. Concentration of manganese at the top of the stellar envelope (Rosseland-mean optical depth of unity) as a function of time.

Figure 5 show the concentration at the top of the envelope, $\chi(\tau_R = 1)$, as a function of time. A maximum occurs when the peak flux reaches $\tau_R = 1$ and is followed by a sharp drop in χ.

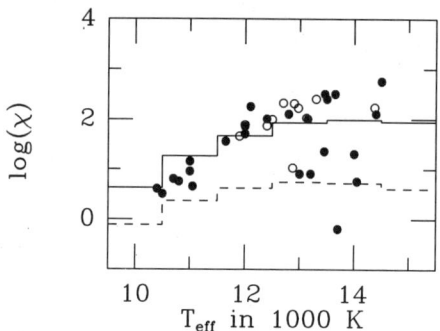

Figure 6. Atmospheric concentrations of manganese for the HgMn stars. Filled circles, results from UV observations (Smith and Dworetsky, 1993): open circles, from optical observation (Adelman *et al.*, 1995). The histograms show calculated concentration at the tops of the envelopes (bottoms of the atmospheres). Upper histogram, maximum calculated concentration: lower histograms, concentrations after diffusion for 10^8 tears.

Figure 6 shows observed atmospheric abundances for Mn in the CP stars with $T_{\rm eff}$ ranging from 10000 to 15000 K (the HgMn stars). Those stars are selected as having enhanced abundances of Mn or Hg or both. The results from UV observations (Smith and Dworetsky, 1993) are shown as filled circles and those from optical observations (Adelman *et al.*, 1995) as open circles. The observed spectrum lines are formed quite high in the atmosphere, at optical depths greater than unity, but some meaningful comparisons can be made with the calculated abundances at $\tau_R = 1$. The upper histogram on Figure 6 shows the maximum calculated abundances and the lower one abundances after diffusion for a time of 10^8 years.

Of the four iron-group elements considered, Cr, Mn, Fe and Ni, it is only for Mn that the diffusion occurs entirely in a Region A, $F < F_{\rm mx}$ for all values of t.

Figure 7 shows F for chromium at $t = 0.0, 0.5, 1.0$ and 1.5. It is seen that at $t = 0.0$ the peak flux is larger than F_{mx} at the bottom of the Ar barrier. The barrier is reached at $t \simeq 1.5$. The CG term is then important.

Figures 8 shows how the flux for chromium works its way around the Ar barrier. Results are shown for time of $t = 1.5$ to 7.5 in steps of 0.5. At $t = 7.5$ the peak flux has passed the barrier. The effect of the CG term is to lead to reductions in the heights of the peak.

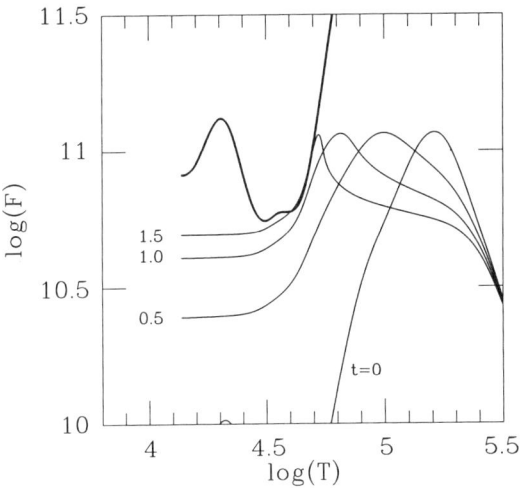

Figure 7. Fluxes for chromium in the 13000 K stellar model at times $t = 0.0, 0.5, 1.0$ and 1.5. At $t \simeq 1.5$ the Ar barrier is reached (shown as a heavy line). If the concentration-gradient (CG) term is neglected, the flux cannot cross the Ar barrier.

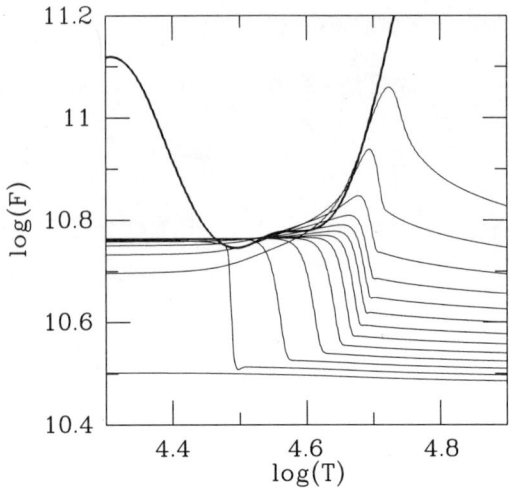

Figure 8. Fluxes for chromium at time $t = 1.5$ to 7.5 in steps of 0.5. The figure shows how the flux passes through the Ar barrier when the CG term is included.

The other two iron-group elements considered, Fe and Ni, are of great interest to astronomers. They differ from Mn and Cr in having smaller values of $g_{\rm rad}$ in outer layers, and in some cases the methods described here cannot be used to obtain solution out as far as $\tau_{\rm R} = 1$. I will not, however, attempt here to give a further description here of work for those two elements.

OTHER RECENT WORK

Richer et al. (1998), to be referred to as RMRITL, have calculated radiative accelerations using monochromatic opacities from the OPAL work (Rogers and Iglesias, 1992), tabulated at 10^4 frequency points. It is noted in my 1979 paper and in their paper that, in some cases, a finer frequency mesh is required. Whereas is my work I have varied the abundance of one element at a time, RMRITL introduce the refinement of considering the effect on $g_{\rm rad}$ for one element of varying the abundance of some other element — for example, the effect on $g_{\rm rad}$ of Mn of varying the abundance of Fe.

RMRITL report some work by Turcotte et al. on effects of diffusion on stellar evolution. That work is restricted to stars somewhat cooler than those considered here. Thus for example they give results for diffusion of Mn in a star of mass 1.4 $M_\odot$. Such a star has an upper convection zone extending out to the stellar surface, and therefore does not present the problems discussed here of escape of Mn at the surface and of the problems which then arise in imposing suitable surface boundary conditions. It is noted by RMRITL, in agreement with the present work, that calculations with a finer frequency mesh are required for the hotter HgMn stars.

FUTURE WORK

Atomic Physics

Data required for the calculation of radiative accelerations have been computed using atomic data obtained by The Opacity Project. Some further atomic-physics work would be of interest.

Line Profiles For electron-impact broadening I have used the Lorentz profile (4) with Γ calculated using the theory of Baranger and RMTRX STGD. That profile is not valid in far line-wings. I am planning to develop RMTRX STGFF, for free-free transitions. Considering radiation by a system of (ion+electron), there are two possible processes: radiation by the electron in the field of the ion; and radiation by the ion perturbed by the colliding electron. That should give results for the line profile which join smoothly to those from the Baranger theory, but which remain valid at larger distances from the line centres.

Momentum Transfer To Photo-Electrons In order to calculate the net momentum given to photo-electrons in photo-ionisation processes it is necessary to make calculations correct to order $1/c$. That requires calculation of amplitudes for both dipole and quadrupole processes. Calculations have been made for hydrogenic systems. It would be of interest to make similar calculations for systems giving resonances in their photo-ionisation cross-sections.

Collisional Ionisation From Excited States It would be of interest to make further studies of collisional ionisation from excited states, including processes which start with collisional excitation to higher excited states.

Diffusion In Stars

The calculations reported here, and similar further work which is in progress, have been done with the intention of developing techniques and attempts to understand some of the main processes. A number of rather drastic approximations have been made. Much further work is required.

Convection Convection will stop diffusion. Normal stars in the temperature range of the HgMn stars considered here will have convection zones in the layers where the ionisation stage of He changes from He^+ to He^{+2}. In the work reported here such convections zone are not included, it being assumed that the He abundances are depleted due to gravitational settling (the HgMn stars show depleted surface abundances of He). It would be of interest to study simultaneously the diffusion of He and of iron-group elements.

Stellar Evolution We have considered diffusion during times of order 10^8 years, comparable with main-sequence life-times, but have not taken into account the normal evolutionary changes in the stellar structures which occur on the main sequence.

Changes In Abundances The iron-group elements make major contributions to Rosseland-mean opacities, particularly in the region of the "Z-bump" (Rogers and Iglesias, 1992; Seaton *et al.*, 1994). Diffusion of those elements can cause major changes in opacities and hence in the structures of the stars. It would be desirable to recalculate the stellar structures taking account of the abundance changes.

References

Adelman S.J., Philip, A.G.D., and Adelman C.J., 1996 *Mon. Not. R., Astr. Soc.*, **282**, 953.
Aller, L.H.and Chapman, S., 1960. *Astrophys. J.*, **132**, 461.
Baranger M., 1958. *Pys. Rev.*, **111**. 494.
Bates D.R., Kingston A.E. and McWhirter R.W.P., 1962. *Proc. R. Soc.*, **A267**, 297.
Burgess, A., 1961. *Mem. Soc. R. Sci. Lieège*.
Burke, P.G., 1975. *Atomic Processes and Applications*, (ed. P.G. Burke and B.L. Moiseiwitsch), North-Holland, Amsterdam.
Burke, V.M., 1992. *J. Phys. B*, **25**, 4917.
Chapman S. and Cowling, T.G., 1052. *Mathematical Theory of non-uniform gases* 2nd. ed., Cambridge Univ. Press.
Griem H.R., 1968. *Phys. Rev.*, **165**, 258.
Hummer D.G. and Mihalas D., 1988. *Phys. Rev.*, **331**, 794.
Massacrier G., 1996. *Astron. Astrophys., Astron. Astrophys.*, **309**, 979.
Michaud, G., 1970. *Astrophys. J.*, /bf 160, 641.
Michaud G., Montmerle T., Cox A., Magee N.H., Hodson S.W. and Martel A., 1979, *Astrophys. J.*, **234**, 206.
Paquette C., Pelletier C., Fontaine G. and Michaud G., 1986. *Astrophys. J. Supp.*, **61**, 197.
Press W.H., Teukolsky S.A., Vettering W.T. and Flannery B.P., 1992. *Numerical recipes in FORTRAN* , Cambridge University Press.

Richer J., Michaud G., Rogers F.R., Iglesias C.A., Turcotte S. and LeBlanc F., 1998. *Astrophys. J.*, **492**, 833.

Rogers, F.J. and Iglesias C.A., 1992. *Astrophys. J. Supp.*, **79**, 507. Seaton, M.J., 1962 *Atomic and molecular processes*, (ed. D R Bates), p375, Academic Press, New York.

Seaton, M.J., 1988. *J. Phys. B*, **21**, 3033.

Seaton M.J., 1995. *J. Phys, B*, **28**, 3185 (and *ibid* 1996, **29**, 2373).

Seaton, M.J., 1997. *Mon. Not. R. Astr. Soc.*, **289**, 700.

Seaton M.J., Yu Yan, Pradhan A.K. and Mihalas D., 1994. *Mon. Not. R. Astr. Soc.*, **266**, 805.

Smith, K.K., 1996. *Astrophys. Sp. Sci.*, **237**, 77.

Smith K.C. and Dworetsky M.M., 1993 *Astr. Astrophys.*, **274**, 335.

Sommerfeld, A., 1930. *Wave Mechanics*, Methuen, London.

The Opacity Project Team. 1995. *The Opacity Project*, Vol. 1, Inst. Phys., Bristol.

Turcotte S., Richer J., Michaud G., Iglesias C.A. and Rogers F.R., 1998, *Astrophys. J. Supp.*, Sept., 1998.

Van Regemorter, H., 1962. *Astro. Phys. J.*, **136**, 906.

ATOMIC PHYSICS OF MUON-CATALYZED FUSION

Isao Shimamura

The Institute of Physical and Chemical Research (RIKEN)
Wako, Saitama 351-0198, Japan

INTRODUCTION

In 1956, Alvarez et al. found the phenomenon of proton-deuteron (p-d) fusion, not in a hot plasma but in a liquid hydrogen bubble chamber when there are muons (μ) in it.[1] At the end of about 0.6% of the stopped μ tracks, secondary μ appeared with an energy of 5.4 MeV, which was released in the p-d fusion reaction. This was interpreted as due to μ-catalyzed fusion (μCF), and was reported sensationally in the New York Times as a possible new source of energy. This article stimulated Jackson to evaluate quickly the parameters relevant to power production.[2] Unfortunately, the rate of μ-catalyzed p-d fusion was too low for efficient energy production, which cooled down the μCF fever. However, the interest in the physics of μCF continued among physicists. Thus, collision processes relevant to μCF, such as a μ-transfer reaction $p\mu + d \to p + d\mu$, were investigated, e.g., by Burke et al.[3]

It was realized later that μCF had already been proposed in 1947 by Frank[4] in the consideration of an alternative explanation of the experimentally known existence of two different kinds of mesons, later known as muons (which turned out not to be mesons) and π mesons (or pions). Sakharov promptly amplified on this idea and discussed the possibility of energy production in μ-catalyzed d-d fusion.[5] The interest in μCF, once faded away because of the low rate of μ-catalyzed p-d fusion, was revived due to Vesman's interpretation of the measured μ-catalyzed d-d fusion rate in terms of a resonance mechanism for the formation of the three-body system $dd\mu$,[6] and especially due to the prediction by Gershtein and Ponomarev that this mechanism would enhance the rate of μ-catalyzed deuteron-triton (d-t) fusion by two orders of magnitude.[7] Indeed, this prediction was verified in 1983 by the measurements by Jones et al.[8] Recent developments in the studies of μCF have been reviewed in, e.g., Refs. 9 and 10.

THE MECHANISM OF MUON-CATALYZED FUSION

The muon is a lepton, like the electron, but its lifetime in vacuum is 2.20 μs. The negatively charged μ has the same charge as the electron and a mass larger than the electron by a factor of 206.8, i.e., a mass smaller than the proton by a factor of 8.88. The Bohr radius of μ moving around a very heavy nucleus (called the muonic atomic unit, m.a.u.) is 2.56×10^{-4} nm or $1/206.8$ a.u., a.u. standing for the electronic atomic unit. This small size of muonic orbits is essential in the mechanism of μCF.

Consider the vibrational wave function $\chi(R_{pd})$ of the hydrogen molecule isotope HD. It has a finite value at the origin $R_{pd} = 0$. In other words, there is a finite probability of finding the two nuclei p and d nearly at the same position, i.e., within the range of the nuclear force. Therefore, p-d fusion occurs in this molecule with a finite probability in the absence of any interaction with external particles or fields. The mechanism of this intramolecular fusion is quite different from that of thermonuclear fusion in hot plasmas in which the Coulomb barrier between the two nuclei at short distances is to be overcome by the kinetic energy of the thermal motion of the nuclei; the tunneling effect in a bound state is the clue to the intramolecular fusion.

The rate of the intramolecular fusion, however, is extremely low. Prepare 1 km^3 of the HD gas at 0 °C and 1 atm (and disregard the reaction 2HD $\to$ H$_2$ + D$_2$). Then, the rate of this fusion in this volume of HD will be once in 3×10^{14} years, which is four orders of magnitude longer than the age of the present universe. It will be once in 10^{25} years for the molecular ion HD$^+$ with a larger equilibrium internuclear distance R_{eq}. If the electron in HD$^+$ ($= pde$) is replaced by a heavier muon μ, however, R_{eq} decreases by a factor of ~ 200, and the value of $|\chi(0)|^2$, and hence the p-d fusion rate, increases by a factor of 3×10^{71}! The rate is even higher by a factor of 3×10^6 for d-t fusion in the system $dt\mu$, namely, 3×10^{30} events per cm^{3}s in principle. The question is whether the system $dt\mu$ can be formed efficiently in some way or other.

The basic experimental scheme is fairly simple. Just introduce a μ beam into the mixture of deuterium D$_2$ and tritium T$_2$, in which there usually exist also the molecules DT formed by the reaction D$_2$ + T$_2$ $\to$ 2DT. Then molecules $dt\mu$ are automatically formed with some probability, and the d-t fusion reaction $d + t \to \alpha + n + 17.6$ MeV occurs in these molecules. Not only do the fusion neutrons n and α particles come out, but also the muons are released. These free muons can further be used to form new $dt\mu$. In this way the muons act as catalysis. A single μ was found to catalyze ~ 150 fusion events on the average before its death[8] in the D$_2$/T$_2$ mixture of the liquid-hydrogen density (LHD), which is 4.25×10^{22} cm^{-3}.

In fact, the elementary processes occurring in the D$_2$/T$_2$ mixture are complicated. First, (i) each μ slows down by ionizing the molecules in the mixture, and (ii) is eventually captured by D$_2$, DT, or T$_2$ into a highly excited muonic orbital by kicking out one of the molecular electrons. (iii) The resultant neutral molecule $dd\mu e$, $dt\mu e$, or $tt\mu e$ will be de-excited by radiative decay, by Auger decay, or in collisions with ambient molecules, and (iv) a muonic atom $d\mu(n)$ or $t\mu(n)$ is formed by molecular dissociation. (v) It will also be de-excited by radiative decay or in collisions with molecules, in which Stark mixing may also occur. The atoms $d\mu(n)$ and $t\mu(n)$ are neutral and very small if they are in a low bound state n. Therefore, they act nearly like neutrons and can easily approach close to a nucleus in a molecule, hardly interacting with the electron cloud. Thus (vi) μ in $d\mu$ can be transferred to t in DT or T$_2$ in a collision $d\mu(n) + t \to d + t\mu(n')$. (vii) When $t\mu(n=1)$ collides with d in D$_2$ or DT, a molecular system $dt\mu$ may be formed in an excited state having a very small binding energy, if the excess energy is absorbed by the ro-vibrational motion of the six-body electronic molecule $[(dt\mu) - d]ee$ or $[(dt\mu) - t]ee$, in which the small system $dt\mu$ may be regarded as a quasi-nucleus. (viii) The muonic molecule $dt\mu$ is de-excited down to the ground state by emission of an Auger electron. (ix) The d-t fusion occurs in the ground-state $dt\mu$, and a neutron and an α particle are emitted with high kinetic energies. Most μ are released and can be captured by D$_2$, DT, or T$_2$ again in process (ii), thus repeating the whole μCF cycle. However, (x) a small fraction of muons can be captured by the fast α particles, forming bound hydrogenic ions αµ. This is called a sticking process. (xi) These fast α particles can release the bound µ in collisions with molecules during their slowing down. This is called a stripping process. The released µ can come back into the µCF cycle. On the other hand, µ cannot be released once the ions αµ become slow. These bound µ are lost from the µCF cycle and no more act as catalysis.

Thus, many kinds of atomic processes are involved in the µCF cycle. The complete understanding of the cycle requires the knowledge of the cross sections for all these processes for many quantum mechanical states. In some of the processes, the collision partners need not be in thermal equilibrium; the velocity distributions need to be known. Furthermore, the actual experiments have been carried out mainly in the liquid phase. Thus the condensed-phase effects also need to be studied. These are the reasons why the theory still cannot explain all the details of experimentally known facts, though many aspects of µCF have been clarified theoretically.

The rates of various processes vary with the density of the hydrogen isotopes. It is customary to talk about the rates (or the corresponding lifetimes) referred to LHD. This custom is followed in this article unless otherwise specified.

FAVORABILITY OF MUON-CATALYZED d-t FUSION

The μ-catalyzed d-t fusion cycle explained in the preceding section can be easily generalized for other μCF cycles. The following summarizes different intramolecular fusion processes for different isotopes:

$$pd\mu \to {}^3\text{He}(0.005\,\text{MeV}) + \gamma(5.49\,\text{MeV}) + \mu \quad (85\%)$$
$$pd\mu \to {}^3\text{He}(0.20\,\text{MeV}) + \mu(5.29\,\text{MeV}) \quad (15\%)$$
$$dd\mu \to t(1.01\,\text{MeV}) + p(3.02\,\text{MeV}) + \mu \quad (42\%)$$
$$dd\mu \to {}^3\text{He}(0.82\,\text{MeV}) + n(2.45\,\text{MeV}) + \mu \quad (58\%)$$
$$pt\mu \to {}^4\text{He}(0.05\,\text{MeV}) + \gamma(19.76\,\text{MeV}) + \mu \quad (\sim 90\%)$$
$$pt\mu \to {}^4\text{He}(0.59\,\text{MeV}) + \mu(19.22\,\text{MeV}) \quad (\sim 10\%)$$
$$dt\mu \to {}^4\text{He}(3.54\,\text{MeV}) + n(14.05\,\text{MeV}) + \mu$$
$$tt\mu \to {}^4\text{He} + 2n(11.33\,\text{MeV}) + \mu$$

This section explains the three main reasons why μ-catalyzed d-t fusion has the highest cycle rate among all kinds of μCF cycles.

Weakly Bound State of $dt\mu$

The quantum states of the three-body systems $pd\mu$, $dd\mu$, $dt\mu$, etc., may be specified by their total-orbital-angular-momentum quantum number J and the label v ($=0,1,2,\cdots$), corresponding to the vibrational quantum number of electronic molecules, as $(pd\mu)_{Jv}$, etc. The system $(dt\mu)_{11}$ happens to have a very small binding energy of 0.660 eV with respect to the energy -2.71 keV of $t\mu(n=1)$, whereas the binding energy of $(dd\mu)_{11}$ with respect to $d\mu(n=1)$ is 1.975 eV. All the other states of these systems and all the states of their isotopes have a binding energy larger than 45 eV. The formation of $dt\mu$ in a six-body muonic molecule occurs efficiently in resonance processes

$$t\mu(n=1) + \text{D}_2 \to [(dt\mu) - d]ee \tag{1}$$

$$t\mu(n=1) + \text{DT} \to [(dt\mu) - t]ee, \tag{2}$$

if the energy of the six-body molecule matches the total energy of the left-hand side, i.e., the sum of the internal energies of the two collision partners and the relative kinetic energy. Since the energies of the vibrational motion $d-d$ in D_2, $d-t$ in DT, $(dt\mu)-d$ in $[(dt\mu)-d]ee$, and $(dt\mu)-t$ in $[(dt\mu)-t]ee$ are of the order of a fraction of an eV, the energy matching condition can be satisfied quite well for thermal collisions only for $(dt\mu)_{11}$ and fairly well for $(dd\mu)_{11}$, but not at all for the other states or for the other systems. Thus the formation rate for $(dt\mu)_{11}$ is $4\times10^8\,\text{s}^{-1}$ at 300 K and is the highest, although even this rate is quite low in the μ-catalyzed d-t fusion cycle and $dt\mu$ formation is a bottleneck in the cycle. The rate of formation of $(dd\mu)_{11}$ is $4\times10^6\,\text{s}^{-1}$ at 300 K. For the other isotopes only the nonresonant mechanism of Auger electron emission is possible, e.g.,

$$d\mu(n=1) + H_2 \rightarrow [(pd\mu)-p]e + e, \tag{3}$$

and the formation rates are $\sim 10^6\,\text{s}^{-1}$ or lower.

Note that $dt\mu$ is not formed efficiently in collisions of $d\mu(n=1)$ with molecules; $d\mu(n=1)$ lies 48.7 eV above $(dt\mu)_{11}$, and the excess energy in the formation of $dt\mu$ cannot be absorbed completely by the ro-vibrational motion of the six-body molecule.

High Fusion Rate in $dt\mu$

The rate $\lambda_{dt}^{(f)}(Jv)$ of intramolecular fusion in a state $(dt\mu)_{Jv}$ is expressible well by the product $P_{dt}(Jv)K_{dt}$, the first factor being the probability density of finding the two nuclei in the state $(dt\mu)_{Jv}$ within the nuclear-force range $R_{dt} \simeq 0$, and K_{dt} being the rate coefficient of d-t fusion per unit probability density in this range.[8, 11] The value of K, determined by nuclear physics, is $1.1\times 10^{-14}\,\text{cm}^3\text{s}^{-1}$ for d-t fusion and is larger than for the other combinations; e.g., $K_{dd} = 1.5 \times 10^{-16}\,\text{cm}^3\text{s}^{-1}$, $K_{tt} = 1.3 \times 10^{-16}\,\text{cm}^3\text{s}^{-1}$, and $K_{pd} = 2.5 \times 10^{-22}\,\text{cm}^3\text{s}^{-1}$.

The atomic physics property P, which is essentially the tunneling probability, is higher for a lighter reduced mass between the two nuclei. Thus $P_{pd}(00)$ is larger than $P_{dt}(00)$. The product $\lambda^{(f)} = PK$, however, is smaller for $(pd\mu)_{00}$ than for $(dt\mu)_{00}$ by nearly seven orders of magnitude. On the other hand, $\lambda^{(f)}$ for the electronic molecule $(pde)_{00}$ is *larger* than that for $(dte)_{00}$ by 17 orders of magnitude, the switching in the order occurring for a lepton mass of about ten times the electron mass.[11]

The transition $J=1\rightarrow 0$ is forbidden for systems with identical nuclei, such as $dd\mu$. Hence the d-d fusion occurs mostly in the $J=1$ states since most $dd\mu$ are formed in the state $(J,v)=(1,1)$. The probability density $P_{dd}(Jv)$ is much smaller for $J=1$ than for $J=0$ because of the centrifugal barrier between the two nuclei for $J=1$. Thus the rate $\lambda^{(f)}$ relevant to actual experiments is much lower for $dd\mu$ than for $dt\mu$. The same argument holds true for $tt\mu$, but the rate of formation of $tt\mu$ is very low, anyway.

Small Sticking Probability

The intramolecular fusion processes listed at the outset of this section indicate that the fastest helium nuclei are produced in the $dt\mu$ fusion. The μ sticking to a fast α particle is unlikely. Thus the sticking probability ω_0 is as small as about 0.7% for $dt\mu$. Even if $\alpha\mu$ is formed, μ may be stripped off again in subsequent fast collisions with a probability R ($\simeq 0.4$) called reactivation fraction, and the net sticking probability ω ($=\omega_0(1-R)$) is only about 0.4%. However, its cumulative effect throughout the 150 cycles is a serious problem. Indeed, ω is one of the crucial quantities that limit the rate of μ-catalyzed d-t fusion cycle. If no sticking occurred, μ could catalyze three times more d-t fusion events.

DYNAMIC PROCESSES INVOLVING MUONIC ATOMS

Some recent work of our group on three-body dynamic processes, relevant to the μCF cycle, will be reported in this section. The hyperspherical coordinates are used.

Hyperspherical Close-Coupling Equations

We start from a set of Jacobi coordinates $(\mathbf{r}_i, \mathbf{R}_{jk})$ appropriate for the arrangement channel $i+(jk)$, where $\mathbf{R}_{jk}$ is the position vector of particle j with respect to particle k, and where $\mathbf{r}_i$ is the position vector of i with respect to the center of mass of j and k. We define the hyperradius ρ_{ijk} by $M\rho_{ijk}^2 = \mu_{i,jk}r_i^2 + \mu_{jk}R_{jk}^2$, where M is the total mass of the three-particle system, and $\mu_{i,jk}$ (μ_{jk}) is the reduced mass for the relative motion between i and the system jk (between j and k). It turns out that $\rho_{ijk} = \rho_{jki} = \rho_{kij}$, which is denoted by ρ. This suggests the convenience of this hyperradius in the treatment of rearrangement collisions. Asymptotic values of ρ correspond to asymptotic separation in any of the three arrangement channels or three-body breakup, while small ρ represents a three-body complex confined in a small region. We calculate quantum states φ_s^{ad} of the total three-body system for fixed values of ρ. They are adiabatic hyperspherical states, which are functions of five angular coordinates that form a set of hyperspherical coordinates together with ρ. The adiabatic hyperspherical states can be interpreted in analogy to adiabatic states of usual diatomic molecules. Expansion of the wave function of the total system in terms of the set $\{\varphi_s^{\text{ad}}\}$ and substitution into the Schrödinger equation lead to coupled differential equations in ρ, called hyperspherical close-coupling (HSCC) equations.[12] Both inelastic and rearrangement collisions occur as transitions between hyperspherical channels in HSCC equations.

The coupling potentials representing the nonadiabatic effect involve no nonlocal potentials even for rearrangement collisions, which is convenient for numerical purposes. Furthermore, the convergence with the increase of the number of coupled channels turns out empirically to be quite rapid. This results from the efficient account of the coupling between the motion in $\mathbf{r}_i$ and that in $\mathbf{R}_{jk}$ by the use of ρ. Also, the adiabatic potential curves provide a visual aid for the physical understanding of the dynamic process, e.g., resonance mechanisms, as well as three-body bound states. In actual numerical calculations, the HSCC equations in terms of $\{\varphi_s^{\text{ad}}\}$ have been switched into coupled equations in piecewise-diabatic states when sharply peaked coupling potentials appear in the former;[12] this technique leads to smooth couplings and to numerical stability.

Muon-Transfer Processes

The muonic molecules $dt\mu$ are formed in reactions (1) and (2), i.e., in collisions of $t\mu(n=1)$ with d in D_2 or DT. A muon bound to d must be first transferred to t for forming $dt\mu$ and contributing to μCF. This μ transfer occurs for $d\mu(n)$ in any state n:

$$d\mu(n) + t \to d + t\mu(n) + \Delta E(= 48.0\,\text{eV}/n^2). \tag{4}$$

This process is almost irreversible because of the large energy difference ΔE between $d\mu(n)$ and $t\mu(n)$ compared with the collision energy.

Similar μ-transfer processes are important in understanding the μ-catalyzed p-d and p-t fusion. The pioneering work of Burke et al.[3] on p-to-d transfer for $n=1$ using the strong-coupling approximation yielded a rate coefficient of 5×10^{-14} cm^{3}s^{-1} at low temperature, more than two orders of magnitude smaller than the Born and weak-coupling results obtained by earlier authors, proving the strong effect of the coupling between the initial and final channels. Recent, more elaborate close-coupling calculations[9, 10] produced a rate coefficient of $4\times10^{-13}\,\mathrm{cm^3 s^{-1}}$, which is closer to Burke et al. than to the too large Born and weak-coupling values.

The rates of d-to-t transfer for $n \geq 2$ for LHD are higher than $10^{11}\,\mathrm{s^{-1}}$ according to semiclassical theories.[13, 14] The rate for $n=1$ is known from several sophisticated quantum mechanical calculations[10] to be $3\times 10^8\,\mathrm{s^{-1}}$, which is by far the lowest among all n because of the largest ΔE for $n=1$. This rate is slightly lower than that of $dt\mu$ formation, which is a bottleneck in the μCF cycle. Therefore, the atoms $d\mu$ affect the formation of $dt\mu$ strongly, once they have cascaded down to the ground state. The fraction of $d\mu(n=1)$, often denoted by q_{1s}, is an important quantity in the μCF cycle. It is determined by the net effect of many elementary processes experienced by $d\mu(n)$ during the cascade down. Muon transfer for $n \geq 2$ is important in the cascade down process because of the competition with other de-excitation processes.

Fully quantum mechanical calculations for general $n \geq 2$ would be quite difficult. We chose the case of $n=2$ as prototype μ transfer from an excited to an excited state at collision energies lower than $0.1\,\mathrm{eV}$.[15] In the HSCC equations we included all hyperspherical channels that dissociate into $d\mu(n=1\text{-}4) + t$ or $d + t\mu(n=1\text{-}4)$, but the inclusion of the $n=4$ channels had only a slight effect on the $n=2$ cross sections. The splitting between the $2s$ and $2p$ states of $d\mu$ in the initial channel due to vacuum polarization is $\Delta E_{\mathrm{vp}} = 0.2\,\mathrm{eV}$, which is higher than the collision energies and is not negligible. Therefore, we added a phenomenological potential $V_{\mathrm{vp}} = -\Delta E_{\mathrm{vp}}|d\mu(2s)><d\mu(2s)|$, which is negligible at small values of ρ in comparison with the strong channel potentials. Note that the splittings in the intermediate and final channels with a common value of n are unimportant because of the large absolute values of the channel kinetic energies. The asymptotic dipole potential ($\propto \rho^{-2}$) in the degenerate $d\mu(n=2)$ channels in the nonrelativistic approximation disappears in the presence of V_{vp}, and the polarization potential ($\propto \rho^{-4}$) is of the longest range.

Partial waves up to $J=8$ were included for convergence in the μ-transfer cross sections. Elastic cross sections were also calculated with inclusion of J up to 35. The latter cross sections are important in determining the non-Maxwellian kinetic-energy distribution of atoms $d\mu(n)$. This distribution would affect the rates of various cascade processes, and hence, the crucial quantity q_{1s}.

The absolute magnitude and the energy dependence (nearly $\propto E^{-1}$) of the cross sections are in general agreement with the recent semiclassical results.[14] An interesting feature in the present quantal results, however, is a resonance structure found in both the μ-transfer and elastic cross sections around $0.02\,\mathrm{eV}$. Partial-wave analysis of this resonance has been made in connection with the adiabatic hyperspherical potentials.[15]

A remark is due here on the presence of a third nucleus and, especially, two electrons in process (4). This is called the screening effect, which may be important at energies lower than $0.1\,\mathrm{eV}$.[14] What really occurs is either of the processes

$$d\mu(n) + \mathrm{DT} \to \mathrm{D_2} + t\mu(n), \tag{5}$$

$$d\mu(n) + \mathrm{T_2} \to \mathrm{DT} + t\mu(n). \tag{6}$$

The product molecule $\mathrm{D_2}$ or DT may break up into fragments. The screening effect was taken into account approximately in semiclassical calculations,[13, 14] but would be difficult to consider in fully quantal treatment.

Hyperfine Transitions

The muonic atoms $p\mu(1s)$ and $t\mu(1s)$ have two different hyperfine states $F=0$ and 1 arising from the interaction of the nuclear spin $\mathbf{s}_1$ with the muonic spin $\mathbf{s}_\mu$. The hyperfine splitting $\Delta E_{\text{hf}} = E(F{=}1) - E(F{=}0)$ is 0.182 eV for $p\mu(1s)$ and 0.237 eV for $t\mu(1s)$. The hyperfine states $F = 1/2$ and $3/2$ of $d\mu(1s)$ are split by 0.048 eV. Because of these splittings the resonance condition in the molecular formation reactions such as (1) and (2) is satisfied at different collision energies for different hyperfine states. Thus the temperature dependence of the molecular formation rate is different for different F. Therefore, the μCF cycle rate is influenced by spin-flip processes such as

$$t\mu(F) + d \rightarrow t\mu(F') + d, \qquad (7)$$

$$t\mu(F) + t \rightarrow t + t\mu(F'). \qquad (8)$$

The elastic processes $F = F'$ are also important in determining the kinetic-energy distribution of $t\mu(F)$, etc., and hence the molecular formation rate.

We took up the process $p\mu(F{=}0,1) + p$ as an example and solved HSCC equations paying special attention to collision energies around and lower than the threshold energy.[16] Though the target p is embedded in H_2 or HD, the screening effect was neglected here again and the pure three-body system was treated. The Hamiltonian of the total system involved the hyperfine interaction $V_{\text{hf}} = c\{\delta(\mathbf{r}_1)\mathbf{s}_1 \cdot \mathbf{s}_\mu + \delta(\mathbf{r}_2)\mathbf{s}_2 \cdot \mathbf{s}_\mu\}$, i.e., the sum of the interactions of the muonic spin $\mathbf{s}_\mu$ with the nuclear spins $\mathbf{s}_i$ ($i = 1$, 2); $\mathbf{r}_i$ is the position vector of μ measured from nucleus i, and c is a constant depending on the muonic atom. The nuclear spin-spin interaction ($\propto \mathbf{s}_1 \cdot \mathbf{s}_2$) was neglected; it is weak for large internuclear distances R_{12} and is much weaker than the strong channel potentials for small R_{12}. The spin-orbit interactions are also negligibly weak.

All hyperspherical channels that dissociate into $p\mu(n{=}1{-}3; F{=}0,1) + p$ were included, but the $n = 3$ channels affected the cross sections by less than 10%. The neglect of V_{hf} reproduced the cross sections fairly well for energies much higher than ΔE_{hf}. A typical Wigner threshold behavior was clearly observed in the cross sections. Previous adiabatic-state expansion calculations in terms of the conventional coordinate system[17] and a modified Faddeev-equation approach[18] yielded results in good agreement with ours. We also treated the spin-flip and elastic processes in collisions $e^-e^+(F) + e^- \rightarrow e^- + e^+e^-(F')$ to study the mass dependence.

CORRECTIONS ON THE ENERGIES OF MUONIC MOLECULES

Since the molecular formation reactions (1) and (2) are resonance processes, the exact *resonance* energies of the six-body systems $[(dt\mu)_{11} - d]ee$ and $[(dt\mu)_{11} - t]ee$ are important in analyzing experimental results. It turned out that an accuracy better than 1 meV is necessary for the interpretation of measured μCF rates.

The systems such as $[(dt\mu)_{Jv} - d]ee$ are treated conveniently and accurately by first regarding the small internal system $dt\mu$ as independent of the rest of the six-body system. Then the total Hamiltonian H may be decomposed into three terms, the first being the Hamiltonian $H_{dt\mu}$ for $dt\mu$, the second being the Hamiltonian H_{Xdee} for the hydrogenlike molecule $Xdee$ in which X is a hypothetical particle formed by compressing $dt\mu$ into a single point, and the last being the remaining interaction V_{fs}.

The ro-vibrational energies of $Xdee$ can be calculated easily from the knowledge of the hydrogen molecule. The nonrelativistic energies of the *bound* states $(dt\mu)_{Jv}$, $(dd\mu)_{Jv}$, etc., were calculated to an extremely high accuracy.[19] (Incidentally, *resonance* states of the three-body systems $dt\mu$, $dd\mu$, etc., were also calculated extensively,[20] and were analyzed in a unified manner[21] using the asymptotic theory of Gailitis and Damburg.[22]) Small corrections on the three-body bound-state energies were also made.[9,10] They include the relativistic effects and the effects of the hyperfine interactions, vacuum polarization, nuclear electromagnetic structure, etc., and the calculated corrections are believed to have more than the desired accuracy.

The interaction V_{fs} may be treated perturbatively. It may be decomposed conveniently into multipole terms. ¿From the viewpoint of the four-body molecule $Xdee$, this interaction is a correction for the finite size of the particle X assumed initially as a point charge. Hence the name finite-size correction. The conventional approach is to begin by solving a simpler problem of the finite-size correction on the energy of $(dt\mu)_{11}e$ up to second order (plus some further effect in some reference) in the finite-size potential similar to V_{fs}.[23, 24] Here, the system turns into a hydrogenlike *atom* when $dt\mu$ is compressed into a point. The finite-size correction for this system is then scaled for $[(dt\mu)_{11} - d]ee$ by the probability-density ratio γ of the wave function of H$_2$ at one of the protons to that of the H atom at the proton. The scaled result is only a small fraction of meV.[23]

We realized that this procedure automatically dropped a strong effect of the first-order correction due to the quadrupole term in V_{fs}; we found this correction to be as large as a few meV and to depend on the rotational angular momenta of $Xdee$ and of the whole six-body system.[25] Direct calculations of the second-order correction for the six-body molecules, though quite difficult, would be highly desirable. We also pointed out that the scaling by the ratio γ has no theoretical ground for the second-order perturbation energy.[25]

This work was recently extended for the system $[(dd\mu)_{11} - d]ee$ and the first-order quadrupole correction was again found to be a few meV and to depend on the rotational angular momenta.[26] For this system a further scaling procedure in a previous work without the quadrupole correction had led to an agreement between the theoretical and experimental resonance energies to within ~0.1 meV. We argued that this scaling procedure is unreliable for several reasons and that the agreement between theory and experiment had been fortuitous.[26]

We also pointed out that wave functions of the total six-body system to be discussed must be formed by correct coupling of the angular momenta of $(dt\mu)_{11}e$ and $Xdee$, although the literature deals with the simple product of wave functions of these subsystems.[25, 26] In fact, we also need to couple spin angular momenta and calculate the fine and hyperfine structure of the six-body resonance energies taking account of the quadrupole finite-size corrections as well, as we have done recently.[27]

CONCLUSIONS

The neutrons obtained from the muon-catalyzed d-t fusion has a high kinetic energy of 14.1 MeV. This energy may be taken out as thermal energy if the fusion neutrons are stopped in matter. A single muon catalyzes ~150 d-t fusion events on the average in the D$_2$/T$_2$ mixture of liquid-hydrogen density. Therefore, an energy of 150×14 MeV $= 2$ GeV can be produced per muon. However, a higher energy of 5–8 GeV is estimated to be consumed for the production of a muon on the average. Muons are produced first by producing high-energy pions in nuclear reactions (which take much input energy), and then by waiting for the pions to decay into high-energy muons, which must slow down before entering the muon-catalyzed fusion (μCF) cycle. Unfortunately, no method of direct production of slow muons is known.

Thus μCF alone seems to be useless as an energy source at this stage. There are proposals of fusion-fission hybrid reactors using μCF, which are claimed by the proposer to be more efficient, cheaper, and safer than the usual breeders. On the other hand, μCF may be used as a source of an intense beam of neutrons with a well-defined energy of 14.1 MeV, or of slow muons of about 10 keV.

The main current interest, however, is in clarifying atomic physics of μCF, which is quite unique compared with the conventional atomic physics problems. This uniqueness stems from the existence of large, medium, and small masses together in the system, namely, proton isotopes, muons, and electrons. A negatively charged particle, much heavier than the electron, can greatly change atomic physics. The uniqueness stems also from the coexistence of Coulomb force and nuclear force, which sometimes contribute to the main physics at the same time. ¿From such an interest we study not only muon-catalyzed d-t fusion but also other combinations, such as p-d fusion and d-d fusion, for better understanding of the physics of μCF. Even the triple $H_2/D_2/T_2$ mixture has attracted recent attention from the viewpoint of possibly high cycle rates. Results of all these studies might pave a way to ideas of increasing the rate of μCF cycle.

Acknowledgments

It is a great pleasure for me to contribute to this volume in honor of Professor Burke, with whom I started working as a visiting postdoctoral fellow at the Queen's University of Belfast in 1975 on leave of absence from University of Tokyo. I would like to express my sincere gratitude to him for all the scientific benefits I got from him through many visits to him.

REFERENCES

1. L. W. Alvarez et al., *Phys. Rev.* **105**, 1127 (1957).
2. J. D. Jackson, *Phys. Rev.* **106**, 330 (1957).
3. P. G. Burke, F. Haas, and I. C. Percival, *Proc. Phys. Soc. London* **73**, 912 (1959).
4. F. C. Frank, *Nature* **160**, 525 (1947).
5. A. D. Sakharov, report written in 1948 [English translation: *Muon Catal. Fusion* **4**, 235 (1989)].
6. E. A. Vesman, *Sov. Phys. JETP Lett.* **5**, 91 (1967).
7. S. S. Gershtein and L. I. Ponomarev, *Phys. Lett.* **72B**, 80 (1977).
8. S. E. Jones et al., *Phys. Rev. Lett.* **51**, 1757 (1983).
9. L. I. Ponomarev, *Contemp. Phys.* **31**, 219 (1991).
10. J. S. Cohen, in "Review of Fundamental Processes and Applications of Atoms and Ions," edited by C. D. Lin (World Scientific, Singapore, 1993), p. 61.
11. I. Shimamura, *Prog. Theor. Phys. (Kyoto)* **82**, 304 (1989).
12. Applications of this method to various kinds of three-body problems have been made. See, e.g., J.-Z. Tang, S. Watanabe, and M. Matsuzawa, *Phys. Rev. A* **46**, 2437 (1992). A. Igarashi and N. Toshima, *Phys. Rev. A* **50**, 232 (1994). J.-Z. Tang and I. Shimamura, *Phys. Rev. A* **50**, 1321 (1994); *ibid.* **51**, R1738 (1995); *ibid.* **52**, R3413 (1995). Y. Zhou and C. D. Lin, *J. Phys. B* **27**, 5065 (1994); *Phys. Rev. Lett.* **75**, 2296 (1995). A. Igarashi and I. Shimamura, *Phys. Rev. A* **56**, 4733 (1997).
13. L. I. Menshikov and L. I. Ponomarev, *Zeit. f. Phys. D* **2**, 1 (1986).
14. W. Czapliński, A. Gula, A. Kravtsov, A. Mikhailov, and N. Popov, *Phys. Rev. A* **50**, 518 (1994).
15. A. Igarashi, T. Shirai, I. Shimamura, and N. Toshima, in preparation.
16. A. Igarashi, I. Shimamura, and N. Toshima, *Phys. Rev. A* **58**, 1166 (1998).
17. L. Bracci, C. Chiccoli, G. Fiorentini, V. S. Melezhik, P. Pasini, L. I. Ponomarev, and J. Wozniak, *Muon Catalyzed Fusion* **4**, 247 (1989).
18. C. Y. Hu, A. A. Kvitsinsky, and J. S. Cohen, *J. Phys. B* **28**, 3629 (1995).
19. For example, M. Kamimura, *Phys. Rev. A* **38**, 621 (1988). H. J. Monkhorst et al., *Phys. Rev. A* **36**, 5494 (1987); *ibid.* **38**, 26, 4859 (1988).
20. For example, S. Hara and T. Ishihara, *Phys. Rev. A* **40**, 4232 (1989).
21. I. Shimamura, *Phys. Rev. A* **40**, 4863 (1989).
22. M. Gailitis and R. Damburg, *Proc. Phys. Soc. London* **82**, 192 (1963).
23. L. I. Menshikov, *Sov. J. Nucl. Phys.* **42**, 918 (1985). D. Bakalov, *Muon Catal. Fusion* **3**, 321 (1988). A. Scrinzi and K. Szalewicz, *Phys. Rev. A* **39**, 4983 (1989).

24. M. R. Harston, I. Shimamura, and M. Kamimura, *Zeit. f. Phys.* D **22**, 635 (1992).
25. M. R. Harston, I. Shimamura, and M. Kamimura, *Phys. Rev. A* **45**, 94 (1992). See also D. D. Bakalov and V. S. Melezhik, *JINR Report* No. P4-81-835 (1981), unpublished.
26. M. R. Harston, S. Hara, Y. Kino, I. Shimamura, H. Sato, and M. Kamimura, *Phys. Rev. A* **56**, 2685 (1997).
27. D. Bakalov, K. Bakalova, V. Korobov, H. J. Monkhorst, and I. Shimamura, *Phys. Rev. A* **57**, 3370 (1998).
28. Y. Kino, M. R. Harston, I. Shimamura, E. A. G. Armour, and M. Kamimura, *Phys. Rev. A* **52**, 870 (1995).

Index

Above threshold
    dissociation, 105,109
    ionization, 78-80, 105
Angle-differential spin-polarization parameters, 46-48
Angle-differential Stokes parameters, 43-46
Angle-integrated cross sections, 39, 40
Angle-integrated Stokes parameters, 41-43
Anisotropic molecular ensembles, 144-149
Atomic
    collisions, JET applications of, 225-247
    collisions, in plasma transport modelling, 227
    collisions, in spectral diagnostics, 233
    and molecular collision theory, fifty years of, 1-7
    physics of muon-catalyzed fusion, 269-278
    processes, 250, 251
Atoms in intense laser fields, 77-103

BERTHA, 4-component relativistic molecular quantum mechanics program, 213-224
Basic Linear Algebra Subprograms (BLAS), 125
Beam stopping coefficients for a neutral deuterium beam, 242
Bloch operator, theorem, 187, 188
Born approximation, 4, 160
    first order for the Ps formation amplitude in hydrogen, 166
Born-Ochkur approximation, 5
Born-Oppenheimer approximation, 4
Breit interaction, interaction integrals, 219

Charge exchange
    cross section, 244

Charge exchange (*cont.*)
    heavy particle, 168
    spectroscopy, 243
Chemically peculiar stars (CP), 262
Chiral molecules, elastic electron collisions with, 137-153
Close coupling approach, 5, 15, 21, 71, 226
    convergent (CCC), 16, 37, 38
    Floquet (FCC), 90
    hyperspherical (HSCC), 21, 273
    pseudostate (PSCC), 21
Collision theory, fifty years of atomic and molecular, 1-7
Collisions of electrons with atoms and molecules: *see also* Electron-impact excitation, ionization
    e^--He, total cross section, 157
    elastic e-Cs collisions, 47, 48
    elastic electron collisions with chiral and oriented molecules, 137-153
    laser-assisted electron-atom collisions, 83, 84
Collisions of positrons with atoms and molecules: *see also* Positron
    e^+-H(1s) collisions, total Ps formation cross section, 162
    e^+-He scattering, total cross section, 157
    Ps formation, 155, 173-175
Continuity equation, 255-257
Continuum distorted wave approximation (CDW), 168
Continuum multiple scattering method, 139
Convergent close coupling: *see* Close coupling approach
Coronal picture, 226
Coulomb barrier in thermonuclear fusion, 271
Coulomb explosion model, field ionization, 106
Coulomb interaction integrals, 219

Coupled channel
    calculations with a hyperspherical basis, 165, 166
    two-centre models, 157
Coupled equation method, 161
Cross section: *see also* Angle-integrated cross sections, Charge exchange, Differential cross section, Differential ionization cross section, Direct ionization, Electron-impact, Photoionization cross section, Positronium, Rosseland mean, Singly differential cross section, Total cross section and Triply differential cross section
    for e^+-H(1s) scattering, 156
    for $e^- + H_2 \, \Sigma_g^+$, 69
    for $e^- + Li_2$ scattering, 73
    for ionization of H(1s) by e^+ impact, 162

Desorption induced by electronic transitions (DIET), 205
Detachment rates, total and partial RMF of H^-, 93
Deuterium, 230, 241, 243
Diabatic potentials and scattering resonances, 57
Differential cross section, 139, 148, 151, 193, 194: *see also* Singly differential cross section, Triply differential cross section
    spin-averaged differential cross section of NiO(001), 194
Differential ionization cross section, 22
Diffusion, 254
Dipole dissociation (DD), 206
Dirac
    equation for a central potential, 215
    equation for a stationary state, 215
    probability density, 98
Direct ionization, 175-178
    cross sections for He and Kr, 176
Direct methods for sparse matrices, 122
    parallelization of, 127
Dissociation of negative ion resonance (NIR) to neutral fragments: CH_3Cl / Graphite, 207
Dissociative electron attachment (DEA), 197, 208

Dissociative ionization, manipulating the process, 108
Distorted wave approaches, 5: *see aslo* CDW, RDW

Effective emission coefficients, 244
Electron attachment to optically excited molecules: O_2 and CO / Graphite, 233
Electron capture to the continuum (ECC), 177
Electron correlation, 219
Electron scattering by small molecules, 67-76
Electron stimulated desorption, 207
    ion angular distributions (ESDIAD), 198
Electron-atom
    collisions, laser assisted, 83, 90
    scattering,
        by argon atoms in the presence of a CO_2 laser, 83
        hydrogen, 10, 16
        krypton, 12, 13
        lithium, 18
        recent progress in, 15-31
        resonances, 9-14
Electron-impact
    cross section, 250
        e-H scattering, 24, 25
    excitation
        of atoms, benchmark studies in, 33-50
        of barium, 21
        of helium, 20, 39-41
        of hydrogen, 16, 17
        of krypton, 43, 46, 47
        of lithium, 18
        of mercury, 46
        of neon, 42
        of sodium, 19, 45
        of xenon, 36, 37
    ionization
        of helium, 26-29
        of hydrogen, 17
    rate coefficient, 250
Electron-ozone scattering, 51-66
Embedding
    calculations of surface electronic structure, 188-190

Embedding (*cont.*)
  method, 184-188
  method for quantum confinement, 190-192
  potential, 186
  and R-matrix methods at surfaces, 183-195
Energy loss spectra; *see* HREELS
  electron (EELS), 53, 200, 204
    from co-adsorbed K and O_2 on graphite, 204
    for helium, 11
    from physisorbed O_2, 200
    R-matrix study of EELS from NiO, 192-195
Exact static exchange approximation (ESE), 54
Experiment
  magnetic angle changing technique, 12
  observation of electron-atom resonances, 9-14
  poloidal section of JET, 229
  pump-probe, 114
  to perform scattered-electron-polarized-photon coincidence studies, 34

Feshbach formalism, 71
Floquet
  close-coupling, 90
  Lipmann - Schwinger equation, 90
  R-matrix-Floquet theory (RMF), 91-94
  theory, 86-91
Fully oriented molecules
  CHBrClF, 144-146
  HBr, 140, 141
  H_2S_2, 140, 142, 143
  numerical procedure and results for, 35-40, 45
Fusion
  intramolecular, 270-272
  muon-catalyzed, atomic physics of, 269-278
  rate, 272
  thermonuclear, 270

G-spinors, 217
Gaussian
  elimination, 123, 127

Gaussian (*cont.*)
  functions, 159
Green's
  functions, 74, 160, 185-187
  theorem, 184

Harmonic
  emission spectra of various noble gases, 81
  generation, 81-83
Hartree-Fock, 5, 18, 54
Harwell
  - Boeing sparse matrix collection, 119, 126
  subroutine library, 133
High resolution electron energy loss spectroscopy (HREELS), 197, 208
Hyperfine transitions, 274, 275
Hyperspherical close-coupling (HSCC), 21, 273
Hyperspherical basis, coupled channel calculations, 165

Integro-differential equations, 159
Intermediate Energy R-Matrix (IERM), 16, 17, 21, 39, 71
Intramolecular fusion, 270-272
Iterative methods, sparse matrices, 128
  parallelization of, 130

Jacobi
  block method, 131
  co-ordinates, 158, 165, 273
Joint European Torus (JET), 225-247
  light impurity simulation, 238
  measurements on molecules, 240
  measuring beam attenuation, 242
  measuring electron temperature and density, 236
  measuring impurity concentrations, 243
  measuring impurity influx, 233
  neutral helium beams, 245
  poloidal section, 229

Keldysch adiabaticity parameter, 79
Kohn method, 70, 72, 74

Laser-assisted electron-atom collisions, 83, 84

Laser-atom interactions, relativistic effects in, 97-99
Laser fields
  bichromatic, 94
  intense, 77-103, 105-118
Laser-induced
  alignment, 106, 112
  charge-asymmetry, 106, 109
  degenerate states (LIDS), 94
  dissociation, 106, 109
  fluorescent emission, 233
  orientation, 106
  stabilization, 106, 111
  trapping, 106
Level populations, 250
Light-induced
  continuum structures (LICS), 94
  states (LIS), 90
Line
  profiles, 265
  ratios for HeI, 236
  saturation, 253
Liquid hydrogen density (LHD), 270, 274
Lorentz profile, 250
LU factorization, 124

Markowitz threshold pivoting, 124
Matrix methods, 119-136
Model adiabatic potential, 52
  to study shapes and locations of single particle resonances, 51-66
Molecules: see also Anisotropic molecular ensembles, BERTHA, Chiral molecules, Collision theory, Collisions of electrons with atoms and molecules, Collisions of positrons with atoms and molecules, Electron attachment to optically excited molecules, Electron scattering by small molecules, Fully oriented molecules, Joint European Torus, Negative ion resonance, Oriented molecules, Partially oriented molecules and Randomly oriented molecules
  on surfaces, negative ion resonance (NIR) of, 197-211
Multiconfiguration Hartree-Fock, 5
Multielectron dissociative ionization (MEDI), 105, 108

Multiphoton
  ionization, 78-80
  processes in atoms and ions, 78-84
Muon-catalyzed fusion, atomic physics of, 269-278
  corrections on the energies of muonic molecules, 275
  dynamic processes involving muonic atoms, 273
  mechanism, 269
  muon transfer process, 273

Negative ion resonance (NIR), 54, 197-211
  dissociation of, to neutral fragments: CH_3Cl / Graphite, 207
  electron scattering by O_2 co-adsorbed with K on graphite, 202
  favourability, 271
  of molecules on surfaces, from spectroscopy to dynamics, 197-211
  scattering by physisorbed O_2 on Ag(110), 199
Neutral beam induced emission, 233

Onion skin model, 231
Opacity project, 5, 249
Optical depth, Rosseland mean, 256
Oriented molecules, elastic electron collisions with chiral and, 137-153
Ozone, electron- scattering, 51-66

Partially oriented molecules, collisions with, 144
Photoionization cross section of Be, 75
Photon stimulated desorption (PSD), 205
Plasma transport modelling, 227
Positron
  atom/molecule collisions, 174
  impact ionization, 177
  scattering by atoms, 155-170
  to positronium scattering, 171-182
Positronium
  beam and scattering, 178-180
  formation, 162-168, 173-175
  scattering cross section for He and H, 179
Proton emission, 116
Pump-probe experiment, 114

Quantum confinement, embedding method for, 190-192
Quantum electro-dynamics (QED), coupling of the electron and photon fields in relativistic QED, 217

R-matrix, 6, 23, 39, 68, 74, 160, 161, 183-195
  and embedding methods at surfaces, 183-195
  embedding and R-matrix theory, 186-188
  Floquet theory (RMF), 91-94
  intermediate energy (IERM), 16, 39, 71
  study of electron energy loss spectroscopy from NiO, 192-195
  with pseudostates (RMPS), 17, 26, 38, 39, 71
  without a box, 73-76
Radiation power loss functions, 227, 239
Radiation pressure and element diffusion in stellar interiors, 249-267
Radiative
  acceleration, calculation of, 252, 253
  flux, 251, 252
  power function for argon, 228
  transfer equation, 252
Randomly oriented molecules, numerical procedure and results for fully oriented and, 139-144: *see also* Fully oriented molecules
Recollision model, 80, 82
Relativistic distorted wave approximation (RDW), 21
Relativistic effects in laser-atom interactions, 97-99
Resonances
  electron-atom, 9-14
  scattering, 57
  single particle, 51-66
Resonant scattering from ozone, 56
Ricatti-Bessel function, 74
Rosseland mean
  cross section, 252
  optical depth, 256

Schrödinger equation, 72, 273
  time-dependent, 84, 91
    numerical solution of 95-97

Schrödinger probability density, 98
Singly differential cross sections, 26
Sparse matrices, 119-136
  application areas, 120
  diagonalization, 132
  direct methods, 122
  frontal and multifrontal methods, 126
  iterative methods, 128
  preconditioning, 129
Spectra: *see also* Energy loss spectra, High resolution electron energy loss spectroscopy, Time-of-flight spectra
  electron energy spectra showing ATI of xenon, 98
  energy spectrum of electrons scattered by argon atoms in the presence of a CO_2 laser, 83
  observed spectra of the 4850-5100 Å region, 240
  plate spectrum from ZETA 1A in the quartz ultraviolet, 226
  spectrum lines of helium and carbon ions, 233
Spectral analysis of
  beam penetrated plasma, 241
  divertor plasma, 233
Static exchange approximation, elastic e^-+H_2 scattering, 69
Stellar interiors, radiation pressure and element diffusion in, 249-267
Steric factors, 148
Sticking probability, 272
Stokes parameters, 34-36: *see also* Angle-differential Stokes parameters, Angle-integrated Stokes parameters
Sturmian functions, 158
Sturmin-Floquet method, 87
Supernova, SN 1987a, 6
Surfaces
  embedding and R-matrix methods at surfaces, 183-195
  embedding calculations of surface electronic structure, 188-190

T-matrix, 22-24, 38, 75, 91, 192-194
Tempkin-Poet model of e-H scattering, 23-25
Thermonuclear fusion, 270
Threshold intensities, 108, 110, 113

Time-dependent Schrödinger equation, 84, 91
   numerical solution of, 95-97
Time-independent, coupled differential equations, 87
Time-of-flight (TOF), spectra, 105, 108
   H_2 TOF spectra, 115
   I^{2+} fragments in the I_2 TOF spectra, 112
Total cross section, 193
   for e^+-He and e^--He scattering, 157
   for e^+-He scattering, 165
   for $e^{\pm}$-He scattering, 172
   for $e^+ + He(1s^2) \to Ps\ (n=1) + He^+(1s)$, 168
   for e^+-Na scattering, 164
   for e^+-Rb scattering, 172
   Ps formation for e^+-H(1s) collisions, 162
   for two states in $e^- + F_2$ scattering, excitation cross section, 252
Total ionization rate versus intensity for H(1s) in a laser field, 89
Transition amplitude for inelastic scattering, 194
Transverse photon interaction, 219
Triply differential cross section for $e^{\pm}$ - H_2, 177
Two-centre coupled channel models, 157-165

Wannier threshold law, 176

Series Publications

Below is a chronological listing of all the published volumes in the *Physics of Atoms and Molecules* series.

ELECTRON AND PHOTON INTERACTIONS WITH ATOMS
Edited by H. Kleinpopper and M. R. C. McDowell

ATOM-MOLECULE COLLISION THEORY: A Guide for the Experimentalist
Edited by Richard B. Bernstein

COHERENCE AND CORRELATION IN ATOMIC COLLISIONS
Edited by H. Kleinpoppen and J. F. Williams

VARIATIONAL METHODS IN ELECTRON-ATOM SCATTERING THEORY
R. K. Nesbet

DENSITY MATRIX THEORY AND APPLICATIONS
Karl Blum

INNER-SHELL AND X-RAYS PHYSICS OF ATOMS AND SOLIDS
Edited by Derek J. Fabian, Hans Kleinpoppen, and Lewis M. Watson

INTRODUCTION TO THE THEORY OF LASER-ATOM INTERACTIONS
Marvin H. Mittleman

ATOMS IN ASTROPHYSICS
Edited by P. G. Burke, W. B. Eissner, D. G. Hummer, and I. C. Percival

ELECTRON-ATOM AND ELECTRON-MOLECULE COLLISIONS
Edited by Juergen Hinze

ELECTRON-MOLECULE COLLISIONS
Edited by Isao Shimamura and Kazuo Takayanagi

ISOTOPE SHIFTS IN ATOMIC SPECTRA
W. H. King

AUTOIONIZATION: Recent Developments and Applications
Edited by Aaron Temkin

ATOMIC INNER-SHELL PHYSICS
Edited by Barnd Crasemann

COLLISIONS OF ELECTRONS WITH ATOMS AND MOLECULES
G. P. Drukarev

THEORY OF MULTIPHOTON PROCESSES
Farhad H. M. Faisal

PROGRESS IN ATOMIC SPECTROSCOPY, Parts A, B, C, and D
Edited by W. Hanle, H. Kleinpoppen, and H. J. Beyer

RECENT STUDIES IN ATOMIC AND MOLECULAR PROCESSES
Edited by Arthur W. Kingston

QUANTUM MECHANICS VERSUS LOCAL REALISM: The Einstein-Podolsky-Rosen Paradox
Edited by Franco Selleri

ZERO-RANGE POTENTIALS AND THEIR APPLICATIONS IN ATOMIC PHYSICS
Yu. N. Demkov and V. N. Ostrovskii

COHERENCE IN ATOMIC COLLISION PHYSICS
Edited by H. J. Beyer, K. Blum, and J. B. West

ELECTRON–MOLECULE SCATTERING AND PHOTOIONIZATION
Edited by P. G. Burke and J. B. West

ATOMIC SPECTRA AND COLLISIONS IN EXTERNAL FIELDS
Edited by K. T. Taylor, M. H. Nayfeh, and C. W. Clark

ATOMIC PHOTOEFFECT
M. Ya. Amusia

MOLECULAR PROCESSES IN SPACE
Edited by Tsutomu Watanabe, Isao Shimamura, Mikio Shimizu, and Yukikazu Itikawa

THE HANLE EFFECT AND LEVEL CROSSING SPECTROSCOPY
Edited by Giovanni Moruzzi and Franco Strumia

ATOMS AND LIGHT: INTERACTIONS
John N. Dodd

POLARIZATION BREMSSTRAHLUNG
Edited by V. N. Tsytovich and I. M. Ojringel

INTRODUCTION TO THE THEORY OF LASER–ATOM INTERACTIONS (Second Edition)
Marvin H. Mittleman

ELECTRON COLLISIONS WITH MOLECULES, CLUSTERS, AND SURFACES
Edited by H. Ehrhardt and L. A. Morgan

THEORY OF ELECTRON–ATOM COLLISIONS, Part 1: Potential Scattering
Philip G. Burke and Charles J. Joachain

POLARIZED ELECTRON/POLARIZED PHOTON PHYSICS
Edited by H. Kleinpoppen and W. R. Newell

INTRODUCTION TO THE THEORY OF X-RAY AND ELECTRONIC SPECTRA OF FREE ATOMS
Romas Karazija

VUV AND SOFT X-RAY PHOTOIONIZATION
Edited by Uwe Becker and David A. Shirley

DENSITY MATRIX THEORY AND APPLICATIONS (Second Edition)
Karl Blum

SELECTED TOPICS ON ELECTRON PHYSICS
Edited by D. Murray Campbell and Hans Kleinpoppen

PHOTON AND ELECTRON COLLISIONS WITH ATOMS AND MOLECULES
Edited by Philip G. Burke and Charles J. Joachain

COINCIDENCE STUDIES OF ELECTRON AND PHOTON IMAPCT IONIZATION
Edited by Colm T. Whelan and H. R. J. Walters

PRACTICAL SPECTROSCOPY OF HIGH-FREQUENCY DISCHARGES
Sergei A. Kazantsev, Vyacheslav K. Khutorshchikov, Günter H. Guthöhrlein, and Laurentius Windholz

IMPACT SPECTROPOLARIMETRIC SENSING
S. A. Kazantsev, A. G. Petrashen, and N. M. Firstova

ELECTRON MOMENTUM SPECTROSCOPY
Erich Weigold and Ian McCarthy

SUPERCOMPUTING, COLLISION PROCESSES, AND APPLICATIONS
Edited by Bell, Berrington, Crothers, Hibbert, and Taylor